Universitext

W0079785

Mathieu Lewin

Spectral Theory and Quantum Mechanics

 Springer

Mathieu Lewin
CNRS & Université Paris-Dauphine
Université PSL
Paris, France

ISSN 0172-5939 ISSN 2191-6675 (electronic)
Universitext
ISBN 978-3-031-66877-7 ISBN 978-3-031-66878-4 (eBook)
https://doi.org/10.1007/978-3-031-66878-4

Mathematics Subject Classification: 81Q10, 47-01

This book is a translation of the original French edition "Théorie spectrale et mécanique quantique" by Mathieu Lewin, published by Springer Nature Switzerland AG in 2022. The translation was done with the help of an artificial intelligence machine translation tool. A subsequent human revision was done primarily in terms of content, so that the book will read stylistically differently from a conventional translation. Springer Nature works continuously to further the development of tools for the production of books and on the related technologies to support the authors.

Translation from the French language edition: "Théorie spectrale et mécanique quantique" by Mathieu Lewin, © The Editor(s) (if applicable) and The Author(s), under exclusive license to Springer Nature Switzerland AG 2022. Published by Springer International Publishing. All Rights Reserved.

This Springer imprint is published by the registered company Springer Nature Switzerland AG
The registered company address is: Gewerbestrasse 11, 6330 Cham, Switzerland

If disposing of this product, please recycle the paper.

Preface

Brief Historical Introduction

In the fall of 1926, John von Neumann arrived in Göttingen, Germany, to work with David Hilbert, the most eminent mathematician of the time. Although he was only 22 years old, the young von Neumann had already completed a thesis in mathematics on set theory in Budapest and also obtained, in parallel, a degree in chemical engineering from ETH Zürich.

He found a highly charged atmosphere in Göttingen. Physicists were in the midst of inventing quantum mechanics, which describes matter at the microscopic scale. Werner Heisenberg, a young assistant to Max Born in Göttingen, had proposed in 1925 a theory based on sorts of infinite matrices [Hei25], later formalized with Born and Pascual Jordan [BJ25, BHJ26]. The following year, the Austrian Erwin Schrödinger had advocated for an entirely different formulation, based on the equation that now bears his name [Sch26] and which Heisenberg had described as "disgusting" [Mac92]. The Briton Paul Dirac had tried to reconcile the two visions in his "theory of transformations" [Dir27], developed simultaneously by Jordan [Jor27]. This uses sorts of matrices with continuous parameters that we would now call integral kernels. For the identity matrix, Dirac had to introduce a strange "function" denoted δ, equal to 0 everywhere except at the origin where it is infinite, so that its integral over all $\mathbb{R}$ is equal to 1.

At nearly 65 years old, Hilbert had been working for many years on spectral problems. It was he who had introduced the term "spectrum," probably in reference to the vibration frequencies of objects like a string or a drum. Influenced by the works of Fredholm around 1900, he had taken a strong interest in linear integral equations, that is, involving a kernel resembling the matrices with continuous parameters of Dirac and Jordan. In a very famous 1906 article [Hil06], Hilbert had proposed adopting a more abstract approach for this type of equations, which he had reformulated for the Fourier coefficients in the space $\ell^2(\mathbb{C})$ of square-summable sequences. Linear operations were then precisely seen as "infinite matrices." In the same article, Hilbert had also discovered that the spectrum of such a "matrix" did not necessarily contain only eigenvalues, and had called the complement "continuous spectrum."

Born was perfectly familiar with Hilbert's work because he had been his student in mathematics and had even become his assistant in 1904. His research with Heisenberg and Jordan was in fact strongly inspired by Hilbert's ideas. According to Heisenberg, "Hilbert indirectly exerted a very strong influence on the development of quantum mechanics," which can "only be fully recognized by those who have studied in Göttingen" [Rei70, p. 182].

For Hilbert, mathematics had to play a major role in these developments. In his famous talk at the International Congress of Mathematics in Paris in 1900, he indeed proposed the axiomatization of Physics as his sixth problem [Hil02]. He had already heavily invested himself in this direction, particularly for the kinetic theory of gases and general relativity. Since 1912, he had employed a full-time personal Physics assistant, tasked with teaching him the latest theories, which he reformulated in his own way and immediately taught to his mathematics students [SM09, Rei70, Sch19]. Lothar Nordheim was this assistant since 1922 and was therefore asked, along with von Neumann, to clarify the situation regarding the "new quantum mechanics."

Von Neumann and Nordheim first focused on the theory of transforma- tions [HvN27] but Dirac's delta function quickly discouraged them. For mathematicians of the time, such an object clearly could not make any sense. It was von Neumann who suddenly understood how to formulate quantum mechanics using the abstract notions of Hilbert space and self-adjoint operators, which he introduced on this occasion. In a few months, von Neumann thus produced about ten foundational works on this subject, which revolutionized both functional analysis and mathematical physics.

At the beginning of the twentieth century, the idea that abstract mathematical structures could be useful for solving concrete problems more effectively was only just beginning to emerge [Die81]. Fréchet had introduced the notion of complete metric space in his thesis in 1906. Riesz and Fischer had shown the following year that the space of Lebesgue square-integrable functions L^2 was complete, and then introduced and studied the L^p spaces in 1910. Hahn and Banach developed the abstract theory of complete normed spaces in the early 1920s.

Von Neumann immediately positioned himself in this trend. After explaining his formulation of quantum mechanics in three foundational articles [von27a, von27c, von27b] in 1927, he set about developing the corresponding mathematical theory. In the paper [von29a] from 1929, he introduced the notion of abstract Hilbert space as we still learn it today, then studied in depth the theory of operators on these spaces, focusing on those that must necessarily be defined on a strict subspace of the total space, which we call "unbounded." His goal was to discuss the differential operators that play a central role in quantum mechanics. He understood the importance of distinguishing the notion of symmetry from that of self-adjointness and, in 1930, managed to prove the spectral theorem [von30] on the diagonalization of self-adjoint operators, drawing inspiration from previous works [Rie13] by Frigyes Riesz on bounded operators. A similar result was independently obtained by the American Marshall Stone [Sto29, Sto30, Sto32a].

The spectral theory of unbounded self-adjoint operators thus resulted from an extremely conducive atmosphere, where physical questions and mathematical reasoning mutually influenced each other. A century after its invention, von Neumann's theory still plays a central role in many branches of mathematics. It is an important tool that will be useful to any applied mathematician. The Laplace operator $\Delta = \sum_{j=1}^{d} \partial_{x_j}^2$ is of course one of the main objects of this theory, which comes into play in many practical situations that it would be impossible to enumerate here. However, von Neumann's theory is far from intuitive and it is good to take the time to study its subtleties in detail. The most disturbing part is undoubtedly the very concept of self-adjoint operator, which is often a source of confusion, even among professional mathematicians. For example, the derivative operator $f \mapsto if'$ has only one possible self-adjoint realization on $\mathbb{R}$, several on the finite interval $(0, 1)$, but none on the semi-infinite interval $(0, +\infty)$. The theory also implies that the "infinite matrices" advocated by Hilbert and his colleagues are really not suitable for unbounded operators. Two self-adjoint operators can very well have the same coefficients $\langle e_n, A e_m \rangle$ in a Hilbert basis (e_n) and yet be very different, for example have completely disjoint spectra. Hilbert did not particularly like this pathology...

Book Content

This book offers an integrated presentation of spectral theory and quantum mechanics, close to the spirit in which these theories emerged. In addition to abstract results, we will present multiple concrete examples and attempt to explain the physical context. However, no prior knowledge of quantum physics is required to read these pages.

Chapter 1 is a quick mathematical introduction to quantum mechanics, which should suit both those who have no previous knowledge of this field and those who already have a good knowledge of quantum physics but wish to learn more about its rigorous aspects.

In **Chaps**. 2–5 we develop the spectral theory according to von Neumann's ideas. We explain what a self-adjoint operator is and how one can show in practice that a particular operator satisfies this property. Then we discuss its diagonalization and we conclude by developing various tools to study its spectrum. These chapters obviously constitute the heart of the book.

The last two chapters are forays into more advanced topics. We state results from more recent research works without always providing all the proofs. Two particular quantum systems are considered. **Chapter** 6 is dedicated to electrons within an atom or a molecule. This is the favorite system of quantum chemists, who seek to understand the spatial configurations of molecules as well as the chemical reactions that can occur between them. In **Chap.** 7 we examine infinite periodic systems that appear in the theory of condensed matter, for example when one wishes to optimize the conduction properties of materials used in electronic components. These are two very important topics from the point of view of applications, but we will only touch on a few theoretical aspects.

Finally, **Appendix A** is a summary of the properties of Sobolev spaces that play a central role throughout the book, and **Appendix B** contains detailed problems on questions that have been overlooked in the body of the text but nevertheless deserve the reader's attention.

Sections marked with a star* are supplements that the hurried reader can skip at first.

The text is mainly derived from a course given in the mathematics department of École Polytechnique starting from 2018, but it has greatly benefited from previous notes [Lew10, Lew17] written on various occasions. This book has first appeared in French [Lew22] in 2022, in a collection of the French Society of Applied and Industrial Mathematics (SMAI). The present English version is a revision of [Lew22], that only contains small corrections and additions compared with the French version.

Prerequisites

The text assumes that the reader has good prior knowledge of

- The elementary theory of Hilbert spaces (Hilbert bases, projections, orthogonal of a subspace, Riesz representation, etc.)
- Functional analysis (completeness, closed graph theorem, duality, weak topologies, compact operators, etc.)
- Measure theory and Lebesgue spaces L^p
- Distribution theory
- Sobolev spaces

For readers who have gaps or wish to deepen their understanding of these topics, we highly recommend the book [LL01] by Elliott H. Lieb and Michael Loss, which has become a standard textbook. The book [Bre11] by Haïm Brezis is another good reference.

Sobolev spaces are probably the least likely to have been previously learned. Appendix A contains everything that is necessary, with proofs of the most important properties. These spaces are introduced in Chap. 1 but begin to play a central role from Chap. 2, where it is shown that the usual differential operators are closed only on these spaces. We advise beginners to read Appendix A after Chap. 1 and refer to it as often as necessary.

Paris, France   Mathieu Lewin
January 2024

Contents

Chapter 1
Introduction to Quantum Mechanics: The Hydrogen Atom

In this chapter, we introduce the basic concepts of quantum mechanics and the link with the spectral theory of operators in infinite dimension. We particularly focus on the simplest physical system, the hydrogen atom, which played an important historical role in the development of quantum mechanics in the early twentieth century. We begin by recalling the classical model and its flaws, before introducing the quantum model. Similar presentations can be found in [Lie76, Lie90, LS10b].

1.1 Classical Mechanics

1.1.1 A Hamiltonian System

Consider a classical particle in $\mathbb{R}^d$ (in an arbitrary dimension $d \geq 1$), which evolves in an external potential $V : \mathbb{R}^d \to \mathbb{R}$. The function V should be thought of as describing a landscape. The places where V is high are more difficult to access because climbing the slope to reach them costs the particle some energy. We will assume here that V is sufficiently regular.

Mathematically, the particle is described by the vector (x, p) in the phase space $\mathbb{R}^d \times \mathbb{R}^d$ where x denotes its position and $p = mv$ its momentum (m is the mass and v the velocity). The dynamics of the particle is a **Hamiltonian system** based on the energy

$$\boxed{E(x, p) = \frac{|p|^2}{2m} + V(x)} \tag{1.1}$$

© The Editor(s) (if applicable) and The Author(s), under exclusive license
to Springer Nature Switzerland AG 2024
M. Lewin, *Spectral Theory and Quantum Mechanics*, Universitext,
https://doi.org/10.1007/978-3-031-66878-4_1

which is the sum of the kinetic energy and the potential energy. This means that we must solve the canonical Hamilton equations

$$
\begin{cases}
\dot{x}(t) = \nabla_p E\big(x(t), p(t)\big) = \dfrac{p(t)}{m}, \\
\dot{p}(t) = -\nabla_x E\big(x(t), p(t)\big) = -\nabla V(x(t)),
\end{cases}
\tag{1.2}
$$

which, after inserting the first into the second, provide Newton's equation

$$
m\,\ddot{x}(t) = -\nabla V(x(t)).
$$

The energy is conserved along the flow

$$
\frac{\mathrm{d}}{\mathrm{d}t} E(x(t), p(t)) = 0,
$$

so that the trajectories are included in the level lines of E in $\mathbb{R}^d \times \mathbb{R}^d$.

A **stationary point** is a solution of (1.2) independent of time,

$$
\begin{cases}
x(t) = x_0, \\
p(t) = p_0,
\end{cases}
$$

which is equivalent to $p_0 = 0$ (no velocity) and $\nabla V(x_0) = 0$. The stationary points of the system are therefore all the pairs $(x_0, 0)$ where x_0 is a critical point of the potential V. Among these critical points, the local minima of V play a particular role, as they are stable points. Indeed, if we assume that the Hessian of V is non-degenerate (positive definite) at such a point x_0, the level lines of E are then deformations of ellipsoids in the vicinity of $(x_0, 0)$ in the phase space, so that the trajectories remain close to this point at all times. Close to a maximum (or a saddle point in dimension $d \geq 2$), the level lines are on the contrary deformations of hyperboloids and the stationary point is unstable. See an example in Fig. 1.1.

The critical values of V are therefore special for the energy of the Hamiltonian system. We like to imagine that the system spends most of its time in the neighborhood of stationary points, with a strong preference for those of minimum energy. To describe this phenomenon precisely, we must make our particle interact with the outside world so that it can exchange energy, and there are many ways to model this behavior. The simplest is probably to add a small friction term in (1.2):

$$
\begin{cases}
\dot{x}(t) = \dfrac{p(t)}{m}, \\
\dot{p}(t) = -\nabla V(x(t)) - \varepsilon\, p(t),
\end{cases}
\tag{1.3}
$$

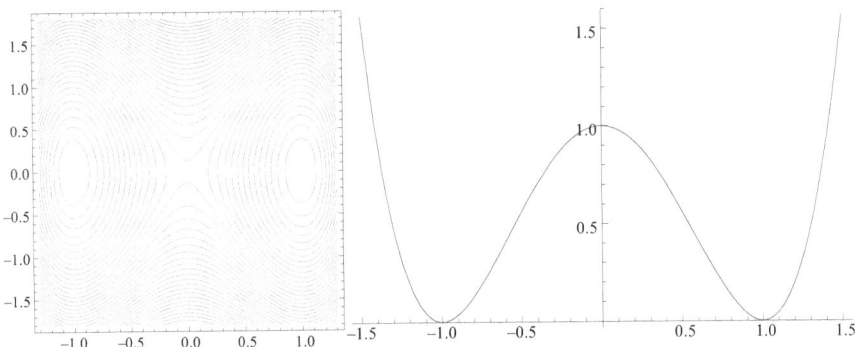

Fig. 1.1 Level lines of $E(x, p) = |p|^2/2 + V(x)$ in the phase space $\mathbb{R} \times \mathbb{R}$ (left) for the potential $V(x) = (x^2 - 1)^2$ with the shape of a Mexican hat (right). In phase space $\mathbb{R} \times \mathbb{R}$, the points $(\pm 1, 0)$ are stable while the point $(0,0)$ is unstable.

with $\varepsilon > 0$. This system is no longer Hamiltonian but it has exactly the same stationary points and now the energy is strictly decreasing along the trajectories:

$$\frac{d}{dt} E(x(t), p(t)) = -\varepsilon |p(t)|^2$$

where $|p|$ denotes the Euclidean norm of the vector $p \in \mathbb{R}^d$, for all $\varepsilon > 0$. Except of course for those that start from a stationary point.

1.1.2 Case of the Hydrogen Atom

Let us now consider the particular case of the classical hydrogen atom. The latter is composed of a proton with charge $+e$ and an electron with charge $-e$, which interact through the Coulomb potential in $\mathbb{R}^3$. The proton is much heavier than the electron (by a factor 1836) and we will assume that it is a pointwise classical particle fixed for all times at the origin of space $0 \in \mathbb{R}^3$, as in Fig. 1.2. This is the **Born-Oppenheimer approximation** [BO27] which consists in focusing, at least initially, on the fast dynamics in the system, that of the electron.

Recall that two charged particles interact with the Coulomb potential

$$\frac{q_1 q_2}{4\pi \varepsilon_0 |x_1 - x_2|} \tag{1.4}$$

where the q_i and $x_i \in \mathbb{R}^3$ are, respectively, the charge and position of the particles in question and ε_0 is the vacuum permittivity. The Coulomb potential (1.4) has two special properties. First, it is singular when $x_1 - x_2 \to 0$, which describes the fact that two particles of opposite charges attract each other enormously when they are

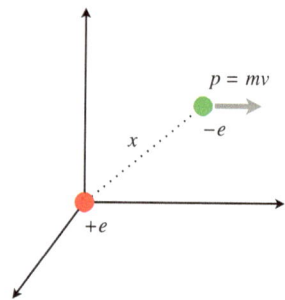

close. As we explain in this section, this divergence at the origin is the main reason
for the instability of the classical hydrogen atom. But the decay at infinity of the
potential is also a significant problem, since the function $x \mapsto 1/|x|$ tends slowly
to 0 and is not integrable. Thus, each particle interacts not only with its closest
neighbors, but also with those that may be very far away. This typically generates
difficulties in the mathematical analysis of systems comprising many particles.

But let us go back to our electron in the hydrogen atom. Its energy is given by
the previous formula (1.1) where V is now its interaction energy with the proton
located at the origin $0 \in \mathbb{R}^3$:

$$E(x, p) := \frac{|p|^2}{2m} - \frac{e^2}{4\pi\varepsilon_0|x|}. \tag{1.5}$$

We see that the associated Hamiltonian system is unstable. The energy is not
bounded below, since we can make x tend to 0 independently of p which itself
can remain fixed:

$$\inf_{\substack{x \in \mathbb{R}^3 \\ p \in \mathbb{R}^3}} E(x, p) = -\infty. \tag{1.6}$$

The property that E is not bounded below physically means that our atom is a kind of
infinite reservoir of energy, which it can exchange with the outside world. Moreover,
the corresponding Hamiltonian system admits no stationary point because

$$|\nabla V(x)| = \frac{e^2}{4\pi\varepsilon_0|x|^2}$$

never vanishes. The solutions of the Hamilton equations (1.2) are conics, like the
Moon revolving around the Earth. A small perturbation of the type (1.3) can cause
the electron to fall on the nucleus, with $E(x(t), p(t)) \to -\infty$.

This instability was a major theoretical problem in physics at the end of the
nineteenth century. But, in addition to these mathematical difficulties, the classical
model also does not reproduce the experimental results. Indeed, if we perform a

Fig. 1.3 Line spectrum observed during a spectroscopy experiment (Balmer series). ©Mathieu Lewin 2021. All rights reserved

spectroscopy experiment and observe the light emitted by an excited hydrogen gas, we find a line spectrum as in Fig. 1.3. This suggests the existence of particular (quantized) energies between which the electron navigates through excitation/de-excitation processes. Such a phenomenon could possibly be described by a potential $V(x)$ having critical points at certain special energies, but certainly not with our Coulomb potential which has no critical point at all.

The unboundedness of the energy of the classical hydrogen atom follows from the possibility of taking x to 0 independently of p which can remain fixed. If there was a link between x and p so that $|p| \to +\infty$ when $|x| \to 0$, the kinetic energy could compensate for the divergence of the potential energy and make the total energy E bounded-below. This is what is achieved by the quantum formalism.

1.2   Quantum Mechanics

1.2.1   A Probabilistic Framework

Quantum mechanics is based on two mathematical axioms, which we present in $\mathbb{R}^d$ for a particle subject to a general potential V.

(i) **Resorting to a probabilistic model.** We must describe our system by two probability measures, say μ that gives the probability that the particle is at $x \in \mathbb{R}^d$ and ν that provides the probability that it has momentum $p \in \mathbb{R}^d$. The (average) energy is then of course given by

$$\frac{1}{2m} \int_{\mathbb{R}^d} |p|^2 d\nu(p) + \int_{\mathbb{R}^d} V(x)\, d\mu(x). \tag{1.7}$$

(ii) **Enforcing of a link between the two probabilities μ and ν.** The latter must be so that $\int_{\mathbb{R}^d} |p|^2 \, d\nu(p)$ diverges when μ is too concentrated at a point, in order to stabilize the system.

Step (i) alone solves nothing at all, since we can always take $\mu = \delta_x$ and $\nu = \delta_p$, which brings us back to the previous classical model. The link (ii) between μ and ν

is by definition given by the **wavefunction**. In the absence of spin (to simplify the presentation), it is a function

$$\psi \in L^2(\mathbb{R}^d, \mathbb{C}) \qquad \text{such that} \qquad \int_{\mathbb{R}^d} |\psi(x)|^2 \, dx = 1.$$

It is then postulated that

- $\mu(x) = |\psi(x)|^2$ is the probability density that the particle is at $x \in \mathbb{R}^d$;
- $v(p) = \hbar^{-d} |\widehat{\psi}(p/\hbar)|^2$ is the probability density that it has momentum $p \in \mathbb{R}^d$.

Here $\widehat{\psi}$ is the Fourier transform of ψ which, throughout this book, is defined by[1]

$$\widehat{\psi}(p) := \frac{1}{(2\pi)^{d/2}} \int_{\mathbb{R}^d} \psi(x) e^{-ix \cdot p} \, dx \tag{1.8}$$

and $\hbar > 0$ is **Planck's constant**, which must be determined experimentally. The definition (1.8) is chosen so that the Fourier transform is an isometry of $L^2(\mathbb{R}^d, \mathbb{C})$, hence v is a probability measure:

$$\int_{\mathbb{R}^d} dv(p) = \hbar^{-d} \int_{\mathbb{R}^d} |\widehat{\psi}(p/\hbar)|^2 \, dp = \int_{\mathbb{R}^d} |\widehat{\psi}(p)|^2 \, dp$$

$$= \int_{\mathbb{R}^d} |\psi(x)|^2 \, dx = \int_{\mathbb{R}^d} d\mu(x) = 1.$$

The two classical variables (x, p) in the phase space $\mathbb{R}^d \times \mathbb{R}^d$ of dimension $2d$ have therefore been replaced by a single variable ψ in the space $L^2(\mathbb{R}^d, \mathbb{C})$ of infinite dimension.

Let us recall that a function very concentrated in space has a very spread out Fourier transform. This implies that if μ is very localized in the vicinity of a point x_0 (we know the position of the particle very well), then v is necessarily very spread out (we poorly know its velocity). This is **Heisenberg's uncertainty principle**, a founding piece of quantum mechanics, which stipulates that position and velocity cannot be known simultaneously. This principle is in practice realized by the Fourier transform, dilated by the factor $\hbar$.

In order to clarify the role of Planck's constant $\hbar$, let us fix $(x_0, p_0) \in \mathbb{R}^d \times \mathbb{R}^d$ and consider the function

$$\psi_\varepsilon(x) = \varepsilon^{-\frac{d}{2}} \varphi\left(\frac{x - x_0}{\varepsilon}\right) e^{i \frac{p_0 \cdot x}{\hbar}}$$

---

[1] Recall that the formula is valid for $\psi \in L^1(\mathbb{R}^d, \mathbb{C}) \cap L^2(\mathbb{R}^d, \mathbb{C})$ and the Fourier transform is then extended by continuity to an isometry on the whole of $L^2(\mathbb{R}^d, \mathbb{C})$.

for some given φ normalized in $L^2(\mathbb{R}^d, \mathbb{C})$, for example with compact support. The factor $\varepsilon^{-d/2}$ was chosen so that ψ_ε is also normalized in $L^2(\mathbb{R}^d, \mathbb{C})$, as it should be. The particle's position is x_0 to an accuracy of order ε and, in the limit $\varepsilon \to 0$, we have the convergence

$$\mu_\varepsilon = |\psi_\varepsilon|^2 \rightharpoonup \delta_{x_0}$$

in the sense of measures. After a calculation, we find that the Fourier transform of ψ_ε, dilated by the factor $\hbar$, reads

$$\hbar^{-\frac{d}{2}}\widehat{\psi_\varepsilon}\left(\frac{p}{\hbar}\right) = \left(\frac{\varepsilon}{\hbar}\right)^{\frac{d}{2}}\widehat{\varphi}\left(\frac{\varepsilon}{\hbar}(p - p_0)\right)e^{-i\frac{x_0 \cdot (p - p_0)}{\hbar}}.$$

The corresponding probability $\nu_\varepsilon(p) = \hbar^{-d}|\widehat{\psi_\varepsilon}(p/\hbar)|^2$ is very spread out when $\varepsilon \to 0$, by a factor $\hbar/\varepsilon \to +\infty$. We therefore lose all information about the momentum when trying to increase our knowledge about the position by taking $\varepsilon \to 0$. The best we can do is to choose $\varepsilon = \sqrt{\hbar}$, which concentrates μ and ν at the same scale $\sqrt{\hbar}$ in the vicinity of x_0 and p_0, respectively. As $\hbar$ is a fixed physical constant, this is the maximum resolution allowed by the theory, when we want to know the position and momentum simultaneously. In the (non-physical) limit $\hbar \to 0$, the link between position and momentum disappears and the measures μ and ν can both converge to a delta.

In quantum mechanics, particles are therefore probabilistic objects. They never have a well-defined position and velocity because of the impossibility of concentrating μ and ν simultaneously. We can possibly imagine them as diffuse objects that are a bit everywhere, like a wave. However, we should not consider ψ or the probabilities μ and ν as "real" waves that we could measure in several places at once.[2] Indeed, the theory specifies that this "wave-like" behavior is only valid if we let the system evolve freely. Particles must necessarily materialize at a point with a certain momentum when we observe them, x and p being drawn at random according to the probabilities μ and ν. The quantum world is therefore quite different from our usual world. Randomness plays a central role, but it seems to reveal itself only under the action of an observer. This strange behavior has generated a lot of discussion but it has been confirmed by all the experiments carried out in the laboratory [Lal19]. For example, if we offer an electron the possibility of passing through two slits, it passes without any problem through both at the same time, which generates interference patterns typical of waves, on a screen placed on the other side [TEM+89]. On the other hand, if we try to find out through which slit it passed, it completely changes its behavior and passes through only one of the two, choosing it at random [FGP10].

---

[2] For N particles we will see later in Sect. 1.5.4 and in Chap. 6 that the system is described by a wavefunction $\Psi(x_1, \ldots, x_N)$ defined on the space $(\mathbb{R}^d)^N$, which is not the physical space in which the particles evolve.

Let us now turn to the mathematical properties of the average energy (1.7) of the system. It is convenient to express this energy using our new variable ψ, which leads to the expression

$$\mathcal{E}(\psi) = \frac{\hbar^2}{2m} \int_{\mathbb{R}^d} |p|^2 |\widehat{\psi}(p)|^2 \mathrm{d}p + \int_{\mathbb{R}^d} V(x) |\psi(x)|^2 \, \mathrm{d}x, \qquad (1.9)$$

after a change of variable in p. Using that $p\widehat{\psi}(p) = -i\widehat{\nabla\psi}(p)$ and Plancherel's theorem $\|\psi\|_{L^2} = \|\widehat{\psi}\|_{L^2}$, we obtain an expression involving only $\psi(x)$:

$$\mathcal{E}(\psi) = \frac{\hbar^2}{2m} \int_{\mathbb{R}^d} |\nabla\psi(x)|^2 \, \mathrm{d}x + \int_{\mathbb{R}^d} V(x) \, |\psi(x)|^2 \, \mathrm{d}x. \qquad (1.10)$$

Multiplying $\mathcal{E}$ by the constant $2m/\hbar^2$ and changing the definition of V, we can remove the constant $\hbar^2/(2m)$ and thus arrive at the simpler form

$$\boxed{\mathcal{E}(\psi) = \int_{\mathbb{R}^d} |\nabla\psi(x)|^2 \, \mathrm{d}x + \int_{\mathbb{R}^d} V(x) \, |\psi(x)|^2 \, \mathrm{d}x.} \qquad (1.11)$$

Now that we have determined the energy of our quantum particle as a function of ψ, several natural questions arise. The first is that of stability. Under what assumptions on the external potential V is the energy $\mathcal{E}$ bounded from below, when we add the constraint $\|\psi\|_{L^2(\mathbb{R}^d, \mathbb{C})} = 1$? If V is itself bounded-below, $V(x) \geq -C$, then we of course have

$$\mathcal{E}(\psi) \geq -C \int_{\mathbb{R}^d} |\psi(x)|^2 \, \mathrm{d}x = -C.$$

But can we allow potentials V that diverge to $-\infty$ at certain places, like the Coulomb potential? We will study this question at length in the following sections.

Another important question is to determine the critical points of the function $\mathcal{E}$ as they must correspond to the stationary states of the system. This question will naturally lead us to spectral theory, as we will see in the next section. Finally, we will explain that the time evolution is still a Hamiltonian system, as in the classical case but in infinite dimension.

Before moving on to the rigorous study of stability, we continue to describe the quantum formalism in an informal way.

1.2.2 Towards Spectral Theory

Assuming that ψ is sufficiently regular, we can integrate by parts the integral involving the gradient, which allows us to express $\mathcal{E}$ in the form

$$\mathcal{E}(\psi) = \left\langle \psi, \left(-\Delta + V(x) \right)\psi \right\rangle \tag{1.12}$$

where $\langle \cdot, \cdot \rangle$ is the usual scalar product of $L^2(\mathbb{R}^d, \mathbb{C})$ (linear on the right),

$$\langle f, g \rangle := \int_{\mathbb{R}^d} \overline{f(x)}g(x)\,dx,$$

and where $\Delta = \sum_{i=1}^{d} \partial_{x_i}^2$ is the Laplacian. Thus, we see that $\mathcal{E}$ is the quadratic form associated with the linear operator

$$H = -\Delta + V(x).$$

Here we simply denote by $V(x)$ the operator $\psi \mapsto V\psi$ of multiplication by the function V. Note that we can find the operator H directly from the classical energy $E(x, p) = |p|^2/2m + V(x)$ by replacing p with $-i\hbar\nabla$, a process called (first) quantization and which is here realized by the Fourier transform. We will come back to this a little later in Sect. 1.5.5. We call H the **Hamiltonian operator**.

We now wish to determine the stationary points of the system, which are the points where the derivative of the energy $\mathcal{E}$ vanishes. Due to Formula (1.12), the derivative is (at least formally) given by $2H\psi$. However, let us recall that ψ must always belong to the unit sphere of $L^2(\mathbb{R}^d, \mathbb{C})$, which means that $\mathcal{E}$ is actually a function defined on a manifold. We must determine its critical points on this manifold and not in the entire space. This amounts to asking that the projection of the gradient onto the tangent space be zero. In the case of a sphere, this simply means that $H\psi$ is parallel to ψ (Fig. 1.4):

$$\boxed{H\psi = \lambda\psi.} \tag{1.13}$$

This is **Schrödinger's equation**, determining the equilibrium points of the system. The critical points ψ of $\mathcal{E}$ on the unit sphere are the eigenfunctions of H and the critical values $\lambda = \mathcal{E}(\psi)$ are the eigenvalues of H.

We have the same interpretation in finite dimension. Consider an $n \times n$ Hermitian matrix A and the corresponding quadratic form $x \in \mathbb{C}^n \mapsto \mathcal{E}(x) = \langle x, Ax \rangle = x^*Ax$, restricted to the unit sphere of $\mathbb{C}^n$. Minimizing $\mathcal{E}$ provides the smallest eigenvalue of A, the corresponding minimizers being exactly the associated eigenvectors. For the largest eigenvalue we need to maximize $\mathcal{E}$ and for the intermediate eigenvalues we need to use the Courant-Fischer min-max formulas.

Fig. 1.4 A critical point x of a functional $\mathcal{E}$ on the unit sphere is such that the gradient $\nabla\mathcal{E}(x)$ is orthogonal to the tangent plane, that is, parallel to x. ©Mathieu Lewin 2021. All rights reserved

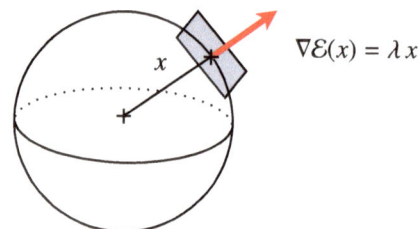

$$\nabla\mathcal{E}(x) = \lambda x$$

In our case, we see after two integrations by parts that the operator H satisfies the same property as Hermitian matrices in finite dimension:

$$\langle \psi_1, H\psi_2 \rangle = \langle H\psi_1, \psi_2 \rangle$$

for all sufficiently regular functions ψ_1, ψ_2. We therefore expect its spectrum to be real, and also that H can be diagonalized in an orthonormal basis, so that the energies of the stationary states are real, as they must be from a physical point of view. Since H is an operator in infinite dimension, the corresponding theory is in fact much more subtle than it appears at first glance. The aim of this book is precisely to explain the spectral theory of operators of the same type as $H = -\Delta + V(x)$. We will see that the **spectrum** of H is not always only composed of eigenvalues. Due to some problems of compactness in infinite dimension, it is conceivable that there exists a sequence (ψ_n) of normalized functions in $L^2(\mathbb{R}^d, \mathbb{C})$ such that $(H - \lambda)\psi_n \to 0$ strongly for a certain λ, without the equation $(H - \lambda)\psi = 0$ admitting a non-trivial solution. Such "quasi eigenvalues" λ's will form what we will call the **continuous spectrum**. These are sorts of "quasi critical values" of the energy $\mathcal{E}$. The presence of such a spectrum immediately suggests that the diagonalization of H will not be an easy task.

We will show in Chap. 5 that the spectrum of H typically has the form given in Fig. 1.5, when V tends to 0 at infinity. The lowest eigenvalue is called the **ground state energy** because it corresponds to minimizing the energy $\mathcal{E}(\psi)$ under the constraint $\|\psi\|_{L^2} = 1$. Indeed, as we have recalled above the first eigenvalue of a Hermitian matrix A satisfies

$$\lambda_1(A) = \min_{\|x\|=1} \langle x, Ax \rangle$$

and the same property will hold in infinite dimension. A corresponding eigenfunction ψ is called a **ground state**. The higher eigenvalues are called **excited energies** and they correspond to unstable stationary states of the system, between which the particle can navigate when it interacts with the outside world. In doing so, it emits light with a wavelength corresponding to the differences between the eigenvalues. This explains the spectral lines seen in spectroscopy experiments. Often, there are no positive eigenvalues at all, the spectrum being made up only of continuous spectrum

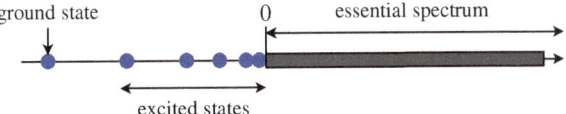

ground state   0   essential spectrum

excited states

Fig. 1.5 Typical shape of the spectrum of the operator $H = -\Delta + V(x)$ describing a quantum particle subjected to an external potential V tending to 0 at infinity.

in $\mathbb{R}^+$. The "quantized" feature of quantum mechanics is therefore mainly seen in the negative part of the spectrum, composed of eigenvalues forming a discrete set.

1.2.3   A Hamiltonian System

At this stage we have defined the energy of a quantum particle in $\mathbb{R}^d$ as well as its stationary points on the unit sphere of $L^2(\mathbb{R}^d, \mathbb{C})$. To complete the picture, we need to introduce the dynamics of the system, which again takes the form of a Hamiltonian system, this time in infinite dimension.

For a Hamiltonian structure we need two variables and a symplectic form. It turns out that we do have two variables because the wavefunction ψ is complex-valued. Writing $\psi = \psi_1 + i\psi_2$ we simply find

$$\mathcal{E}(\psi) = \mathcal{E}(\psi_1) + \mathcal{E}(\psi_2)$$

because V is a real function. We therefore obtain a Hamiltonian system by crossing the derivatives of $\mathcal{E}$ with respect to ψ_1, ψ_2 as follows:

$$\begin{cases} \dfrac{\partial \psi_1}{\partial t} = \dfrac{\partial \mathcal{E}}{\partial \psi_2}(\psi) = 2H\psi_2 \\ \dfrac{\partial \psi_2}{\partial t} = -\dfrac{\partial \mathcal{E}}{\partial \psi_1}(\psi) = -2H\psi_1. \end{cases}$$

This can be rewritten in a single equation as

$$i\frac{\partial \psi}{\partial t} = 2H\psi.$$

In this formalism, the symplectic form is therefore the multiplication by i. We can remove the factor 2 by changing the unit of time, which provides the **time-dependent Schrödinger equation**

$$\boxed{i\frac{\partial \psi}{\partial t} = H\psi.}\tag{1.14}$$

A formal calculation shows that the L^2 norm and the energy of any solution are conserved over time:

$$\frac{\mathrm{d}}{\mathrm{d}t}\mathcal{E}(\psi(t)) = \frac{\mathrm{d}}{\mathrm{d}t}\int_{\mathbb{R}^d} |\psi(t, x)|^2 \, \mathrm{d}x = 0.$$

The conservation of the L^2 norm is obviously crucial for our probabilistic interpretation of $|\psi|^2$ and $|\widehat{\psi}|^2$ to persist at all times along the trajectories. We will study Schrödinger's equation (1.14) rigorously in Sect. 4.7.1 of Chap. 4.

One might wonder if there is a link between the classical Hamilton equations (1.2) and Schrödinger's equation (1.14). Because of the probabilistic interpretation of $|\psi|^2$ and $|\widehat{\psi}|^2$, the average position of the particle is given by

$$\langle X \rangle_t := \int_{\mathbb{R}^d} x \, |\psi(t, x)|^2 \, \mathrm{d}x$$

and its average momentum is

$$\langle P \rangle_t := \int_{\mathbb{R}^d} p \, |\widehat{\psi}(t, p)|^2 \, \mathrm{d}p = -i \int_{\mathbb{R}^d} \overline{\psi(t, x)} \, \nabla_x \psi(t, x) \, \mathrm{d}x.$$

Here we have used the abbreviated notation $\langle A \rangle_t = \langle \psi(t), A\psi(t) \rangle$ for the average value of an operator A in the state $\psi(t, x)$, as well as $P := -i\nabla_x$ and $(X\psi)(x) := x\psi(x)$. A (formal) calculation then shows that

$$\begin{cases} \dfrac{\mathrm{d}}{\mathrm{d}t}\langle X \rangle_t = \langle P \rangle_t, \\ \dfrac{\mathrm{d}}{\mathrm{d}t}\langle P \rangle_t = -\langle \nabla V(X) \rangle_t. \end{cases} \tag{1.15}$$

These equations are usually called the **Ehrenfest's relations** [Ehr27]. The Hamilton equations remain true in some way, *on average*, but the reader should remember that, in general,

$$\left\langle \nabla V(X) \right\rangle_t \neq \nabla V\big(\langle X \rangle_t\big)$$

otherwise we would find exactly the classical equations (1.2). We ask the reader to verify (1.15), assuming the existence of a smooth solution $\psi(t)$ that decays rapidly enough for everything to make sense.

We conclude this section with a remark about the link between the eigenfunctions of the operator H and the time-dependent equation (1.14). The term "stationary state" may be a source of confusion, because if we assume that ψ is time-independent, we find the equation $H\psi = 0$ and not $H\psi = \lambda\psi$. The reason is that our model is invariant under "phase changes", that is, multiplication by a complex of modulus 1 in the form $\psi \mapsto e^{i\theta}\psi$. All physical quantities calculated from ψ are invariant under this action of the group $\mathbb{S}^1$, as we will explain in more detail in

Sect. 1.5. We should therefore rather work *modulo* phases, that is, in the quotient of the unit sphere of $L^2(\mathbb{R}^d, \mathbb{C})$ by the equivalence relation

$$\psi_1 \sim \psi_2 \quad \Longleftrightarrow \quad \exists \theta \in \mathbb{R} : \psi_2 = e^{i\theta} \psi_1.$$

Thus, when we say that ψ is time-independent, we must always think "modulo a phase". This amounts to considering functions of the form $e^{i\theta(t)}\psi(x)$ where ψ does not depend on time and θ does not depend on x. Inserting into Eq. (1.14) we find $H\psi = -\theta'(t)\psi$ for all t. Since ψ cannot be an eigenfunction associated with several eigenvalues at once, we deduce that $\theta'(t) = -\lambda$ is constant. Thus the stationary solutions are precisely those of the form

$$\psi(t) = e^{-i\lambda t}\psi_0, \qquad \text{with} \quad H\psi_0 = \lambda\psi_0. \tag{1.16}$$

These are indeed the critical points of $\mathcal{E}$ on the unit sphere, discussed in Sect. 1.2.2. The phase depends linearly on time, but this plays no role in our modeling, all important quantities calculated from ψ will indeed be time-independent.

We have completed our informal description of the quantum formalism, which takes the form of a Hamiltonian system in infinite dimension. We have not explained the winding historical path that led physicists to the postulates introduced in this section. There are in fact several equivalent presentations of the same theory. Instead of focusing on the wavefunction, Heisenberg [Hei25] started from the axiom that, in dimension $d = 1$, the position and momentum are described by two self-adjoint operators X and P satisfying the commutation relation.

$$[X, P] = i\hbar.$$

Such operators do not exist in finite dimension (to see this, take the trace) and this is what led him to "infinite matrices". Stone and von Neumann later proved that, up to isomorphism, the only possibility is to take X equal to the operator of multiplication by x and $P = -i\hbar d/dx$ as in our presentation (see [RS72, Thm. VIII.14] and [RS79, Thm. XI.84]). Schrödinger [Sch26] introduced Eq. (1.14) based on the wavefunction ψ but he did not identified the probabilistic interpretation of $|\psi|^2$ and $|\widehat{\psi}|^2$, which is due to Born [Bor26].

In the following two sections, we study the stability of our quantum model and the existence of a ground state, starting with the hydrogen atom before turning to the general case.

1.3   The Quantum Hydrogen Atom

After this description of quantum mechanics, let us return to the case of the quantum electron in the hydrogen atom, whose energy is written with the physical constants

$$\mathcal{E}(\psi) = \frac{\hbar^2}{2m} \int_{\mathbb{R}^3} |\nabla \psi(x)|^2 \, dx - \frac{e^2}{4\pi \varepsilon_0} \int_{\mathbb{R}^3} \frac{|\psi(x)|^2}{|x|} \, dx. \qquad (1.17)$$

To study the stability of this system it is more convenient to work in **atomic units**. We make the change of variable

$$x = \frac{4\pi \varepsilon_0 \hbar^2}{me^2} x', \qquad \mathcal{E} = \frac{m}{\hbar^2} \left(\frac{e^2}{4\pi \varepsilon_0} \right)^2 \mathcal{E}'$$

(which also changes the function ψ) and obtain the following energy, still denoted $\mathcal{E}$ for simplicity:

$$\mathcal{E}(\psi) = \frac{1}{2} \int_{\mathbb{R}^3} |\nabla \psi(x)|^2 \, dx - \int_{\mathbb{R}^3} \frac{|\psi(x)|^2}{|x|} \, dx. \qquad (1.18)$$

We therefore have, in a certain way, $e^2/4\pi \varepsilon_0 = \hbar^2/m = 1$. It is possible to replace the factor $1/2$ in front of the first integral by 1, but we prefer to keep it in this section for better consistency with the physical literature.

We could remove all the physical parameters because the two terms of energy have a different homogeneity. The term $\int_{\mathbb{R}^3} |\nabla \psi|^2$ is the inverse of the square of a distance[3] while the potential energy is the inverse of a distance. If we replace $1/|x|$ by $1/|x|^2$ (or the Laplacian by a differential operator of order one) we would not be able to eliminate all the physical constants.

To have a finite kinetic energy, it seems natural to work in the Sobolev space

$$H^1(\mathbb{R}^3, \mathbb{C}) := \left\{ \psi \in L^2(\mathbb{R}^3, \mathbb{C}) \ : \ \nabla \psi \in L^2(\mathbb{R}^3, \mathbb{C})^3 \right\}$$

where we recall that $\nabla \psi$ is here understood in the sense of distributions. We refer to Appendix A, which contains a reminder of the most important properties of Sobolev spaces. Throughout the rest of the book, we will frequently use the notation $L^2(\mathbb{R}^d)$, $H^1(\mathbb{R}^d)$, etc., without specifying that the functions are complex-valued, which we will always assume. When the context is clear, we will even sometimes write L^2, H^1, etc. We will write $L^2(\mathbb{R}^d, \mathbb{R})$, $H^1(\mathbb{R}^d, \mathbb{R})$, etc., if the functions are assumed to be real-valued.

---

[3] As $\int_{\mathbb{R}^3} |\psi(x)|^2 \, dx = 1$, $|\psi(x)|^2$ behaves like the inverse of the cube of a distance to compensate for the dx.

1.3.1 Stability

We will now prove that the hydrogen atom is stable in quantum mechanics, using classical functional inequalities.

Proposition 1.1 (Stability of the Quantum Hydrogen Atom) *The functional $\mathcal{E}$ in* (1.18) *is well defined and continuous on* $H^1(\mathbb{R}^3)$. *There exists a universal constant* $C < 0$ *such that*

$$\mathcal{E}(\psi) \geq C$$

for all $\psi \in H^1(\mathbb{R}^3)$ *such that* $\int_{\mathbb{R}^3} |\psi(x)|^2 \, \mathrm{d}x = 1$.

There are many possible proofs of Proposition 1.1. Later in Theorem 1.3 we will give a simple one that even allows us to precisely identify the value of the infimum and the corresponding minimizer, but which only works for the potential $1/|x|$. Here we provide a more general proof that adapts to multiple situations, for example when V is not exactly equal to $1/|x|$. We will use the following **Sobolev inequality**, recalled and proved in Theorem A.13 of Appendix A.

Theorem 1.2 (Sobolev Inequality in Dimension $d = 3$) *Let* $\psi \in L^1_{\mathrm{loc}}(\mathbb{R}^3)$ *be such that* $\nabla\psi \in L^2(\mathbb{R}^3)$ *and the set* $\{x \in \mathbb{R}^3 \,:\, |\psi(x)| \geq M\}$ *has finite measure for all* $M > 0$. *Then* $\psi \in L^6(\mathbb{R}^3)$ *and we have*

$$\left(\int_{\mathbb{R}^3} |\psi(x)|^6 \, \mathrm{d}x \right)^{\frac{1}{3}} \leq S_3 \int_{\mathbb{R}^3} |\nabla\psi(x)|^2 \, \mathrm{d}x \tag{1.19}$$

where $S_3 = (4/3)2^{-\frac{2}{3}}\pi^{-\frac{4}{3}}$ *is the best possible constant in this inequality.*

We have stated the Sobolev inequality with very weak assumptions on ψ because it is not very natural to assume that $\psi \in L^2(\mathbb{R}^3)$, since the $L^2(\mathbb{R}^3)$ norm does not appear in (1.19). For $\psi \in L^2(\mathbb{R}^3)$ as is our case, the sets $\{|\psi| \geq M\}$ are of finite measure since

$$\left| \{|\psi| \geq M\} \right| = \int_{\{|\psi| \geq M\}} \mathrm{d}x \leq \frac{1}{M^2} \int_{\mathbb{R}^3} |\psi(x)|^2 \, \mathrm{d}x < \infty.$$

The $L^6(\mathbb{R}^3)$ norm is very natural in (1.19), as it has the same homogeneity as the kinetic energy (when it is raised to the power $1/3$). To see this, we can insert a rescaled function $\psi(x/\varepsilon)$ and verify that the two terms behave the same, proportional to ε. Equivalently, we can just count the units of length (the "$\mathrm{d}x$") in each side of the inequality. As $\mathrm{d}x \sim L^3$ and $\nabla \sim L^{-1}$, the term on the right behaves in $|\psi|^2 L$ and the one on the left in $(|\psi|^6 L^3)^{1/3} = |\psi|^2 L$. In fact, an inequality like (1.19) could not be true for another power $p \neq 6$ as we would arrive at a contradiction by dilating the function ψ.

The Sobolev inequality (1.19) allows us to quantify how the kinetic energy blows up when the wavefunction concentrates in space. It is therefore a **quantitative form of Heisenberg's uncertainty principle**. Indeed, if $\psi \in H^1(\mathbb{R}^d)$ has its support in a set $\Omega \subset \mathbb{R}^3$ and is normalized in $L^2(\mathbb{R}^3)$, we have by Hölder's inequality

$$1 = \int_\Omega |\psi(x)|^2 \, dx \leq |\Omega|^{\frac{2}{3}} \left(\int_{\mathbb{R}^3} |\psi(x)|^6 \, dx \right)^{\frac{1}{3}}.$$

We thus obtain an inequality due to Poincaré:

$$\int_{\mathbb{R}^3} |\nabla \psi(x)|^2 \, dx \geq \frac{1}{S_3 |\Omega|^{\frac{2}{3}}}. \tag{1.20}$$

If Ω is small, the kinetic energy must therefore be large.

The physical literature often highlights another quantitative version of the uncertainty principle, called **Heisenberg's inequality**, which states that

$$\left(\int_{\mathbb{R}^3} |\nabla \psi(x)|^2 \, dx \right)^{\frac{1}{2}} \left(\int_{\mathbb{R}^3} |x|^2 |\psi(x)|^2 \, dx \right)^{\frac{1}{2}} \geq \frac{3}{2} \int_{\mathbb{R}^3} |\psi(x)|^2 \, dx \tag{1.21}$$

(or a similar version with a variance). Unfortunately, there does not seem to exist a mathematical proof of the stability of the hydrogen atom solely based on this inequality.

Let us now return to the proof of the stability of the hydrogen atom based on the Sobolev inequality.

***Proof (of Proposition 1.1 with the Sobolev Inequality** (1.19))* The idea is to treat the singularity at the origin separately, by splitting the integral in the following way, for $r > 0$:

$$\int_{\mathbb{R}^3} \frac{|\psi(x)|^2}{|x|} \, dx = \int_{|x| \leq r} \frac{|\psi(x)|^2}{|x|} \, dx + \int_{|x| > r} \frac{|\psi(x)|^2}{|x|} \, dx.$$

Outside the ball of radius r, we simply estimate $1/|x|$ by $1/r$ as follows:

$$\int_{|x| > r} \frac{|\psi(x)|^2}{|x|} \, dx \leq r^{-1} \int_{|x| > r} |\psi(x)|^2 \, dx \leq r^{-1} \int_{\mathbb{R}^3} |\psi(x)|^2 \, dx.$$

We then bound the potential energy inside the ball using Hölder's inequality (we take $1/|x|$ to the power $3/2$ and $|\psi|^2$ to the power 3 to make the L^6 norm appear):

$$\int_{|x| \leq r} \frac{|\psi(x)|^2}{|x|} \, dx \leq \left(\int_{|x| \leq r} \frac{dx}{|x|^{\frac{3}{2}}} \right)^{\frac{2}{3}} \left(\int_{|x| \leq r} |\psi(x)|^6 \, dx \right)^{\frac{1}{3}}$$

$$\leq \left(4\pi \int_0^r \sqrt{s}\, ds\right)^{\frac{2}{3}} \left(\int_{\mathbb{R}^3} |\psi(x)|^6\, dx\right)^{\frac{1}{3}}$$

$$\leq \left(\frac{8\pi}{3}\right)^{\frac{2}{3}} S_3\, r \int_{\mathbb{R}^3} |\nabla\psi(x)|^2\, dx.$$

In the last line, we used the Sobolev inequality (1.19). In conclusion, we have shown that

$$\int_{\mathbb{R}^3} \frac{|\psi(x)|^2}{|x|}\, dx \leq \kappa r \int_{\mathbb{R}^3} |\nabla\psi(x)|^2\, dx + r^{-1} \int_{\mathbb{R}^3} |\psi(x)|^2\, dx \qquad (1.22)$$

with $\kappa := (8\pi/3)^{2/3} S_3 \simeq 0.753$. This proves in particular that $|x|^{-1/2}\psi$ belongs to $L^2(\mathbb{R}^3)$ when ψ is in $H^1(\mathbb{R}^3)$. Now we take $r = 1/(2\kappa)$ and obtain

$$\int_{\mathbb{R}^3} \frac{|\psi(x)|^2}{|x|}\, dx \leq \frac{1}{2} \int_{\mathbb{R}^3} |\nabla\psi(x)|^2\, dx + 2\kappa \int_{\mathbb{R}^3} |\psi(x)|^2\, dx. \qquad (1.23)$$

The estimate (1.23) shows that the linear map $\psi \in H^1(\mathbb{R}^3) \mapsto \psi|x|^{-1/2} \in L^2(\mathbb{R}^3)$ is bounded, hence continuous. Thus, the energy $\mathcal{E}$ is continuous on $H^1(\mathbb{R}^3)$ and

$$\mathcal{E}(\psi) \geq -2\kappa \int_{\mathbb{R}^3} |\psi(x)|^2\, dx \simeq -1.506 \quad \text{for} \int_{\mathbb{R}^3} |\psi(x)|^2\, dx = 1,$$

which concludes the proof of Proposition 1.1. $\square$

1.3.2   Ground State

We have seen that the energy of the quantum electron in the hydrogen atom was bounded-below (for wavefunctions ψ normalized in L^2). The next step is to study the existence of a minimizer, called the **ground state**. The following theorem provides the exact value of the minimum of $\mathcal{E}$ as well as the uniqueness of the associated minimizer. We also obtain the information that the corresponding function is an eigenfunction of the operator $H = -\Delta/2 - 1/|x|$, as was announced in Sect. 1.2.2.

Theorem 1.3 (Ground State of the Hydrogen Atom) *The infimum of $\mathcal{E}$ on the sphere of $L^2(\mathbb{R}^3)$ is a minimum, which equals*

$$\min_{\substack{\psi \in H^1(\mathbb{R}^3) \\ \int_{\mathbb{R}^3} |\psi|^2 = 1}} \mathcal{E}(\psi) = -\frac{1}{2}. \qquad (1.24)$$

The corresponding minimizers are all of the form

$$\psi(x) = \frac{e^{i\theta}}{\sqrt{\pi}} \, e^{-|x|}, \qquad \theta \in \mathbb{R}. \tag{1.25}$$

So we have uniqueness up to a phase. These minimizers belong to $H^2(\mathbb{R}^3)$ and solve Schrödinger's equation

$$\left(-\frac{\Delta}{2} - \frac{1}{|x|}\right)\psi = -\frac{1}{2}\psi \tag{1.26}$$

where each term is in $L^2(\mathbb{R}^3)$.

Recall that $\mu(x) = |\psi(x)|^2 = \pi^{-1}e^{-2|x|}$ is the probability distribution for the position x of the electron. We deduce that the electron spends most of its time close to the nucleus, attracted by its opposite charge. Its average position is even centered on the nucleus since ψ is radial: $\int_{\mathbb{R}^3} x|\psi(x)|^2 \, dx = 0$. Nevertheless, as ψ is strictly positive on $\mathbb{R}^3$, the electron has a non-zero probability of escaping far from the nucleus. It is exponentially small in the distance, though. Note that ψ is C^∞ (even real-analytic) everywhere except at the origin, where it is Lipschitz but not C^1. This is a reminiscence of the divergence of the potential.

The following proof is based on the same method that is usually employed for Hardy's inequality (studied in Exercise 1.27 below).

Proof For $\psi \in C_c^\infty(\mathbb{R}^3)$ we compute

$$\int_{\mathbb{R}^3} \left|\nabla\psi(x) + \frac{x}{|x|}\psi(x)\right|^2 \, dx$$

$$= \int_{\mathbb{R}^3} |\nabla\psi(x)|^2 \, dx + 2\Re \int_{\mathbb{R}^3} \overline{\psi(x)}\,\frac{x}{|x|} \cdot \nabla\psi(x) \, dx + \int_{\mathbb{R}^3} |\psi(x)|^2 \, dx.$$

Using that $\nabla|\psi|^2 = 2\Re\overline{\psi}\nabla\psi$ and integrating by parts, we find

$$2\Re \int_{\mathbb{R}^3} \overline{\psi(x)}\,\frac{x}{|x|} \cdot \nabla\psi(x) \, dx = \int_{\mathbb{R}^3} \frac{x}{|x|} \cdot \nabla|\psi|^2(x) \, dx$$

$$= -\sum_{k=1}^{3} \int_{\mathbb{R}^3} |\psi(x)|^2 \partial_{x_k}\left(\frac{x_k}{|x|}\right) \, dx$$

$$= -2\int_{\mathbb{R}^3} \frac{|\psi(x)|^2}{|x|} \, dx.$$

Thus, we have shown that

$$0 \leq \frac{1}{2} \int_{\mathbb{R}^3} \left| \nabla \psi(x) + \frac{x}{|x|} \psi(x) \right|^2 \mathrm{d}x$$

$$= \frac{1}{2} \int_{\mathbb{R}^3} |\nabla \psi(x)|^2 \, \mathrm{d}x - \int_{\mathbb{R}^3} \frac{|\psi(x)|^2}{|x|} \, \mathrm{d}x + \frac{1}{2} \int_{\mathbb{R}^3} |\psi(x)|^2 \, \mathrm{d}x. \tag{1.27}$$

Since $C_c^\infty(\mathbb{R}^3)$ is dense in $H^1(\mathbb{R}^3)$ (Theorem A.8), the relation (1.27) remains valid in $H^1(\mathbb{R}^3)$. We can rewrite (1.27) as follows:

$$\mathcal{E}(\psi) = \frac{1}{2} \int_{\mathbb{R}^3} \left| \nabla \psi(x) + \frac{x}{|x|} \psi(x) \right|^2 \mathrm{d}x - \frac{1}{2} \int_{\mathbb{R}^3} |\psi(x)|^2 \, \mathrm{d}x. \tag{1.28}$$

This proves that the infimum of $\mathcal{E}$ (with the normalization constraint on ψ) is at least equal to $-1/2$. We can make $\mathcal{E}(\psi)$ equal to $-1/2$ by canceling the first term, that is, by determining all the functions $\psi \in H^1(\mathbb{R}^3)$ such that

$$\nabla \psi(x) = -\frac{x}{|x|} \psi(x) \tag{1.29}$$

for almost every $x \in \mathbb{R}^3$. The function $\psi(x) = \pi^{-1/2} e^{-|x|}$ satisfies this property and belongs to $H^2(\mathbb{R}^3)$ (therefore to $H^1(\mathbb{R}^3)$). Moreover, the constant $\pi^{-1/2}$ was precisely chosen so that it is normalized in $L^2(\mathbb{R}^3)$. We therefore conclude that the infimum is a minimum and that it is exactly $-1/2$. A calculation shows that $\psi = \pi^{-1/2} e^{-|x|}$ is a solution of Schrödinger's Equation (1.26).

For uniqueness, we use that if $\psi \in H^1(\mathbb{R}^3)$ satisfies (1.29), then the function $\eta(x) := e^{|x|} \psi(x)$ belongs to $H^1_{\mathrm{loc}}(\mathbb{R}^3)$ (it is not necessarily square integrable at infinity) and satisfies $\nabla \eta(x) = 0$ almost everywhere, therefore in the sense of distributions. The only solutions to this equation on $\mathbb{R}^3$ are the constant solutions, $\eta(x) = c$, which indeed shows that the solutions of (1.29) in $H^1(\mathbb{R}^3)$ are all in the form $ce^{-|x|}$. As we only consider normalized functions in $L^2(\mathbb{R}^3)$, we must have $|c| = \pi^{-1/2}$, that is $c = \pi^{-1/2} e^{i\theta}$. $\qquad \square$

We can rephrase our information about the value of the ground state energy of the hydrogen atom into an inequality, following the ideas of Kato [Kat51].

Corollary 1.4 (Kato's Inequality)   *We have the inequality*

$$\int_{\mathbb{R}^3} \frac{|\psi(x)|^2}{|x - R|} \, \mathrm{d}x \leq \eta \int_{\mathbb{R}^3} |\nabla \psi(x)|^2 \, \mathrm{d}x + \frac{1}{4\eta} \int_{\mathbb{R}^3} |\psi(x)|^2 \, \mathrm{d}x, \tag{1.30}$$

for all $\psi \in H^1(\mathbb{R}^3)$, all $R \in \mathbb{R}^3$ and all $\eta > 0$.

Proof For $R = 0$ and $\eta = 1/2$, this is exactly our inequality on the ground state energy of the hydrogen atom

$$\frac{1}{2}\int_{\mathbb{R}^3}|\nabla\psi(x)|^2\,dx - \int_{\mathbb{R}^3}\frac{|\psi(x)|^2}{|x|}\,dx \geq -\frac{1}{2}\int_{\mathbb{R}^3}|\psi(x)|^2\,dx.$$

We can obtain (1.30) for all $\eta > 0$ and all $R \in \mathbb{R}^3$ by applying this inequality to the dilated and translated function $(2\eta)^{3/2}\psi(2\eta x + R)$. Performing the appropriate changes of variable concludes the proof.   □

1.3.3 Spectrum

As discussed in Sect. 1.2.2, the eigenfunctions of the operator $H = -\Delta/2 - 1/|x|$ play a central role in our theory as they correspond to the stationary states of the system. The result goes as follows.

Theorem 1.5 (Spectrum of the Hydrogen Atom)   *The non-zero solutions $\psi \in H^1(\mathbb{R}^3)$ of the equation*

$$\left(-\frac{\Delta}{2} - \frac{1}{|x|}\right)\psi = \lambda\,\psi \tag{1.31}$$

in the sense of distributions on $\mathbb{R}^3$ are all in $H^2(\mathbb{R}^3)$ and they exist if and only if

$$\lambda = -\frac{1}{2n^2}, \qquad n \in \mathbb{N}. \tag{1.32}$$

Here we find the famous negative spectrum of the hydrogen atom. The very special, quantized, negative energies correspond to the stationary states between which the electron can navigate depending on the energy it exchanges with the outside world.

We will not provide the complete proof of Theorem 1.5 in this book. The spectral theory that we will develop in the following chapters allows us to prove that the possible $\lambda < 0$ form a countable set and tend to 0 (this is Theorem 5.46 in Chap. 5). However, to show (1.32) it is necessary to compute these eigenvalues explicitly, which is a bit tedious. The detailed calculation can be read in [Tes09, Hal13]. It is based on the invariance of the model by rotations around the origin $0 \in \mathbb{R}^3$, which allows one to work in the eigenspaces of the angular momentum (Sect. 4.8.3). In the end, the computation is reduced to studying an ordinary differential equation for the radial part f of ψ, in the form

$$-\frac{1}{2}f''(r) - \frac{1}{r}f'(r) + \frac{\ell(\ell+1)}{2r^2}f(r) - \frac{1}{r}f(r) = \lambda\,f(r)$$

where $\ell \in \mathbb{N} \cup \{0\}$, an equation that must then be solved explicitly. Here it is necessary to determine the possible values of $f(0)$ or $f'(0)$ so that the unique solution, given by the Cauchy-Lipschitz theorem, tends to 0 at infinity, hence the corresponding function is in $L^2(\mathbb{R}^3)$. The solutions are expressed using Laguerre polynomials [Tes09, Hal13].

The absence of non-negative eigenvalues follows from the **Virial identity**[4] which stipulates that any $H^1(\mathbb{R}^3)$ solution of Eq. (1.31) must necessarily satisfy

$$\int_{\mathbb{R}^3} |\nabla \psi(x)|^2 \, dx = \int_{\mathbb{R}^3} \frac{|\psi(x)|^2}{|x|} \, dx, \tag{1.33}$$

which immediately implies

$$\lambda = \frac{1}{2} \int_{\mathbb{R}^3} |\nabla \psi(x)|^2 \, dx - \int_{\mathbb{R}^3} \frac{|\psi(x)|^2}{|x|} \, dx = -\frac{1}{2} \int_{\mathbb{R}^3} |\nabla \psi(x)|^2 \, dx < 0.$$

Therefore, there cannot exist non-negative eigenvalues. The Virial identity (1.33) can be obtained by multiplying Eq. (1.31) by $(3/2)\overline{\psi(x)} + x \cdot \nabla \overline{\psi(x)}$ and integrating by parts. This requires verifying that all the terms make sense in $H^1(\mathbb{R}^3)$, which we will discuss later in Sect. 4.8.4.

Exercise 1.6 (Continuous Spectrum) As we mentioned in Sect. 1.2.2, the spectrum of H also contains a "continuous spectrum" over the entire interval $[0, +\infty)$, typical of the infinite dimension, which corresponds to "quasi eigenvalues". The existence of this spectrum will be shown in Corollary 5.38 of Chap. 5. For all $\lambda \geq 0$, we can construct a sequence ψ_n normalized in $L^2(\mathbb{R}^3)$ such that $(H - \lambda)\psi_n \to 0$ as follows. Let $\chi \in C_c^\infty(\mathbb{R}^3)$ be such that $\int_{\mathbb{R}^3} |\chi|^2 = 1$. We consider the sequence of functions $\psi_n(x) = e^{ik \cdot x} n^{-3/2} \chi(x/n)$, for some $k \in \mathbb{R}^3$. Show that

$$\left(-\frac{\Delta}{2} - \frac{1}{|x|} - \frac{|k|^2}{2} \right) \psi_n \to 0$$

strongly in $L^2(\mathbb{R}^3)$ and that $\psi_n \rightharpoonup 0$ weakly in $H^2(\mathbb{R}^3)$.

1.4 One Particle in $\mathbb{R}^d$ with a General Potential V

After having studied the hydrogen atom in detail, we now discuss the more general case of a quantum particle in $\mathbb{R}^d$ (with $d \geq 1$), subject to any external potential V, satisfying the condition that it tends in some way to 0 at infinity. That is, we consider

---

[4] Also called Pohožaev's identity, named after the one who seems to have been one of the first to use it to study nonlinear equations [Poh65].

the functional

$$\mathcal{E}(\psi) = \int_{\mathbb{R}^d} |\nabla\psi(x)|^2 \, dx + \int_{\mathbb{R}^d} V(x)|\psi(x)|^2 \, dx.$$

The study carried out in this section is in the same spirit as [LL01, Chap. 11].

1.4.1   Spaces $L^p(\mathbb{R}^d) + L^q(\mathbb{R}^d)$

We wish to work with conditions on the potential V that are sufficiently general without being too complicated. They must allow V to diverge locally at certain points of $\mathbb{R}^d$, like the Coulomb potential in dimension $d = 3$ which diverges at the origin. A natural framework would be to work with the spaces $L^p(\mathbb{R}^d)$, because these marry well with Sobolev embeddings. However, it is not very reasonable physically to suppose that V belongs to a single space $L^p(\mathbb{R}^d)$. Indeed, local divergences have nothing to do with the speed of convergence to 0 at infinity and the same space may not cover these two regions. For example, the Coulomb potential $V(x) = -1/|x|$ does not belong to any $L^p(\mathbb{R}^3)$. Its power p is integrable close to the origin for $p < 3$ and at infinity for $p > 3$. For this reason, we will rather work with *sums of spaces L^p*. Those can properly handle different behaviors in several regions of space. We therefore begin by recalling the definition and elementary properties of these spaces.

Definition 1.7 (Sums of Lebesgue Spaces) Let $1 \le p, q \le \infty$. We call $L^p(\mathbb{R}^d) + L^q(\mathbb{R}^d)$ the vector space composed of functions $f \in L^1_{loc}(\mathbb{R}^d)$ that can be written $f = f_p + f_q$ with $f_p \in L^p(\mathbb{R}^d)$ and $f_q \in L^q(\mathbb{R}^d)$.

The space $L^p(\mathbb{R}^d) + L^q(\mathbb{R}^d)$ is a Banach space when it is equipped with the norm

$$\|f\|_{L^p(\mathbb{R}^d)+L^q(\mathbb{R}^d)} = \inf\left\{\|f_p\|_{L^p(\mathbb{R}^d)} + \|f_q\|_{L^q(\mathbb{R}^d)} : f = f_p + f_q\right\} \qquad (1.34)$$

but we will rarely use this norm because V will often be given once and for all. See Exercise 1.26. It is important to remember that the decomposition $f = f_p + f_q$ is **not unique**, which complicates things a bit. Indeed, we notice that any $g \in L^p(\mathbb{R}^d)$ can be written

$$g = \underbrace{g\,\mathbb{1}(|g| \ge M)}_{\in L^1(\mathbb{R}^d)\cap L^p(\mathbb{R}^d)} + \underbrace{g\,\mathbb{1}(|g| < M)}_{\in L^p(\mathbb{R}^d)\cap L^\infty(\mathbb{R}^d)}. \qquad (1.35)$$

The second function is bounded in modulus by $|g| \in L^p(\mathbb{R}^d)$ and by M, so it belongs to the mentioned intersection. For the first function, we note that

$$\int_{|g(x)| \geq M} |g(x)| \, dx \leq M^{1-p} \int_{\mathbb{R}^d} |g(x)|^p \, dx,$$

so that $g \, \mathbb{1}(|g| \geq M)$ indeed belongs to $L^1(\mathbb{R}^d)$. Thus, any function $g \in L^p(\mathbb{R}^d)$ can be decomposed in the form $g = g_< + g_>$ where $g_<$ belongs to all the spaces $L^r(\mathbb{R}^d)$ with $1 \leq r \leq p$ and $g_>$ to all those with $p \leq r \leq +\infty$. If $f = f_p + f_q \in L^p(\mathbb{R}^d) + L^q(\mathbb{R}^d)$ with for example $p \leq q$, we can apply this decomposition to f_p and add $f_p \mathbb{1}(|f_p| \leq M) \in L^q(\mathbb{R}^d)$ to f_q, or apply it to f_q and add $f_q \mathbb{1}(|f_q| \geq M) \in L^p(\mathbb{R}^d)$ to f_p, hence the non-unique character of the decomposition. The same argument also allows us to see that

- $L^r(\mathbb{R}^d) \subset L^p(\mathbb{R}^d) + L^q(\mathbb{R}^d)$ for all $p \leq r \leq q$;
- $L^{p_1}(\mathbb{R}^d) + L^{q_1}(\mathbb{R}^d) \subset L^{p_2}(\mathbb{R}^d) + L^{q_2}(\mathbb{R}^d)$ for all $p_2 \leq p_1 \leq q_1 \leq q_2$, that is, the space increases when we decrease the smallest index and increase the largest;
- it is unnecessary to consider sums of more than two spaces because

$$L^{p_1}(\mathbb{R}^d) + L^{p_2}(\mathbb{R}^d) + L^{p_3}(\mathbb{R}^d) = L^{\min(p_1, p_2, p_3)}(\mathbb{R}^d) + L^{\max(p_1, p_2, p_3)}(\mathbb{R}^d).$$

We will always work with the assumption that V belongs to a space of the form $L^p(\mathbb{R}^d) + L^q(\mathbb{R}^d)$, where p is the *minimal* exponent allowed by our theory (which controls local singularities) and q is the *maximal* exponent (which will usually be $q = +\infty$), so as to work with the most general assumptions.

Often, we will need to assume in addition that the potential V becomes negligible at infinity, so that our quantum particle is free far from the origin. We could work with the simple assumption that $V \to 0$ at infinity, as is the case for the Coulomb potential, but we will rather use a more general viewpoint, which will not complicate the proofs at all. As $C_c^\infty(\mathbb{R}^d)$ is dense in $L^p(\mathbb{R}^d)$ for all $1 \leq p < +\infty$ (but not for $p = +\infty$), we deduce that $C_c^\infty(\mathbb{R}^d)$ is dense in $L^p(\mathbb{R}^d) + L^q(\mathbb{R}^d)$ for the norm (1.34), provided that p and q are both finite. This naturally leads to the question of what is the closure of $C_c^\infty(\mathbb{R}^d)$ in $L^p(\mathbb{R}^d) + L^\infty(\mathbb{R}^d)$ for the norm (1.34), when $p < +\infty$. This is the space introduced in the following definition.

Definition 1.8 (Negligibility at Infinity)   Let $1 \leq p < +\infty$. We call

$$L^p(\mathbb{R}^d) + L_\varepsilon^\infty(\mathbb{R}^d) \tag{1.36}$$

the space of functions $f \in L^p(\mathbb{R}^d) + L^\infty(\mathbb{R}^d)$ such that for all $\varepsilon > 0$ there exists $f_p \in L^p(\mathbb{R}^d)$ and $f_\infty \in L^\infty(\mathbb{R}^d)$ with $f = f_p + f_\infty$ and $\|f_\infty\|_{L^\infty(\mathbb{R}^d)} \leq \varepsilon$. This space is the closure of $C_c^\infty(\mathbb{R}^d)$ in $L^p(\mathbb{R}^d) + L^\infty(\mathbb{R}^d)$ for the norm (1.34). We will say that its elements are **negligible at infinity**.

The notation with the ε as a subscript was introduced in [RS72]. It has the advantage of being short and effective, but the disadvantage of not being very visible. We encourage the reader to be attentive to the presence or absence of this subscript in the statements. For the assertion concerning the closure of $C_c^\infty(\mathbb{R}^d)$, see Exercise 1.26.

We can show that $f \in L^p(\mathbb{R}^d) + L^\infty(\mathbb{R}^d)$ belongs to $L^p(\mathbb{R}^d) + L_\varepsilon^\infty(\mathbb{R}^d)$ if and only if

$$\lim_{R \to \infty} \left\| \mathbb{1}_{\mathbb{R}^d \setminus B_R} f \right\|_{L^p(\mathbb{R}^d) + L^\infty(\mathbb{R}^d)} = 0$$

for the norm introduced in (1.34) (see again Exercise 1.26), where B_R denotes the ball of radius R centered at the origin. It is in this sense that f is negligible at infinity. By writing

$$f = f\mathbb{1}(|f| \geq \varepsilon) + f\mathbb{1}(|f| < \varepsilon)$$

we see that the functions of $L^r(\mathbb{R}^d)$ are all negligible at infinity when $r < \infty$:

$$L^r(\mathbb{R}^d) \subset L^p(\mathbb{R}^d) + L_\varepsilon^\infty(\mathbb{R}^d) \qquad \text{for all } p \leq r < \infty.$$

On the other hand, the constant function $f \equiv 1$ does not belong to $L^p(\mathbb{R}^d) + L_\varepsilon^\infty(\mathbb{R}^d)$, so that

$$L^\infty(\mathbb{R}^d) \not\subset L^p(\mathbb{R}^d) + L_\varepsilon^\infty(\mathbb{R}^d).$$

Remark 1.9 If $f = f_p + f_\infty \in L^p(\mathbb{R}^d) + L^\infty(\mathbb{R}^d)$ with $1 \leq p < \infty$, we can write

$$f_p = f_p\mathbb{1}(|f_p| \geq M) + f_p\mathbb{1}(|f_p| < M)$$

where the second term is in $L^\infty(\mathbb{R}^d)$ and can be added to f_∞, while the first is as small as we want in $L^p(\mathbb{R}^d)$, since by dominated convergence

$$\lim_{M \to \infty} \int_{|f_p| \geq M} |f_p(x)|^p \, dx = 0.$$

Thus, the functions of $f \in L^p(\mathbb{R}^d) + L^\infty(\mathbb{R}^d)$ with $1 \leq p < \infty$ can always be written $f = \tilde{f}_p + \tilde{f}_\infty$ with $\tilde{f}_p$ as small as we want in $L^p(\mathbb{R}^d)$.

1.4.2   Stability

In the rest of this chapter, we will work with the assumption that $V \in L^p(\mathbb{R}^d, \mathbb{R}) + L^\infty(\mathbb{R}^d, \mathbb{R})$ is a real-valued function, with

$$\begin{cases} p = 1 & \text{if } d = 1, \\ p > 1 & \text{if } d = 2, \\ p = \frac{d}{2} & \text{if } d \geq 3, \end{cases} \tag{1.37}$$

and we will add the additional assumption that $V \in L^p(\mathbb{R}^d, \mathbb{R}) + L^\infty_\varepsilon(\mathbb{R}^d, \mathbb{R})$ when necessary. These conditions on p are related to the Sobolev inequality recalled below (see also Theorem A.15 in Appendix A). The following lemma contains the most important properties of the term involving the potential V.

Lemma 1.10 (Potential Energy) *Let $d \geq 1$ and*

$$V \in L^p(\mathbb{R}^d, \mathbb{R}) + L^\infty(\mathbb{R}^d, \mathbb{R})$$

with p satisfying (1.37). Then $|V(x)|^{1/2}\psi$ belongs to $L^2(\mathbb{R}^d)$ for all $\psi \in H^1(\mathbb{R}^d)$. For all $\varepsilon > 0$, there exists a constant C_ε such that

$$\left| \int_{\mathbb{R}^d} V(x)|\psi(x)|^2 \, dx \right| \leq \int_{\mathbb{R}^d} |V(x)| \, |\psi(x)|^2 \, dx$$

$$\leq \varepsilon \int_{\mathbb{R}^d} |\nabla\psi(x)|^2 \, dx + C_\varepsilon \int_{\mathbb{R}^d} |\psi(x)|^2 \, dx \tag{1.38}$$

for all $\psi \in H^1(\mathbb{R}^d)$. The map

$$\psi \in H^1(\mathbb{R}^d) \mapsto \int_{\mathbb{R}^d} V(x)|\psi(x)|^2 \, dx \in \mathbb{R} \tag{1.39}$$

*is therefore **strongly continuous**. If furthermore*

$$V \in L^p(\mathbb{R}^d, \mathbb{R}) + L^\infty_\varepsilon(\mathbb{R}^d, \mathbb{R}),$$

*then the map (1.39) is also **weakly continuous**. That is, for any sequence $\psi_n \rightharpoonup \psi$ converging weakly in $H^1(\mathbb{R}^d)$, we have*

$$\lim_{n \to \infty} \int_{\mathbb{R}^d} V(x)|\psi_n(x)|^2 \, dx = \int_{\mathbb{R}^d} V(x)|\psi(x)|^2 \, dx.$$

Remark 1.11 The potential energy is in general not weakly continuous if V is not negligible at infinity. For example, for $V = 1$ we simply find $\int_{\mathbb{R}^d} |\psi(x)|^2 \, dx$ which is not weakly continuous.

Proof We can directly bound $\int_{\mathbb{R}^d} |V| \, |\psi|^2$. The idea is of course to write $V = V_p + V_\infty$ with $V_p \in L^p(\mathbb{R}^d)$ and $V_\infty \in L^\infty(\mathbb{R}^d)$, and then estimate the two terms separately. However, the reader may have noticed that the chosen exponent for p will lead us to involve the critical exponent of the Sobolev embedding in dimensions $d \geq 3$, which may not allow us to obtain the estimate (1.38) with an ε as small as we want. We thus follow Remark 1.9 and start by writing

$$V_p = V_p \, \mathbb{1}(|V_p| \geq M) + V_p \, \mathbb{1}(|V_p| < M)$$

where the first term is small in $L^p(\mathbb{R}^d)$ for $M \to \infty$, by dominated convergence. We obtain

$$\int_{\mathbb{R}^d} |V(x)| \, |\psi(x)|^2 \, dx$$

$$\leq \int_{|V_p| \geq M} |V_p(x)| \, |\psi(x)|^2 \, dx + \left(M + \|V_\infty\|_{L^\infty(\mathbb{R}^d)} \right) \int_{\mathbb{R}^d} |\psi(x)|^2 \, dx$$

$$\leq \left\| V_p \mathbb{1}(|V_p| \geq M) \right\|_{L^p(\mathbb{R}^d)} \|\psi\|^2_{L^{2p'}(\mathbb{R}^d)} + \left(M + \|V_\infty\|_{L^\infty(\mathbb{R}^d)} \right) \int_{\mathbb{R}^d} |\psi(x)|^2 \, dx,$$

where $p' = p/(p-1)$ is the conjugate exponent of p. In dimension $d \geq 3$, we find $2p' = 2d/(d-2)$ which is the critical Sobolev exponent, for which we have the inequality similar to (1.19)

$$\|\psi\|^2_{L^{\frac{2d}{d-2}}(\mathbb{R}^d)} \leq S_d \int_{\mathbb{R}^d} |\nabla \psi(x)|^2 \, dx \tag{1.40}$$

(see Theorem A.13 in Appendix A). We have therefore proved, as desired, that

$$\int_{\mathbb{R}^d} |V(x)| \, |\psi(x)|^2 \, dx \leq S_d \left\| V_p \mathbb{1}(|V_p| \geq M) \right\|_{L^p(\mathbb{R}^d)} \int_{\mathbb{R}^d} |\nabla \psi(x)|^2 \, dx$$

$$+ \left(M + \|V_\infty\|_{L^\infty(\mathbb{R}^d)} \right) \int_{\mathbb{R}^d} |\psi(x)|^2 \, dx.$$

The constant in front of the gradient can be made as small as we want by taking M very large, which increases the one in front of the L^2 norm.

In dimension $d = 2$, the constraint $p > 1$ implies that $2p' < \infty$ and we can then use the Gagliardo-Nirenberg inequality

$$\|\psi\|_{L^{2p'}(\mathbb{R}^2)} \leq C \|\nabla \psi\|^{\frac{1}{p}}_{L^2(\mathbb{R}^2)} \|\psi\|^{\frac{1}{p'}}_{L^2(\mathbb{R}^2)} \leq C \|\psi\|_{H^1(\mathbb{R}^2)}$$

recalled in Theorem A.15 of Appendix A. In dimension $d = 1$ we have $p = 1$ so that $2p' = \infty$ and we have the inequality

$$\|\psi\|_{L^\infty(\mathbb{R})} \leq \sqrt{2}\,\|\psi'\|_{L^2(\mathbb{R})}^{\frac{1}{2}}\,\|\psi\|_{L^2(\mathbb{R})}^{\frac{1}{2}} \leq \|\psi\|_{H^1(\mathbb{R})}\,,$$

as explained in (A.23) in Appendix A. We find in these two cases

$$\int_{\mathbb{R}^d} |V(x)|\,|\psi(x)|^2\,\mathrm{d}x \leq C\,\big\|V_p \mathbb{1}(|V_p| \geq M)\big\|_{L^p(\mathbb{R}^d)} \int_{\mathbb{R}^d} |\nabla\psi(x)|^2\,\mathrm{d}x$$
$$+ \Big(M + \|V_\infty\|_{L^\infty(\mathbb{R}^d)} + C\,\big\|V_p \mathbb{1}(|V_p| \geq M)\big\|_{L^p(\mathbb{R}^d)}\Big) \int_{\mathbb{R}^d} |\psi(x)|^2\,\mathrm{d}x,$$

for a constant C depending on the dimension, which allows us to conclude in the same way.

It remains to prove the weak continuity when $V \in L^p(\mathbb{R}^d) + L^\infty_\varepsilon(\mathbb{R}^d)$ is negligible at infinity. Let $\psi_n \rightharpoonup \psi$ be a sequence that converges weakly in $H^1(\mathbb{R}^d)$. We give ourselves a $\delta > 0$ and, this time, we write $V = V_1 + V_2 \in L^p(\mathbb{R}^d) + L^\infty(\mathbb{R}^d)$ with $\|V_2\|_{L^\infty} \leq \delta$. We claim that the potential energy associated with V_1 is weakly continuous:

$$\lim_{n\to\infty} \int_{\mathbb{R}^d} V_1(x)|\psi_n(x)|^2\,\mathrm{d}x = \int_{\mathbb{R}^d} V_1(x)|\psi(x)|^2\,\mathrm{d}x. \tag{1.41}$$

This follows immediately from the fact that

$$|\psi_n|^2 \rightharpoonup |\psi|^2 \qquad \text{weakly in } L^{p'}(\mathbb{R}^d). \tag{1.42}$$

By the Sobolev embeddings, we know that (ψ_n) is bounded in $L^{2p'}(\mathbb{R}^d)$ and therefore that $(|\psi_n|^2)$ is bounded in $L^{p'}(\mathbb{R}^d)$. In general, it is *false* that the weak limit of a square is the square of the weak limit. But this is true for a sequence that converges weakly in a Sobolev-type space! Indeed, the Rellich-Kondrachov Theorem A.18 recalled in Appendix A implies the strong *local* convergence of ψ_n to ψ in $L^2(\mathbb{R}^d)$. This means that $\psi_n \to \psi$ strongly in $L^q(B_R)$ for all $2 \leq q < 2p'$ and any $R > 0$, and therefore that $|\psi_n|^2 \to |\psi|^2$ strongly in $L^r(B_R)$ for all $1 \leq r < p'$ and any $R > 0$. In dimensions $d = 1, 2$ $r = p'$ is even included because the exponent is subcritical. As $(|\psi_n|^2)$ is also bounded in $L^{p'}(\mathbb{R}^d)$, its admits weakly convergent subsequences and we deduce, after for example testing against functions of $C^\infty_c(\mathbb{R}^d)$, that the weak limit (1.42) is true, and therefore that (1.41) occurs. To conclude the argument, we use $|V_2| \leq \delta$ to deduce that

$$\left| \int_{\mathbb{R}^d} V(x)|\psi_n(x)|^2\,\mathrm{d}x - \int_{\mathbb{R}^d} V(x)|\psi(x)|^2\,\mathrm{d}x \right|$$
$$\leq \left| \int_{\mathbb{R}^d} V_1(x)|\psi_n(x)|^2\,\mathrm{d}x - \int_{\mathbb{R}^d} V_1(x)|\psi(x)|^2\,\mathrm{d}x \right| + 2C\delta$$

for large enough n, where

$$\|\psi\|^2_{L^2(\mathbb{R}^d)} \le C := \limsup_{n\to\infty} \|\psi_n\|^2_{L^2(\mathbb{R}^d)} < \infty.$$

The previous reasoning thus shows that

$$\limsup_{n\to\infty} \left| \int_{\mathbb{R}^d} V(x)|\psi_n(x)|^2 \, dx - \int_{\mathbb{R}^d} V(x)|\psi(x)|^2 \, dx \right| \le 2C\delta$$

and by taking $\delta \to 0$ we obtain the desired limit. $\square$

By taking $\varepsilon = 1/2$ in (1.38), we find that the model is stable.

Corollary 1.12 (Stability) *Let $d \ge 1$ and $V \in L^p(\mathbb{R}^d, \mathbb{R}) + L^\infty(\mathbb{R}^d, \mathbb{R})$ with p satisfying (1.37). The energy*

$$\mathcal{E}(\psi) = \int_{\mathbb{R}^d} |\nabla\psi(x)|^2 \, dx + \int_{\mathbb{R}^d} V(x)|\psi(x)|^2 \, dx. \tag{1.43}$$

is well defined and continuous on $H^1(\mathbb{R}^d)$, with the estimate

$$\mathcal{E}(\psi) \ge \frac{1}{2} \int_{\mathbb{R}^d} |\nabla\psi(x)|^2 \, dx - C \int_{\mathbb{R}^d} |\psi(x)|^2 \, dx \tag{1.44}$$

where $C = C_{1/2}$ corresponds to $\varepsilon = 1/2$ in (1.38).

The assumptions on the positive part $V_+ = \max(V, 0)$ of the potential are only useful to show that $\mathcal{E}$ is well defined and continuous on $H^1(\mathbb{R}^d)$. The function V_+ plays no role in lower bounds on the energy $\mathcal{E}$ as in (1.44). The constant C only depends on the negative part $V_- = \max(-V, 0)$.

1.4.3 Existence of a Ground State

When $V \in L^p(\mathbb{R}^d, \mathbb{R}) + L^\infty(\mathbb{R}^d, \mathbb{R})$ with p as in (1.37), Corollary 1.12 leads us to define

$$\boxed{I := \inf_{\substack{\psi \in H^1(\mathbb{R}^d) \\ \int_{\mathbb{R}^d} |\psi|^2 = 1}} \mathcal{E}(\psi).} \tag{1.45}$$

where $\mathcal{E}$ is given by (1.43). We now study when I is a minimum, when V is negligible at infinity.

Inequality (1.44) means that $\mathcal{E}$ is **coercive** for the norm of $H^1(\mathbb{R}^d)$ on the sphere of $L^2(\mathbb{R}^d)$, that is, the sets $\{\psi \in H^1(\mathbb{R}^d) : \|\psi\|_{L^2(\mathbb{R}^d)} = 1, \mathcal{E}(\psi) \le C\}$ are bounded in $H^1(\mathbb{R}^d)$ for every C. Moreover, Lemma 1.10 implies that $\mathcal{E}$ is weakly lower semi-continuous (*wlsc*) when V is negligible at infinity (the proof is provided below). A coercive function that is weakly *wlsc* always reaches its minimum on a weakly closed set. Unfortunately, the set $\{\psi \in H^1(\mathbb{R}^d) \ : \ \|\psi\|_{L^2(\mathbb{R}^d)} = 1\}$ on which $\mathcal{E}$ is minimized is *not* weakly closed due to the constraint $\|\psi\|_{L^2(\mathbb{R}^d)} = 1$ which does not pass to the weak limit. The existence of a minimum is therefore not always guaranteed for I. In fact, if for example $V \equiv 0$ it is clear that $I = 0$ and is not reached. The following theorem shows that the infimum of $\mathcal{E}$ is always reached when $I < 0$.

Theorem 1.13 (One Quantum Particle: Existence) *Let $d \geq 1$ and $V \in L^p(\mathbb{R}^d, \mathbb{R}) + L^\infty_\varepsilon(\mathbb{R}^d, \mathbb{R})$ negligible at infinity, with p satisfying (1.37). Then we always have $I \leq 0$.*

- *If $I = 0$, there exists a minimizing sequence $(\psi_n) \subset H^1(\mathbb{R}^d)$ such that $\|\psi_n\|_{L^2(\mathbb{R}^d)} = 1$ and $\psi_n \rightharpoonup 0$ weakly. The infimum I in (1.45) may or may not be attained.*
- *If $I < 0$, then all the minimizing sequences admit a subsequence that converges strongly in $H^1(\mathbb{R}^d)$. In particular, the infimum I in (1.45) is a minimum and there always exists at least one minimizer.*

The condition $I < 0$ is physically very natural. A particle that escapes to infinity no longer feels the potential V and its minimum energy is then zero, because only the kinetic energy remains. Therefore, the assumption $I < 0$ is used to suppress compactness problems at infinity, by rendering them energetically unfavorable.

A simpler system, also important from a physical point of view, is when the particle is *confined*, which corresponds to taking a potential V that tends to $+\infty$ at infinity. The existence of a minimizer for this problem is always guaranteed because escaping to infinity now costs an infinite amount of energy and therefore becomes totally impossible. This important but much easier situation is studied in detail in Exercise 1.28 below. For example, $V(x) = |x|^2$ corresponds to attaching our quantum particle to a spring nailed at the origin; this is the harmonic oscillator.

Remark 1.14 (Existence if $I = 0$) It is entirely possible that $I = 0$ and is nevertheless attained. This is for example the case for $V(x) = -d(2-d)(1+|x|^2)^{-2}$ in dimensions $d \geq 5$, with the minimizer $\psi_0(x) = c(1 + |x|^2)^{\frac{2-d}{2}}$.

Proof Let χ be any function of $C^\infty_c(\mathbb{R}^d)$ such that $\int_{\mathbb{R}^d} |\chi|^2 = 1$. Consider the sequence of dilated functions $\chi_n(x) = n^{-d/2} \chi(x/n)$. Then we have $\chi_n \to 0$ uniformly and, as

$$\int_{\mathbb{R}^d} |\nabla \chi_n(x)|^2 \, dx = \frac{1}{n^2} \int_{\mathbb{R}^d} |\nabla \chi(x)|^2 \, dx, \qquad \int_{\mathbb{R}^d} |\chi_n(x)|^2 \, dx = \int_{\mathbb{R}^d} |\chi(x)|^2 \, dx,$$

the sequence (χ_n) is bounded in $H^1(\mathbb{R}^d)$. Therefore, we must have $\chi_n \rightharpoonup 0$ weakly in $H^1(\mathbb{R}^d)$. As V is negligible at infinity, we deduce from Lemma 1.10 that

$$\lim_{n \to \infty} \int_{\mathbb{R}^d} V(x)|\chi_n(x)|^2 \, dx = 0$$

and therefore that $\mathcal{E}(\chi_n) \to 0$. By definition of the infimum, this shows that $I \leq 0$. If $I = 0$, this sequence is suitable for the first part of the statement.

Now suppose that $I < 0$ and consider a minimizing sequence (ψ_n). We first note that (ψ_n) is bounded in $H^1(\mathbb{R}^d)$ from (1.44) and since $\|\psi_n\|_{L^2} = 1$. Therefore, extracting a subsequence if necessary, we can assume that $\psi_n \rightharpoonup \psi$ weakly in $H^1(\mathbb{R}^d)$. From Lemma 1.10 we obtain

$$\lim_{n \to \infty} \int_{\mathbb{R}^d} V(x)|\psi_n(x)|^2 \, dx = \int_{\mathbb{R}^d} V(x)|\psi(x)|^2 \, dx.$$

In particular

$$\mathcal{E}(\psi_n) = \int_{\mathbb{R}^d} |\nabla \psi_n(x)|^2 \, dx + \int_{\mathbb{R}^d} V(x)|\psi(x)|^2 \, dx + o(1)_{n \to \infty}$$

and since $\mathcal{E}(\psi_n)$ converges to I, the gradient term $\int_{\mathbb{R}^d} |\nabla \psi_n(x)|^2 \, dx$ converges. Because of the weak convergence we always have

$$\liminf_{n \to \infty} \int_{\mathbb{R}^d} |\nabla \psi_n(x)|^2 \, dx = \lim_{n \to \infty} \int_{\mathbb{R}^d} |\nabla \psi_n(x)|^2 \, dx \geq \int_{\mathbb{R}^d} |\nabla \psi(x)|^2 \, dx,$$

and

$$1 = \liminf_{n \to \infty} \int_{\mathbb{R}^d} |\psi_n(x)|^2 \, dx \geq \int_{\mathbb{R}^d} |\psi(x)|^2 \, dx.$$

The inequality on the gradient limit provides

$$I = \lim_{n \to \infty} \mathcal{E}(\psi_n) \geq \int_{\mathbb{R}^d} |\nabla \psi(x)|^2 \, dx + \int_{\mathbb{R}^d} V(x)|\psi(x)|^2 \, dx = \mathcal{E}(\psi).$$

This is the *wlsc* character of $\mathcal{E}$ mentioned above. We can immediately note that the weak limit ψ of the sequence (ψ_n) cannot be zero because otherwise we would have $I \geq \mathcal{E}(0) = 0$, which contradicts the assumption that $I < 0$. Since $\psi \neq 0$ and $\mathcal{E}$ is quadratic, we can then write

$$0 > I \geq \mathcal{E}(\psi) = \|\psi\|_{L^2(\mathbb{R}^d)}^2 \, \mathcal{E}\left(\frac{\psi}{\|\psi\|_{L^2(\mathbb{R}^d)}}\right) \geq \|\psi\|_{L^2(\mathbb{R}^d)}^2 \, I.$$

As $\|\psi\|_{L^2(\mathbb{R}^d)} \leq 1$ and $I < 0$ this shows that $\|\psi\|_{L^2(\mathbb{R}^d)} = 1$, so that $\psi_n \to \psi$ strongly in $L^2(\mathbb{R}^d)$, with $\mathcal{E}(\psi) = I$. The limit ψ is therefore a minimizer. Returning to the above limits, we also find that

$$\lim_{n \to \infty} \int_{\mathbb{R}^d} |\nabla \psi_n(x)|^2 \, dx = \int_{\mathbb{R}^d} |\nabla \psi(x)|^2 \, dx,$$

which implies strong convergence in $H^1(\mathbb{R}^d)$. To summarize, we have indeed shown that, when $I < 0$, any minimizing sequence has a subsequence that converges strongly in $H^1(\mathbb{R}^d)$, its limit being a minimizer.   □

The following proposition provides a relatively simple condition on V that implies $I < 0$, and which also applies to the hydrogen atom.

Proposition 1.15 (Existence If V Decreases Slowly at Infinity) *If the function* $V \in L^p(\mathbb{R}^d, \mathbb{R}) + L_\varepsilon^\infty(\mathbb{R}^d, \mathbb{R})$ *satisfies*

$$V(x) \leq -\frac{c}{|x|^\alpha}, \qquad for \ |x| \geq R$$

with $c, R > 0$ and $0 < \alpha < 2$, then $I < 0$ and therefore a minimizer exists.

Proof Let $\chi \in C_c^\infty(B_2 \setminus B_1)$ be supported in the annulus located between the balls of radii 1 and 2, and such that $\int_{\mathbb{R}^d} |\chi|^2 = 1$. We dilate χ as before by defining $\chi_n(x) = n^{-d/2}\chi(x/n)$. After changing variables we obtain

$$\mathcal{E}(\chi_n) = \frac{1}{n^2} \int_{\mathbb{R}^d} |\nabla \chi(x)|^2 \, dx + \int_{B_2 \setminus B_1} V(nx)|\chi(x)|^2 \, dx$$

$$\leq \frac{1}{n^2} \int_{\mathbb{R}^d} |\nabla \chi(x)|^2 \, dx - \frac{c}{n^\alpha} \int_{B_2 \setminus B_1} \frac{|\chi(x)|^2}{|x|^\alpha} \, dx.$$

This is negative for large enough n, since $0 < \alpha < 2$ so the term in $n^{-\alpha}$ wins over the first.   □

The condition that $\alpha < 2$ is optimal. If the potential decreases faster than $1/|x|^2$ at infinity, it is perfectly possible to have $I = 0$. In fact, it is enough that V is small enough in $L^{d/2}(\mathbb{R}^d)$, in dimension $d \geq 3$.

Proposition 1.16 (Non-existence If V Is Small in $L^{d/2}(\mathbb{R}^d)$) *In dimension $d \geq 3$, if*

$$\|V\|_{L^{d/2}(\mathbb{R}^d)} < (S_d)^{-1}$$

where S_d is the best Sobolev constant appearing in (1.40), then we have $I = 0$ and the infimum is not attained.

Proof As before, we use the Hölder and Sobolev inequalities (1.40) to estimate

$$\mathcal{E}(\psi) \geq \int_{\mathbb{R}^d} |\nabla \psi(x)|^2 \, dx - \|V\|_{L^{d/2}(\mathbb{R}^d)} \|\psi\|^2_{L^{\frac{2d}{d-2}}(\mathbb{R}^d)}$$

$$\geq \left(1 - \|V\|_{L^{d/2}(\mathbb{R}^d)} S_d\right) \int_{\mathbb{R}^d} |\nabla \psi(x)|^2 \, dx.$$

This shows that $\mathcal{E} \geq 0$ and thus $I = 0$. If $\mathcal{E}(\psi) = 0$, then we must have from the previous inequality $\nabla \psi = 0$ therefore $\psi = 0$, which contradicts the constraint $\int_{\mathbb{R}^d} |\psi|^2 = 1$. Hence there is no minimizer. □

The situation is more complicated in dimensions $d \in \{1, 2\}$ where existence is not at all related to the size of the potential V in an L^p space. In particular, the minimum is always reached if for example $V \leq 0$ almost everywhere, with of course $V \neq 0$ somewhere.

Proposition 1.17 (Existence in Dimensions $d \in \{1, 2\}$) *Let $d \in \{1, 2\}$ and $V \in L^p(\mathbb{R}^d) + L^\infty_\varepsilon(\mathbb{R}^d)$ with p as in (1.37). If $d = 2$, we further assume that $V_+ \in L^1(\mathbb{R}^2)$. If*

$$\int_{\mathbb{R}^d} V(x) \, dx < 0$$

then $I < 0$ and a minimizer exists.

In dimension $d = 2$, we allow here $\int_{\mathbb{R}^d} V(x) \, dx = -\infty$ when V_- is not integrable.

Proof Let us start with the dimension $d = 1$, where the proof is very similar to that of Proposition 1.15. We again take $\chi \in C^\infty_c(\mathbb{R})$ such that $0 \leq \chi \leq 1$, this time with $\chi \equiv 1$ on $[-1, 1]$. We then introduce $\chi_n = n^{-1/2} \chi(x/n)$ and obtain

$$\mathcal{E}(\chi_n) = \frac{1}{n^2} \int_{-\infty}^{\infty} |\chi'(x)|^2 \, dx + \frac{1}{n} \int_{\mathbb{R}} V(x) \chi(nx)^2 \, dx$$

$$\leq \frac{1}{n^2} \int_{-\infty}^{\infty} |\chi'(x)|^2 \, dx + \frac{1}{n} \left(\int_{\mathbb{R}} V(x)_+ \, dx - \int_{-n}^{n} V(x)_- \, dx\right).$$

Here we have used that $\chi(nx)^2 \leq 1$ for the term involving V_+ and that $\chi(nx)^2 \geq \mathbb{1}_{[-n,n]}(x)$ for the one involving V_-. By monotone convergence we have $\int_{-n}^{n} V_- \rightarrow \int_{\mathbb{R}} V_-$, which is strictly greater than $\int_{\mathbb{R}} V_+$ by assumption. Thus the second term is negative and it wins over the first. We have therefore proved that $\mathcal{E}(\chi_n) < 0$ for n large enough. The result follows after division by $\|\chi_n\|^2_{L^2}$.

The previous argument is just inoperative in dimension $d = 2$, since the second term is then multiplied by $1/n^2$ instead of $1/n$, hence becomes of the same order as the kinetic energy. We need to localize in a more subtle way and instead take

$$\chi_n(x) = \begin{cases} n^{-1} & \text{if } |x| \le n, \\ \dfrac{\log\left(\frac{n+n^2}{|x|}\right)}{n\log(1+n)} & \text{if } n \le |x| \le n+n^2, \\ 0 & \text{if } |x| \ge n+n^2. \end{cases}$$

The logarithm is natural in dimension $d = 2$ because we have $\Delta \log|x| = 0$ on $\mathbb{R}^2 \setminus \{0\}$, which allows us to minimize the error on the kinetic energy. A calculation provides

$$\int_{\mathbb{R}^2} |\nabla \chi_n|^2 dx = \frac{2\pi}{n^2 \log(1+n)}$$

so that we obtain by arguing as before

$$\mathcal{E}(\chi_n) \le \frac{2\pi}{n^2 \log(1+n)} + \frac{1}{n^2}\left(\int_{\mathbb{R}^2} V(x)_+ \, dx - \int_{|x|\le n} V(x)_- \, dx\right).$$

The potential energy therefore again wins over the kinetic energy, though this time only by a logarithm. $\square$

1.4.4   Uniqueness of the Ground State

The following statement provides the uniqueness of the minimizer, when it exists.

Theorem 1.18 (One Quantum Particle: Uniqueness) *Let $d \ge 1$ and $V \in L^p(\mathbb{R}^d, \mathbb{R}) + L^\infty(\mathbb{R}^d, \mathbb{R})$ with p satisfying (1.37). If I in (1.45) is attained, then the minimizers are unique up to a phase. They all take the form $e^{i\theta}\psi$ with $\theta \in \mathbb{R}$ and $\psi \in H^1(\mathbb{R}^d)$ a strictly positive function on $\mathbb{R}^d$, that is a solution of Schrödinger's equation*

$$\left(-\Delta + V(x)\right)\psi(x) = I\,\psi(x) \tag{1.46}$$

in the sense of distributions and in $H^{-1}(\mathbb{R}^d)$.

The theorem and its proof can be found in [RS78, Sec. XIII]. A different demonstration with additional assumptions on V can be read in Sect. 1.6 below.

The uniqueness of the minimizer in the quantum case should be compared with the classical case, for which there is no uniqueness in general. Indeed, if a function V reaches its global minimum at several points, the function $E(x, p) = |p|^2 + V(x)$

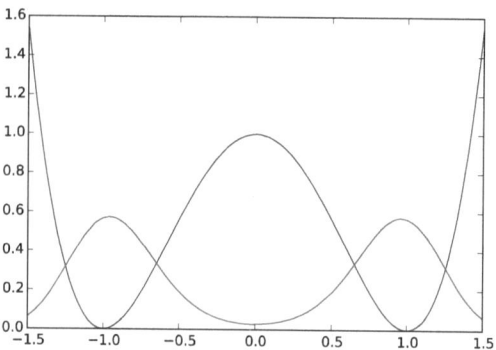

Fig. 1.6 Numerical calculation of the first eigenfunction ψ of Theorem 1.18 (in green on the figure) in the case where $d = 1$ and the potential is $V(x) = (x^2 - 1)^2$ (represented in blue on the figure). The particle is localized in both wells at the same time. ©Mathieu Lewin 2021. All rights reserved

then has several distinct minima. The quantum model provides a unique minimizer ψ, but which is typically localized in all the wells at once. See Fig. 1.6 for a numerical example.

In this section, we have studied a quantum particle in a general external potential V. We have seen that there is always a minimizer when the infimum of the energy I is negative and V is negligible at infinity (Theorem 1.13). The minimizer is always unique, when it exists (Theorem 1.18). We refer to Exercise 1.28 for the simpler case where V tends to $+\infty$ at infinity.

1.5 Abstract Formalism of Quantum Mechanics

In this section, we present the abstract formalism of quantum mechanics, due to von Neumann [von27a, von27c, von27b, von32a], without going into technical details since the rigorous formulation of the principles below requires the notions of spectral theory that we will develop later. The reader can always think of the finite-dimensional case $\mathfrak{H} = \mathbb{C}^d$ for simplicity.

1.5.1 Physical System, States

A finite physical system is described by a **complex Hilbert space** $\mathfrak{H}$, which we always assume to be separable. The **states** of the system are by definition the vectors of the unit sphere $S\mathfrak{H} = \{v \in \mathfrak{H} : \|v\| = 1\}$ of $\mathfrak{H}$, modulo phases.[5] The set of physical states is therefore the quotient $S\mathfrak{H}/\sim$ of the unit sphere by the equivalence

---

[5] Our definition corresponds to the "pure" states which are not necessarily suitable for all practical situations. A **mixed state** is by definition a collection (n_i, v_i) with v_i a Hilbert basis of $\mathfrak{H}$, $n_i \geq 0$ and $\sum_i n_i = 1$ which forms a sort of probability on the pure states. Mixed states are frequently represented by the operator Γ, defined by $\Gamma f = \sum n_i \langle v_i, f \rangle v_i$ called a "density

relation defined by $v \sim v'$ if and only if $v = e^{i\theta}v'$. It is often more convenient to work in the sphere $S\mathfrak{H}$ rather than in the quotient, provided that at all times it is verified that the considered quantities are indeed independent of the phases.

Example 1.19 For the electron of the hydrogen atom, we have $\mathfrak{H} = L^2(\mathbb{R}^3, \mathbb{C})$ when neglecting the spin and $\mathfrak{H} = L^2(\mathbb{R}^3 \times \{\uparrow, \downarrow\}, \mathbb{C}) \simeq L^2(\mathbb{R}^3, \mathbb{C}^2)$ if we take it into account. We will see other examples in Chap. 6.

1.5.2   Observables

The **physical observables** are the quantities that can in principle be measured in experiments: energy, position, speed, etc. In the quantum formalism, they are represented by **self-adjoint operators**. Most of the important practical examples are unbounded operators, which greatly complicates the mathematical definition of self-adjointness. We will talk about this at length in the next chapter.

Example 1.20 For a quantum particle evolving in $\mathbb{R}^d$, the "position" observable is the operator $X : f \mapsto xf$, which is actually a vector containing d distinct operators $X_j : f \mapsto x_j f$ corresponding to the d axes of $\mathbb{R}^d$ in the chosen frame. Similarly, the "momentum" observable is the collection $P = (P_1, \ldots, P_d)$ of the d operators $P_j : f \mapsto -i\partial_{x_j} f$. Finally, we have seen that the "energy" observable is the operator $H = -\Delta + V(x)$ when the particle is subject to an external potential V.

For an observable described by the self-adjoint operator A, the average value of this quantity when the system is in the state $v \in S\mathfrak{H}$ is by definition $\langle v, Av \rangle$ (the vector v must satisfy appropriate conditions for this scalar product to make sense). This is only the average value. The quantities that can be obtained by experiment are actually random, given by a probability measure $\mu_{A,v}$ on $\mathbb{R}$ (depending on A and the state v), called the **spectral measure**. It is of course such that the average value is given by

$$\langle v, Av \rangle = \int_{\mathbb{R}} a \, d\mu_{A,v}(a).$$

The probability $\mu_{A,v}$ is always concentrated on the spectrum of A and its rigorous definition is a bit subtle, we will see it in Chap. 4.

For the convenience of the reader, let us explain the definition of $\mu_{A,v}$ in finite dimension $\mathfrak{H} = \mathbb{C}^d$, in which case A is just a hermitian $d \times d$ matrix. The matrix A can be diagonalized in an orthonormal basis $v_1, \ldots, v_d$ of $\mathbb{C}^d$, with eigenvalues

---

matrix". Moreover, we only study here systems comprising a finite number of particles; infinite systems are rather described by operator algebras [BR02a, BR02b].

$\lambda_1, \ldots, \lambda_d$ (some λ_j may coincide in case of degeneracy). The measure $\mu_{A,v}$ is in this case defined on $\{\lambda_1, \ldots, \lambda_d\} \subset \mathbb{R}$ by the probabilities $|\langle v, v_j \rangle|^2$, that is

$$\mu_{A,v} = \sum_{j=1}^{d} |\langle v, v_j \rangle|^2 \delta_{\lambda_j}. \tag{1.47}$$

As the v_j form an orthonormal basis, we have

$$\int_{\mathbb{R}} d\mu_{A,v}(a) = \sum_{j=1}^{d} |\langle v, v_j \rangle|^2 = \|v\|^2 = 1$$

as well as

$$\int_{\mathbb{R}} a \, d\mu_{A,v}(a) = \sum_{j=1}^{d} \lambda_j |\langle v, v_j \rangle|^2 = \langle v, Av \rangle$$

as we wanted. Experimental measurements can therefore only give the values $\lambda_1, \ldots, \lambda_d$, with probabilities that depend on the state v of the system. If it turns out that v is parallel to one of the v_j, then we will necessarily obtain λ_j. But when v varies over the unit sphere we can obtain all the possible probabilities on $\{\lambda_1, \ldots, \lambda_d\}$.

The definition (1.47) easily extends to operators A in infinite dimension that can be diagonalized in an orthonormal basis (for example the compact self-adjoint operators recalled later in Sect. 5.3). On the other hand, the definition for a general self-adjoint operator is more difficult, because of the continuous spectrum. We will see it in Chap. 4.

Remark 1.21 An important postulate is the one that expresses the future of a quantum system after an experimental measurement (the "wavefunction collapse"). While this axiom is traditionally included in the quantum formalism, we have chosen to ignore it here. We will indeed only study isolated quantum systems, whose behavior is not perturbed by any observer.

1.5.3   Dynamics

The dynamics of an isolated quantum system is described by Schrödinger's equation

$$\begin{cases} i \dfrac{d}{dt} v(t) = H v(t) \\ v(0) = v_0, \end{cases}$$

where H is the "energy" observable called the **Hamiltonian**. The operator H could also depend on time, when the system is subject to external forces that vary. If H is independent of time, we obtain a Hamiltonian system, as we explained in Sect. 1.2.3. In finite dimension, the solution of the equation with $v(0) = v_0$ is then given by the formula

$$v(t) = e^{-itH} v_0 \qquad (1.48)$$

where e^{-itH} is by definition the unitary matrix whose eigenvalues are equal to $e^{-it\lambda_j}$ in the basis of the v_j's that diagonalize H. The average energy

$$\langle v(t), Hv(t) \rangle = \left\langle e^{-iHt} v_0, He^{-iHt} v_0 \right\rangle = \left\langle v_0, e^{iHt} He^{-iHt} v_0 \right\rangle = \langle v_0, Hv_0 \rangle$$

is conserved over time. In fact, the spectral measure μ_{H,v_0} is invariant by the dynamics because

$$\left| \left\langle e^{-iHt} v_0, v_j \right\rangle \right|^2 = \left| \left\langle v_0, e^{iHt} v_j \right\rangle \right|^2 = \left| \left\langle v_0, e^{i\lambda_j t} v_j \right\rangle \right|^2 = \left| \langle v_0, v_j \rangle \right|^2$$

for every eigenvector v_j of H. Moreover, the stationary states (modulo phases) are the eigenvectors of H. These properties will remain true in infinite dimension for any self-adjoint operator H.

1.5.4   Union of Quantum Systems

When we put together two physical systems represented by the spaces $\mathfrak{H}_1$ and $\mathfrak{H}_2$ and Hamiltonians H_1 and H_2, the union of the two systems is always described by the tensor product $\mathfrak{H} = \mathfrak{H}_1 \otimes \mathfrak{H}_2$. The Hamiltonian of the total system is often in the form

$$H = H_1 \otimes \mathbb{1} + \mathbb{1} \otimes H_2 + I_{12} \qquad (1.49)$$

where $H_1 \otimes \mathbb{1}$ means that the operator only acts on the first part of the tensor product and is the identity on the other part: $(H_1 \otimes \mathbb{1})v_1 \otimes v_2 = (H_1 v_1) \otimes v_2$. The operator I_{12} which involves the two components then describes the interaction between the two systems.

Thus, a system comprising N quantum particles will have to be described by a tensor product of N Hilbert spaces. For $\mathfrak{H} = L^2(\Omega)$ we get $\mathfrak{H}^{\otimes N} \simeq L^2(\Omega^N)$, which therefore generates a growth in the number of variables of wavefunctions with N. This makes the numerical approximation of the solutions of Schrödinger's equation very difficult as soon as N becomes too large. This is the "curse of dimensionality". We will come back to this when we study atoms and molecules in Chap. 6.

Example 1.22 (Complete Hydrogen Atom) If we describe the proton of the hydrogen atom as a quantum particle, instead of a fixed classical particle as we did at the beginning of this chapter, we must work in

$$\mathfrak{H} = L^2(\mathbb{R}^3, \mathbb{C}) \otimes L^2(\mathbb{R}^3, \mathbb{C}) = L^2(\mathbb{R}^3 \times \mathbb{R}^3, \mathbb{C}).$$

Here, the tensor product means that we consider the closure of the space of finite linear combinations of functions in the form $f(x)g(y)$ with $x, y \in \mathbb{R}^3$, which provides all the square integrable functions on $\mathbb{R}^3 \times \mathbb{R}^3$. The Hamiltonian describing the fully quantized hydrogen atom (we have again neglected spin for simplicity) is then given by

$$H = -\frac{\Delta_x}{2m} - \frac{\Delta_y}{2M} - \frac{e^2}{4\pi\varepsilon_0|x - y|}$$

which is indeed in the form (1.49). The terms are interpreted from left to right as the kinetic energy of the electron (of mass m), that of the proton (of mass $M \approx 1836\,m$) and the Coulomb interaction between them. The state of the system is represented by wavefunctions $\Psi(x, y) \in L^2(\mathbb{R}^3 \times \mathbb{R}^3, \mathbb{C})$, where $|\Psi(x, y)|^2$ is the probability density that the electron is at $x \in \mathbb{R}^3$ and the proton is at $y \in \mathbb{R}^3$, with a similar interpretation for the momenta in Fourier space. The corresponding energy is, of course,

$$\mathcal{E}(\Psi) = \frac{1}{2m} \int_{\mathbb{R}^3} \int_{\mathbb{R}^3} |\nabla_x \Psi(x, y)|^2 \, dx \, dy + \frac{1}{2M} \int_{\mathbb{R}^3} \int_{\mathbb{R}^3} |\nabla_y \Psi(x, y)|^2 \, dx \, dy$$
$$- \frac{e^2}{4\pi\varepsilon_0} \int_{\mathbb{R}^3} \int_{\mathbb{R}^3} \frac{|\Psi(x, y)|^2}{|x - y|} \, dx \, dy. \qquad (1.50)$$

The Born-Oppenheimer approximation (consisting of treating the proton as a classical particle) can be justified in the limit $M \to \infty$.

Exercise 1.23 Show that the infimum of the energy (1.50) in the set $\{\Psi \in H^1(\mathbb{R}^6) : \|\Psi\|_{L^2(\mathbb{R}^6)} = 1\}$ is finite and determine its value. Is there a minimizer? You can perform the change of variable $u = (mx + My)/(m + M)$ and $v = x - y$.

1.5.5 Quantization*

The abstract formalism presented so far does not specify how to choose the space $\mathfrak{H}$ and the observables to describe a particular system. We briefly discuss this issue here.

We often already have a *classical* description of the physical system under study and the task is to deduce the corresponding *quantum* model, a process called

quantization.[6] In principle, the rule is just to replace p by $-i\hbar\nabla$, which transforms the classical energy E of the system into an operator H. However, this procedure is not clearly defined in all cases. Imagine a function $a(x, p)$ defined on the phase space $\mathbb{R}^d \times \mathbb{R}^d$ of a particle evolving in $\mathbb{R}^d$. We would like to be able to associate with each such function an operator A denoted by $A = \mathrm{Op}_\hbar(a)$ on $L^2(\mathbb{R}^d)$, so that the (quantization) map

$$a \mapsto \mathrm{Op}_\hbar(a)$$

satisfies good mathematical properties. For example, it seems natural to ask that it is an algebra homomorphism if we restrict ourselves to a set of functions a forming an algebra. More precisely we can ask

- that the quantization map be linear;
- that if $a(x, p) = f(x)$, $\mathrm{Op}_\hbar(a)$ is the operator $\psi(x) \mapsto f(x)\psi(x)$;
- that if $a(x, p) = g(p)$, $\mathrm{Op}_\hbar(a)$ is the operator $\psi \mapsto g(-i\hbar\nabla)\psi$ defined in Fourier by $\mathcal{F}\{g(-i\hbar\nabla)\psi\}(k) = g(\hbar k)\widehat{\psi}(k)$;
- that $\mathrm{Op}_\hbar\big(F(a)\big) = F\big(\mathrm{Op}_\hbar(a)\big)$ where the term on the right is understood in the sense of functional calculus (see Sect. 4.3);
- that if $a \geq 0$ almost everywhere on $\mathbb{R}^d \times \mathbb{R}^d$, then the operator $\mathrm{Op}_\hbar(a)$ is self-adjoint and positive, in the sense that $\langle v, \mathrm{Op}_\hbar(a)v \rangle \geq 0$ for all v;
- that the quantization map preserves the trace norm:

$$\frac{1}{(2\pi\hbar)^d} \int_{\mathbb{R}^d} \int_{\mathbb{R}^d} a(x, p)\, dx\, dp = \mathrm{tr}\left(\mathrm{Op}_\hbar(a)\right)$$

where $\mathrm{tr}(A) = \sum_j \langle v_j, Av_j \rangle$ is the trace of A in a Hilbert basis.

Unfortunately, there exist no quantization procedure that satisfies all these natural properties. There are many possible solutions, each with its advantages and disadvantages, and only verifying a part of the above properties. They are all close in an appropriate sense in the limit $\hbar \to 0$.

To understand the difficulty of the quantization process, consider the example of a function in the form $a(x, p) = f(x)g(p)$. One might think of defining its quantization as

$$A_1 := f(x)g(-i\hbar\nabla).$$

Recall that $g(-i\hbar\nabla)$ is the operator that multiplies by $g(\hbar k)$ in Fourier. In space coordinates this corresponds to a convolution with the function $(2\pi)^{d/2}\hbar^{-d}\check{g}(x/\hbar)$

---

[6] Of course, it is the quantum model that is essential. The classical model should in principle be deduced from the latter in an appropriate limit, and not the other way around.

where $\check{g}$ is the inverse Fourier transform of g. Thus A_1 is the operator defined on $L^2(\mathbb{R}^d)$ by

$$\left(A_1\psi\right)(x) = (2\pi)^{\frac{d}{2}}\hbar^{-d}\, f(x) \int_{\mathbb{R}^d} \check{g}\left(\frac{x-y}{\hbar}\right)\psi(y)\,dy.$$

The one defined in the other direction

$$A_2 := g(-i\hbar\nabla)f(x)$$

is different and acts as

$$\left(A_2\psi\right)(x) = (2\pi)^{\frac{d}{2}}\hbar^{-d} \int_{\mathbb{R}^d} \check{g}\left(\frac{x-y}{\hbar}\right) f(y)\psi(y)\,dy,$$

namely with $f(x)$ replaced by $f(y)$. It can be shown that A_1 and A_2 are well defined and bounded when for example $f, g \in L^p(\mathbb{R}^d)$ with $2 \le p \le +\infty$ (see Sect. 5.3). A third possibility, more symmetric, is

$$A_3 = \frac{A_1 + A_2}{2},$$

that is

$$\left(A_3\psi\right)(x) = (2\pi)^{\frac{d}{2}}\hbar^{-d} \int_{\mathbb{R}^d} \check{g}\left(\frac{x-y}{\hbar}\right) \frac{f(x) + f(y)}{2}\psi(y)\,dy.$$

The operator A_3 is symmetric if $\check{g}$ is even, which seems more natural given that we should normally work with self-adjoint operators. But one can also introduce the operator defined by

$$\left(A_4\psi\right)(x) = (2\pi)^{\frac{d}{2}}\hbar^{-d} \int_{\mathbb{R}^d} \check{g}\left(\frac{x-y}{\hbar}\right) f\left(\frac{x+y}{2}\right)\psi(y)\,dy \tag{1.51}$$

which is called the **Weyl quantization** and is also symmetric. As

$$\hbar^{-d}\check{g}(\hbar^{-1}\cdot) \rightharpoonup \delta_0 \int_{\mathbb{R}^d}\check{g}(x)\,dx = \frac{g(0)}{(2\pi)^{d/2}}\delta_0$$

in the sense of distributions, the four solutions have a similar behavior in the limit $\hbar \to 0$.

In practice, we rarely encounter such quantization problems, since classical Hamiltonians are very often sums of functions of x and functions of p, as we saw for the hydrogen atom. However, there are some important operators that mix x and p. Being aware of the mathematical limitations of the quantization process is also necessary for a good understanding of the quantum formalism.

1.6   Proof of Theorem 1.18 *

In this section we write the proof of Theorem 1.18 concerning the uniqueness (up to a phase) of a minimizer for the variational problem

$$I = \inf_{\substack{\psi \in H^1(\mathbb{R}^d) \\ \int_{\mathbb{R}^d} |\psi|^2 = 1}} \mathcal{E}(\psi) \tag{1.52}$$

seen in (1.45), where

$$\mathcal{E}(\psi) = \int_{\mathbb{R}^d} |\nabla \psi(x)|^2 \, dx + \int_{\mathbb{R}^d} V(x) |\psi(x)|^2 \, dx,$$

assuming that such a minimizer exists. There are two classical methods to show uniqueness. The first is a spectral method based on a generalization to unbounded self-adjoint operators of the Perron-Frobenius theory for matrices, which can be read in [RS78, Sec. XIII] and in Problem B.5 in Appendix B. The second method uses tools coming from the theory of partial differential equations and this is the one we present in this section. The argument proceeds in four steps:

1. If there are minimizers, there is one that is non-negative.
2. Every minimizer solves the equation

$$\left(-\Delta + V(x)\right)\psi(x) = I\,\psi(x) \tag{1.53}$$

   in the sense of distributions and in $H^{-1}(\mathbb{R}^d)$.
3. Any non-negative solution of Eq. (1.53) is in fact strictly positive.
4. A strictly positive solution of the equation is necessarily a minimizer and is unique up to phase.

In order to simplify the discussion as much as possible and to focus on the main ideas, we will write the proof only in the case where the potential is uniformly bounded:

$$\boxed{V \in L^\infty(\mathbb{R}^d, \mathbb{R}).}$$

Our argument can easily be generalized to more singular potentials, as we will indicate. However, the proof under the exact assumptions of the theorem is more difficult and we will only give the corresponding references.

Step 1: There Exists a Non-negative Minimizer This step is based on the following important lemma, taken from [LL01, Thm 6.17 & Thm. 7.8] where the reader can find the detailed proof.

Lemma 1.24 (Convexity of Gradients) *For all $F \in H^1(\mathbb{R}^d, \mathbb{C})$, we have $|F| \in H^1(\mathbb{R}^d, \mathbb{R})$ with the inequality*

$$|\nabla |F|(x)|^2 \leq |\nabla F(x)|^2 \tag{1.54}$$

almost everywhere. In particular,

$$\int_{\mathbb{R}^d} |\nabla |F|(x)|^2 \, dx \leq \int_{\mathbb{R}^d} |\nabla F(x)|^2 \, dx. \tag{1.55}$$

By introducing the real and imaginary parts, $F = f + ig$, the inequality (1.54) can also be written in the form

$$\left| \nabla \sqrt{f^2 + g^2}(x) \right|^2 \leq |\nabla f(x)|^2 + |\nabla g(x)|^2. \tag{1.56}$$

By setting $f = \sqrt{t\rho_1}$ and $g = \sqrt{(1-t)\rho_2}$ with $t \in [0, 1]$, we find that the functional $0 \leq \rho \mapsto \int_{\mathbb{R}^d} |\nabla \sqrt{\rho}|^2$ is convex, hence the title of the lemma.

If f and g are sufficiently smooth and do not vanish, the proof of (1.56) is very simple. We write

$$\nabla \sqrt{f^2 + g^2}(x) = \frac{f(x)\nabla f(x) + g(x)\nabla g(x)}{\sqrt{f(x)^2 + g(x)^2}},$$

whose square is estimated by

$$\left| \nabla \sqrt{f^2 + g^2}(x) \right|^2 = \frac{|f(x)\nabla f(x) + g(x)\nabla g(x)|^2}{f(x)^2 + g(x)^2}$$

$$= |\nabla f(x)|^2 + |\nabla g(x)|^2 - \frac{|g(x)\nabla f(x) - f(x)\nabla g(x)|^2}{f(x)^2 + g(x)^2}$$

$$\leq |\nabla f(x)|^2 + |\nabla g(x)|^2.$$

The proof of the lemma in $H^1(\mathbb{R}^d)$ is done by regularization and can be read in [LL01]. If $|F| = \sqrt{f^2 + g^2} > 0$ on $\mathbb{R}^d$, one can show that there is equality if and only if $g\nabla f = f\nabla g$, which means that $f = cg$.

Since the potential energy depends only on $|\psi|$, Lemma 1.24 implies

$$\mathcal{E}(\psi) \geq \mathcal{E}(|\psi|).$$

This inequality shows that if ψ is a minimizer for I, then $|\psi|$ also is. Thus when minimizers exist we can always choose one that is non-negative, which we wanted to prove in Step 1.

Step 2: Every Minimizer Solves Schrödinger's Equation Let $\psi_0 \in H^1(\mathbb{R}^d)$ with $\|\psi_0\|_{L^2(\mathbb{R}^d)} = 1$ be any minimizer for problem I. We show in this step that ψ_0 is a solution of Eq. (1.53). This is a very general argument which simply states that $\mathcal{E}$ must be stationary at ψ_0 on the unit sphere $\{\psi \in H^1(\mathbb{R}^d) : \|\psi\|_{L^2} = 1\}$. For this, it is not necessary to talk about derivatives in infinite dimension, we can just look at the derivative in an arbitrary direction, which amounts to calculating a one-dimensional derivative. So let $0 \neq \chi \in H^1(\mathbb{R}^d)$ be any function (the chosen direction for the derivative of the energy). We define

$$\psi_\varepsilon = \frac{\psi_0 + \varepsilon\chi}{\|\psi_0 + \varepsilon\chi\|_{L^2(\mathbb{R}^d)}}.$$

We have

$$\|\psi_0 + \varepsilon\chi\|^2_{L^2(\mathbb{R}^d)} = 1 + 2\varepsilon\Re\int_{\mathbb{R}^d} \overline{\psi_0(x)}\chi(x)\,dx + \varepsilon^2\int_{\mathbb{R}^d} |\chi(x)|^2\,dx,$$

which does not vanish for sufficiently small ε. Thus ψ_ε is a well-defined function in $H^1(\mathbb{R}^d)$, normalized in $L^2(\mathbb{R}^d)$ as it should be. Now, we write that the function $\varepsilon \mapsto \mathcal{E}(\psi_\varepsilon)$ reaches its minimum at $\varepsilon = 0$ and therefore its derivative vanishes at $\varepsilon = 0$. The result of the calculation is:

$$\Re\left(\int_{\mathbb{R}^d} \nabla\overline{\psi_0(x)} \cdot \nabla\chi(x)\,dx + \int_{\mathbb{R}^d} V(x)\overline{\psi_0(x)}\chi(x)\,dx\right.$$

$$\left. - I\int_{\mathbb{R}^d} \overline{\psi_0(x)}\chi(x)\,dx\right) = 0. \tag{1.57}$$

Since the equation is valid for all $\chi \in H^1(\mathbb{R}^d)$, we can replace χ with $i\chi$ and thus remove the real part:

$$\int_{\mathbb{R}^d} \nabla\overline{\psi_0(x)} \cdot \nabla\chi(x)\,dx + \int_{\mathbb{R}^d} V(x)\overline{\psi_0(x)}\chi(x)\,dx - I\int_{\mathbb{R}^d} \overline{\psi_0(x)}\chi(x)\,dx = 0. \tag{1.58}$$

This is a good time to recall that if $\psi_0 \in H^1(\mathbb{R}^d)$, then $\Delta\psi_0$ is a distribution that is also in the dual $H^{-1}(\mathbb{R}^d)$ of the space $H^1(\mathbb{R}^d)$, via the duality formula

$$_{H^{-1}(\mathbb{R}^d)}\langle -\Delta\psi_0, \chi\rangle_{H^1(\mathbb{R}^d)} := \langle\nabla\psi_0, \nabla\chi\rangle_{L^2(\mathbb{R}^d)}.$$

Similarly, $V\psi_0$ belongs to $H^{-1}(\mathbb{R}^d)$ with the relation

$$_{H^{-1}(\mathbb{R}^d)}\langle V\psi_0, \chi\rangle_{H^1(\mathbb{R}^d)} := \int_{\mathbb{R}^d} V(x)\overline{\psi_0(x)}\chi(x)\,dx.$$

Let us recall that $|V|^{1/2}\psi$ belongs to $L^2(\mathbb{R}^d)$ when $\psi \in H^1(\mathbb{R}^d)$ by Lemma 1.10. Thus, Formula (1.58) means exactly that ψ_0 solves (1.53) in $H^{-1}(\mathbb{R}^d)$. If we restrict ourselves to $\chi \in C_c^\infty(\mathbb{R}^d)$, we find the equation in the sense of distributions.

The proof we have provided is valid with the general assumptions of Theorem 1.18. In our case where $V \in L^\infty(\mathbb{R}^d, \mathbb{R})$, we know that $V\psi_0 \in L^2(\mathbb{R}^d)$ and therefore immediately deduce from Eq. (1.53) that $\Delta\psi_0 \in L^2(\mathbb{R}^d)$, that is to say $\psi_0 \in H^2(\mathbb{R}^d)$, according to Lemma A.6 (elliptic regularity) in Appendix A. The terms of the equation are therefore all in $L^2(\mathbb{R}^d)$.

Step 3: Any Non-negative Solution of the Equation Is Strictly Positive At this step we know that there exists a non-negative minimizer $\psi_0 \in H^2(\mathbb{R}^d)$, that solves Eq. (1.53). We now claim that ψ_0 is strictly positive, in the sense that for any radius $R > 0$, there exists a constant $c_R > 0$ (depending on R) such that

$$\psi_0(x) \geq c_R > 0, \qquad \text{for almost all } x \in B_R. \tag{1.59}$$

This step is the most delicate of the proof. The argument we are going to use relies heavily on the fact that V is bounded and that $\psi_0 \in H^2(\mathbb{R}^d)$.

We rewrite Eq. (1.53) in the form

$$(-\Delta + C)\psi_0 = (C + I - V)\psi_0 =: g \tag{1.60}$$

and choose C large enough so that g is non-negative. As V is bounded, we can take for example $C = \|V\|_{L^\infty(\mathbb{R}^d)} + |I| + 1$ so that $g \geq \psi_0 \geq 0$. As ψ_0 is non-zero we also find that $g \in L^2(\mathbb{R}^d)$ is non-zero. Equation (1.60) can be explicitly solved. By going to the Fourier domain, we find $(C + |k|^2)\widehat{\psi_0}(k) = \widehat{g}(k)$ and therefore

$$\psi_0(x) = (2\pi)^{-\frac{d}{2}} \Phi * g(x) = (2\pi)^{-\frac{d}{2}} \int_{\mathbb{R}^d} \Phi(x - y)g(y)\,dy \tag{1.61}$$

where Φ is the inverse Fourier transform of the function $k \mapsto (C + |k|^2)^{-1}$. In dimension $d = 3$ this function is explicit,

$$\Phi(x) = \sqrt{\frac{\pi}{2}} \frac{e^{-\sqrt{C}|x|}}{|x|},$$

and is called the Yukawa potential. In dimension $d = 1$ we have $\Phi(x) = \sqrt{\pi/2}\, e^{-\sqrt{C}|x|}$. For the other dimensions we can for example use the integral formula

$$\frac{1}{C + |k|^2} = \int_0^\infty e^{-t(C+|k|^2)}\,dt$$

and the Fourier transform of Gaussians, to deduce after a change of variables that

$$\Phi(x) = \frac{2^{-d/2}}{|x|^{d-2}} \int_0^\infty e^{-sC|x|^2 - \frac{1}{4s}} \frac{ds}{s^{\frac{d}{2}}}. \tag{1.62}$$

This allows us to conclude that the function Φ is strictly positive everywhere and decreasing with respect to $|x|$. In fact, Φ is even C^∞ on $\mathbb{R}^d \setminus \{0\}$ and belongs to $L^1(\mathbb{R}^d, \mathbb{R})$.

We can now show (1.59) using (1.61). Since g is positive but not identically zero, we can choose R large enough so that $\int_{B_R} g > 0$. We then estimate, for all $x \in B_R$,

$$\psi_0(x) \geq (2\pi)^{-\frac{d}{2}} \int_{B_R} \Phi(x - y) g(y) \, dy$$

$$\geq (2\pi)^{-\frac{d}{2}} \Phi(2R) \int_{B_R} g(y) \, dy =: c_R > 0, \tag{1.63}$$

since $|x - y| \leq |x| + |y| \leq 2R$ and Φ is radial decreasing. This concludes the proof of (1.59), under our assumption that V is bounded.

The reader can verify that the previous proof works the same if we assume that V is bounded from above and we replace condition (1.37) by the stronger assumption that $p = \max(2, d)$ in dimension $d \neq 2$ and $p > 2$ in dimension $d = 2$, so that $V\psi_0 \in L^2(\mathbb{R}^d)$. This last condition on p can be further improved using the Rellich-Kato theorem that we will see in Chap. 3.

The proof of the strict positivity of ψ_0 is much more complicated in the general case, especially when V is not bounded from above. The property (1.59) is called the strong maximum principle and it was proven by Trudinger [Tru73] assuming however that $p > d/2$ if $d \geq 3$ (see also [AS82] for a probabilistic approach). For $p = d/2$ in dimension $d \geq 3$, we unfortunately get the weaker conclusion that $\psi_0 > 0$ almost everywhere, that is, $\{\psi_0 = 0\}$ has zero Lebesgue measure [Tru77], which complicates a bit the rest of the proof.

Step 4: Any Strictly Positive Solution of the Equation Is the Unique Minimizer Up to a Phase Let $\psi_0 \in H^2(\mathbb{R}^d)$ be a strictly positive function in the sense of (1.59), normalized in $L^2(\mathbb{R}^d)$, which is a solution of the equation $(-\Delta + V)\psi_0 = \lambda \psi_0$, for some $\lambda \in \mathbb{R}$. We do not assume *a priori* that ψ_0 is a minimizer. We will show that $\lambda = I$ and ψ_0 is the unique minimizer of I, up to a phase. We will then have proven more than necessary since in our case we already know that $\lambda = I = \mathcal{E}(\psi_0)$. Taking the scalar product against ψ_0 we already find that $\lambda = \mathcal{E}(\psi_0)$.

For all $\varphi \in C_c^\infty(\mathbb{R}^d)$, the function φ/ψ_0 belongs to $H^1(\mathbb{R}^d)$, because ψ_0 satisfies (1.59) and we have

$$\nabla \frac{\varphi}{\psi_0} = \frac{\nabla \varphi}{\psi_0} - \varphi \frac{\nabla \psi_0}{\psi_0^2}.$$

We then calculate

$$
\int_{\mathbb{R}^d} \psi_0(x)^2 \left| \nabla \frac{\varphi}{\psi_0}(x) \right|^2 dx = \int_{\mathbb{R}^d} \left| \nabla \varphi(x) - \frac{\varphi(x) \nabla \psi_0(x)}{\psi_0(x)} \right|^2 dx
$$

$$
= \int_{\mathbb{R}^d} |\nabla \varphi(x)|^2 \, dx + \int_{\mathbb{R}^d} |\varphi(x)|^2 \frac{|\nabla \psi_0(x)|^2}{\psi_0(x)^2} \, dx
$$

$$
- 2 \Re \int_{\mathbb{R}^d} \varphi(x) \overline{\nabla \varphi(x)} \cdot \frac{\nabla \psi_0(x)}{\psi_0(x)} \, dx
$$

$$
= \int_{\mathbb{R}^d} |\nabla \varphi(x)|^2 \, dx + \int_{\mathbb{R}^d} |\varphi(x)|^2 \frac{\Delta \psi_0(x)}{\psi_0(x)} \, dx.
$$

Here we have integrated by parts the cross term, using that $2\Re(\varphi \overline{\nabla \varphi}) = \nabla |\varphi|^2$. We then insert the equation $\Delta \psi_0 = (V - \lambda) \psi_0$ with $\lambda = \mathcal{E}(\psi_0)$ and finally obtain the relation

$$
\mathcal{E}(\varphi) = \int_{\mathbb{R}^d} \psi_0(x)^2 \left| \nabla \frac{\varphi}{\psi_0}(x) \right|^2 dx + \mathcal{E}(\psi_0) \int_{\mathbb{R}^d} |\varphi(x)|^2 \, dx. \tag{1.64}
$$

This formula is called the **ground state resolution** and it is the generalization to an arbitrary potential V of the relation (1.28) that we used for the hydrogen atom. As the first term is non-negative, this proves that $\mathcal{E}(\varphi) \geq \mathcal{E}(\psi_0) \int_{\mathbb{R}^d} |\varphi|^2$ for all $\varphi \in C_c^\infty(\mathbb{R}^d)$ and therefore, by density, also for all $\varphi \in H^1(\mathbb{R}^d)$. In other words, ψ_0 is a minimizer and thus $\mathcal{E}(\psi_0) = I = \lambda$. The same density argument also shows that $\psi_0 \nabla(\varphi/\psi_0)$ belongs to $L^2(\mathbb{R}^d)$ for all $\varphi \in H^1(\mathbb{R}^d)$, with the relation (1.64). So φ is a minimizer if and only if the first term vanishes, that is

$$
\nabla \frac{\varphi}{\psi_0}(x) = 0
$$

almost everywhere on $\mathbb{R}^d$. This means that $\varphi = c \psi_0$, as in the proof of uniqueness for the hydrogen atom in Theorem 1.3. This concludes the proof of Theorem 1.18 in the case where $V \in L^\infty(\mathbb{R}^d, \mathbb{R})$. $\qquad \square$

Remark 1.25 When $p = d/2$ in dimensions $d \geq 3$ we only know that $\psi_0 > 0$ almost everywhere according to [Tru77]. We should rather expand

$$
\int_{\mathbb{R}^d} (\psi_0(x) + \varepsilon)^2 \left| \nabla \frac{\varphi}{\psi_0 + \varepsilon}(x) \right|^2 dx
$$

and take the limit $\varepsilon \to 0$ at the end.

Complementary Exercises

Exercise 1.26 ($L^p(\mathbb{R}^d) + L^q(\mathbb{R}^d)$)   Let $1 \leq p, q \leq \infty$. We endow the space $L^p(\mathbb{R}^d) + L^q(\mathbb{R}^d)$ seen in Definition 1.7 with the norm (1.34).

1. Show that $L^p(\mathbb{R}^d) + L^q(\mathbb{R}^d)$ is a Banach space.
2. Show that for $1 < p, q \leq \infty$, the space $L^p(\mathbb{R}^d) + L^q(\mathbb{R}^d)$ identifies with the dual of $L^{p'}(\mathbb{R}^d) \cap L^{q'}(\mathbb{R}^d)$, the latter being equipped with the norm
$$\|g\|_{L^{p'}(\mathbb{R}^d) \cap L^{q'}(\mathbb{R}^d)} := \|g\|_{L^{p'}(\mathbb{R}^d)} + \|g\|_{L^{q'}(\mathbb{R}^d)}, \text{ with } p' = p/(p-1) \text{ and}$$
$q' = q/(q-1)$.
3. Let $1 \leq p < \infty$. We recall that $f \in L^p(\mathbb{R}^d) + L^\infty_\varepsilon(\mathbb{R}^d)$ when for all $\varepsilon > 0$ there exists $g \in L^p(\mathbb{R}^d)$ and $h \in L^\infty(\mathbb{R}^d)$ such that $f = g + h$ and $\|h\|_{L^\infty(\mathbb{R}^d)} \leq \varepsilon$ (Definition 1.8). Let $f \in L^p(\mathbb{R}^d) + L^\infty(\mathbb{R}^d)$. Show that the following assertions are equivalent:

   (i) $f \in L^p(\mathbb{R}^d) + L^\infty_\varepsilon(\mathbb{R}^d)$,
   (ii) $\{|f| \geq \eta\}$ has finite measure for all $\eta > 0$,
   (iii) $\lim_{R \to \infty} \|\mathbb{1}_{\mathbb{R}^d \setminus B_R} f\|_{L^p(\mathbb{R}^d) + L^\infty(\mathbb{R}^d)} = 0$.

4. Let $(h_n)_{n \geq 1}$ be a sequence of $L^\infty(\mathbb{R}^d)$ that converges to h. Show that if $|\{|h_n| \geq \eta\}| < \infty$ for all $\eta > 0$ and all $n \geq 1$, then the limit also satisfies $|\{|h| \geq \eta\}| < \infty$ for all $\eta > 0$.
5. Show that for all $1 \leq p < \infty$, $L^p(\mathbb{R}^d) + L^\infty_\varepsilon(\mathbb{R}^d)$ is a closed subspace of $L^p(\mathbb{R}^d) + L^\infty(\mathbb{R}^d)$.
6. Show that the closure of $C_c^\infty(\mathbb{R}^d)$ in $L^p(\mathbb{R}^d) + L^\infty(\mathbb{R}^d)$ is precisely $L^p(\mathbb{R}^d) + L^\infty_\varepsilon(\mathbb{R}^d)$.

Exercise 1.27 (Hardy's Inequality)

1. Let $u \in C_c^\infty(\mathbb{R}^d)$, in dimension $d \geq 3$. Expand

$$\int_{\mathbb{R}^d} \left| \nabla u(x) + \alpha \frac{x}{|x|^2} u(x) \right|^2 dx$$

and optimize over α to find **Hardy's inequality**

$$\int_{\mathbb{R}^d} \frac{|u(x)|^2}{|x|^2} dx \leq \left(\frac{2}{d-2} \right)^2 \int_{\mathbb{R}^d} |\nabla u(x)|^2 dx. \tag{1.65}$$

2. Show that if $u \in H^1(\mathbb{R}^d)$ with $d \geq 3$, then $x \mapsto |x|^{-1} u(x)$ is in $L^2(\mathbb{R}^d)$ and satisfies (1.65).
3. Using the radial function $u(x) = |x|^{-\alpha} \mathbb{1}(|x| \leq 1) + (2 - |x|) \mathbb{1}(1 \leq |x| \leq 2)$, show that the constant $2/(d-2)$ is the best possible in (1.65).
4. Provide a proof of Proposition 1.1 on the stability of the hydrogen atom, which uses Hardy's inequality (1.65) in place of the Sobolev inequality.

Exercise 1.28 (A Particle Confined in $\mathbb{R}^d$) Let V be a real-valued measurable function such that $V_- = \max(-V, 0) \in L^p(\mathbb{R}^d, \mathbb{R}) + L^\infty(\mathbb{R}^d, \mathbb{R})$ where p satisfies (1.37) and $V_+ = \max(V, 0) \in L^1_{\text{loc}}(\mathbb{R}^d)$ with $V_+(x) \to +\infty$ when $|x| \to +\infty$. We consider the energy

$$\mathcal{E}(\psi) = \int_{\mathbb{R}^d} |\nabla \psi(x)|^2 \, dx + \int_{\mathbb{R}^d} V(x)|\psi(x)|^2 \, dx$$

and the space $\mathcal{V} := \{\psi \in H^1(\mathbb{R}^d) : \sqrt{V_+}\psi \in L^2(\mathbb{R}^d)\}$ equipped with the norm

$$\|\psi\|^2_{\mathcal{V}} = \|\psi\|^2_{H^1(\mathbb{R}^d)} + \int_{\mathbb{R}^d} V_+(x)|\psi(x)|^2 \, dx.$$

1. Show that $\mathcal{V}$ is complete.
2. Show that $\mathcal{E}$ is well defined and continuous on $\mathcal{V}$.
3. Show that $I = \inf\{\mathcal{E}(\psi) : \psi \in \mathcal{V}, \int_{\mathbb{R}^d} |\psi|^2 = 1\}$ is finite.
4. Show that $\mathcal{V}$ is compactly embedded into $L^2(\mathbb{R}^d)$.
5. Deduce that I is attained and write the equation satisfied by any minimizer.
6. Show the uniqueness of the minimizer up to a phase when $V_+ \in L^\infty_{\text{loc}}(\mathbb{R}^d)$. You can follow the arguments of Sect. 1.6 and use [LL01, Thm. 9.10] instead of (1.63) to show the strict positivity of ψ, locally.
7. We now consider the harmonic oscillator $V(x) = \omega^2 |x|^2$ which corresponds to attaching our quantum particle to a spring nailed at the origin, ω being related to the stiffness of the spring.

   a. Show that

   $$\mathcal{E}(\psi) = \int_{\mathbb{R}^d} |\nabla \psi(x) + \omega x \psi(x)|^2 \, dx + \omega d \int_{\mathbb{R}^d} |\psi(x)|^2 \, dx \qquad (1.66)$$

   for all $\psi \in \mathcal{V}$. This formula is the equivalent, for the harmonic oscillator, of the relation (1.28) for the hydrogen atom and of the ground state resolution (1.64).

   b. Deduce that $I = \omega d$ and that the minimizers are all of the form

   $$\psi(x) = (\omega/\pi)^{\frac{d}{4}} e^{i\theta} e^{-\omega\frac{|x|^2}{2}}, \qquad \theta \in \mathbb{R}. \qquad (1.67)$$

   c. Show Heisenberg's inequality (1.21) by optimizing over ω.

Chapter 2
Self-adjointness

A $d \times d$ square matrix A with complex coefficients is called **self-adjoint** or **Hermitian** when $A^* = A$ where A^* is the matrix obtained by transposing and taking the complex conjugate of all the coefficients. The property $A^* = A$ is equivalent to $\langle v, Aw \rangle_{\mathbb{C}^d} = \langle Av, w \rangle_{\mathbb{C}^d}$ where $\langle v, w \rangle_{\mathbb{C}^d} = v^* w$ is the scalar product of $\mathbb{C}^d$. Self-adjoint matrices are all diagonalizable in an orthonormal basis and their eigenvalues are real. The generalization of this result to infinite dimension is quite involved. In this chapter we introduce and study the concept of self-adjointness.

2.1 Operators, Graph, Extension

Let $\mathfrak{H}$ be any separable Hilbert space over the field $\mathbb{C}$. It is often necessary to consider linear maps A that are only defined on a subspace $D(A)$ of $\mathfrak{H}$, called the **domain of** A.

Definition 2.1 (Operators in Infinite Dimension) An **operator on** $\mathfrak{H}$ is a linear map $A : D(A) \to \mathfrak{H}$, where $D(A)$ is a dense vector subspace of $\mathfrak{H}$.

We can work without the assumption that $D(A)$ is dense, but adding it greatly simplifies the framework. Density will play an important role when defining the adjoint of A in Sect. 2.4. In finite dimension, the only dense subspace is the entire space. In infinite dimension, it is absolutely necessary to always specify the domain $D(A)$ on which we work. As we will see from examples, the resolution of the eigenvalue equation $Av = \lambda v$ strongly depends on the considered domain.

The simplest example of an operator A is that of a linear map defined on the entire space $D(A) = \mathfrak{H}$, but we will see many examples of operators that cannot be defined on all $\mathfrak{H}$ and whose domain is necessarily a strict subspace of $\mathfrak{H}$. We particularly think of the differentiation $f \mapsto f'$ which is linear but not defined on $\mathfrak{H} = L^2(\mathbb{R})$ with values in the same space. It can be defined on $L^2(\mathbb{R})$ but f' is

© The Editor(s) (if applicable) and The Author(s), under exclusive license
to Springer Nature Switzerland AG 2024
M. Lewin, *Spectral Theory and Quantum Mechanics*, Universitext,
https://doi.org/10.1007/978-3-031-66878-4_2

then a distribution that does not necessarily belong to $L^2(\mathbb{R})$. The differentiation operator is however well defined on $C_c^\infty(\mathbb{R})$, for example, or on the Sobolev space $H^1(\mathbb{R}) \subsetneq L^2(\mathbb{R})$.

Instead of giving A and its domain $D(A)$, it is sometimes convenient to work with the **graph of** A, which contains both the operator and its domain of definition.

Definition 2.2 (Graph) The **graph of an operator** $(A, D(A))$ is the vector subspace of $\mathfrak{H} \times \mathfrak{H}$ given by

$$G(A) = \{(v, Av) \in D(A) \times \mathfrak{H}\}. \tag{2.1}$$

Conversely, it is legitimate to ask under what condition a subspace of $\mathfrak{H} \times \mathfrak{H}$ is the graph of an operator, which is the subject of the following lemma.

Lemma 2.3 (Characterization of Graphs) *A set $G \subset \mathfrak{H} \times \mathfrak{H}$ is the graph of an operator $(A, D(A))$ if and only if*

(i) G is a vector subspace of $\mathfrak{H} \times \mathfrak{H}$;
(ii) $(0, y) \in G$ implies $y = 0$;
(iii) $D = \{x \in \mathfrak{H} \; : \; \exists y \in \mathfrak{H}, \; (x, y) \in G\}$ is dense in $\mathfrak{H}$.

Proof Conditions *(i)–(iii)* are clearly necessary. Conversely, *(i)* and *(ii)* imply that if we have (x, y) and (x, y') in G, then $y = y'$. Hence for any $x \in D$ there is a unique y so that $(x, y) \in G$ and we can let $Ax := y$. The so-defined operator A is linear by *(i)* and its domain is dense by *(iii)*.   □

We can compare different realizations of the same operator on domains included in each other, using the concept of extension and restriction.

Definition 2.4 (Extension, Restriction)   Let A and B be two operators defined respectively on $D(A)$ and $D(B)$. We say that B is an **extension of** A and that A is a **restriction of** B, if $D(A) \subset D(B)$ and if B coincides with A on $D(A)$, that is, if $Bx = Ax$ for all $x \in D(A)$. An equivalent definition is to require that $G(A) \subset G(B)$, which we simply denote as $A \subset B$.

Example 2.5 We can define the derivative operator $f \mapsto f'$ in $D(A_1) = C_c^\infty(\mathbb{R})$ or in $D(A_2) = H^1(\mathbb{R})$, and we then have $A_1 \subset A_2$.

2.2 Spectrum

Throughout this book, we use the simple notation 1 for the identity of the ambient Hilbert space $\mathfrak{H}$. Thus, $A - z$ is the operator

$$\boxed{A - z = A - z \, \mathrm{Id}_{\mathfrak{H}}.}$$

The spectrum of a square matrix is defined as the set of complex numbers z such that $\det(A - z) = 0$, that is, such that $A - z$ is not invertible. Here, invertibility is equivalent to the injectivity or surjectivity of $A - z$. Moreover, the inverse $(A - z)^{-1}$ is always continuous because it is linear. In infinite dimension, the situation is more complex because a linear application can be injective without being surjective, and vice versa. In addition, the inverse may exist without being continuous. The definition of the spectrum is as follows.

Definition 2.6 (Spectrum) Let A be an operator defined on $D(A) \subset \mathfrak{H}$. The **resolvent set** of A is the subset of $\mathbb{C}$

$$\rho(A) := \Big\{ z \in \mathbb{C} \text{ such that } A - z : D(A) \to \mathfrak{H}$$

$$\text{is invertible, with bounded inverse } (A - z)^{-1} : \mathfrak{H} \to D(A) \subset \mathfrak{H} \Big\}.$$

The **spectrum of** A is the set $\sigma(A) = \mathbb{C} \setminus \rho(A)$.

The assumption that $(A - z)^{-1}$ is bounded means that there exists a constant C such that $\|(A - z)^{-1} v\| \le C \|v\|$ for all $v \in \mathfrak{H}$ or, in other words, that $(A - z)^{-1}$ defines a continuous map on $\mathfrak{H}$, but which takes its values in $D(A)$. When $z \in \rho(A)$, the operator $(A - z)^{-1}$ is called the **resolvent**. Note that if $D(A) = \mathfrak{H}$ and A is bounded, then the invertibility of $A - z$ automatically implies that $(A - z)^{-1}$ is bounded, by the open mapping theorem.

The set $\rho(A)$ contains all $z \in \mathbb{C}$ such that the equation $(A - z)v = w$ has a unique solution $v \in D(A)$ for any given $w \in \mathfrak{H}$ (this is the existence of the inverse), this solution v depending continuously on w in $\mathfrak{H}$ (this is the boundedness of $(A - z)^{-1}$). We see that a complex number λ can belong to the spectrum of A for several different reasons. For example, $A - \lambda$ might not be injective, and there then exists a $v \ne 0$ such that $Av = \lambda v$. In this case, λ is called an **eigenvalue of** A and v is an **associated eigenvector**. The (geometric) **multiplicity** of λ is by definition the dimension of $\ker(A - \lambda)$ and it can be finite or infinite. But it is also possible that $A - \lambda$ is injective without being surjective, or even that it is invertible but its inverse is not bounded on $\mathfrak{H}$.

Example 2.7 (Shift) On $\mathfrak{H} = \ell^2(\mathbb{N})$ we introduce the right shift S defined by $S(\mathbf{x}) = (0, x_1, x_2, \ldots)$ for $\mathbf{x} = (x_1, x_2, \ldots) \in \ell^2(\mathbb{N})$. Then S is injective but not surjective. Therefore $0 \in \sigma(S)$ but 0 is not an eigenvalue.

Before going further we begin by proving that the spectrum of an operator is always a closed set. The argument is based on the important property that the inverse $(A - z)^{-1}$ is bounded when $z \in \rho(A)$.

Lemma 2.8 (Holomorphy and Closure of $\sigma(A)$) *Let A be an operator defined on its domain $D(A)$ and $z \in \rho(A)$. Then the open ball of center z and radius*

$$\frac{1}{\|(A - z)^{-1}\|}$$

is included in $\rho(A)$ and the resolvent is given by the normally convergent series

$$(A - z')^{-1} = (A - z)^{-1} \sum_{n \geq 0} (z' - z)^n (A - z)^{-n}, \tag{2.2}$$

for all z' in this ball. In particular, $\rho(A)$ is open and therefore $\sigma(A)$ is closed.

Proof We can write $A - z' = (1 - (z' - z)(A - z)^{-1})(A - z)$ where the operator $A - z$ on the right is a bijection from $D(A)$ to $\mathfrak{H}$ since $z \in \rho(A)$ by assumption, and the operator $1 - (z' - z)(A - z)^{-1}$ is bounded on $\mathfrak{H}$. But we know that for any bounded operator B with norm $\|B\| < 1$, the operator $1 - B$ is invertible with $(1 - B)^{-1} = \sum_{n \geq 0} B^n$. Thus, the operator $1 - (z' - z)(A - z)^{-1}$ is invertible for $|z' - z| \, \|(A - z)^{-1}\| < 1$. As a composition of invertible operators, we then conclude that $A - z'$ is invertible from $D(A)$ to $\mathfrak{H}$ with a bounded inverse, given by Formula (2.2) in the statement. □

Remark 2.9 (Holomorphy and $\sigma(A) \neq \emptyset$) The formula (2.2) means that the map $z \in \rho(A) \mapsto (A - z)^{-1}$ can be expanded in a normally convergent power series in the vicinity of each point of $\rho(A)$. In particular, the function $f(z) := \langle v_1, (A - z)^{-1} v_2 \rangle$ is holomorphic on $\rho(A)$ for all $v_1, v_2 \in \mathfrak{H}$. We can use this information to show that $\sigma(A) \neq \emptyset$ for any *bounded* operator A. As $\|z(A - z)^{-1}\| = \|(A/z - 1)^{-1}\| \leq 2$ for $|z| \geq 2\|A\|$, this implies that $|f(z)| \leq 2\|v_1\| \, \|v_2\|/|z|$ and therefore $f(z) \to 0$ at infinity. However, since a holomorphic function on all the complex plane that tends to 0 at infinity is necessarily zero, we deduce that $\rho(A) \neq \mathbb{C}$, that is $\sigma(A) \neq \emptyset$, when A is bounded. If A is not bounded, it is however quite possible that $\sigma(A) = \emptyset$, see Exercise 2.10 below. We will show later in Theorem 2.23 that *symmetric* operators however always have a non-empty spectrum.

Exercise 2.10 (Empty Spectrum) Here is an example from [RS72]. Let $Pf = -if'$ in $\mathfrak{H} = L^2(I)$ with $I = (0, 1)$, defined on the domain $D(P) := \{f \in H^1(I) : f(0) = 0\}$. Also let S_z be the operator defined by

$$(S_z f)(x) := i \int_0^x e^{iz(x-y)} f(y) \, dy$$

on all $\mathfrak{H}$. Show that $P - z$ is invertible with inverse equal to S_z for all $z \in \mathbb{C}$. Deduce that $\sigma(P) = \emptyset$.

A natural question is to determine the spectrum of $(A - z)^{-1}$ when $z \in \rho(A)$. The following lemma specifies that it is the closure of the image of the spectrum of A by the function $\lambda \mapsto (\lambda - z)^{-1}$.

Lemma 2.11 (Spectrum of the Resolvent) *Let A be an operator defined on its domain $D(A)$ such that $\rho(A) \neq \emptyset$, and let $z \in \rho(A)$. The spectrum of the resolvent is*

$$\sigma\big((A - z)^{-1}\big) = \left\{ \frac{1}{\lambda - z} \; : \; \lambda \in \sigma(A) \right\} \cup \begin{cases} \emptyset & \text{if } A \text{ is bounded on } D(A) = \mathfrak{H}, \\ \{0\} & \text{if } A \text{ is not bounded.} \end{cases}$$

$$(2.3)$$

Proof We will show that for all $z' \neq z$, we have

$$z' \in \rho(A) \quad \text{if and only if} \quad \frac{1}{z' - z} \in \rho\big((A - z)^{-1}\big). \tag{2.4}$$

For this, we write

$$A - z' = A - z - (z' - z) = (z' - z)\left((z' - z)^{-1} - (A - z)^{-1} \right)(A - z). \tag{2.5}$$

If $(z' - z)^{-1} \in \rho((A - z)^{-1})$, the operator in the parenthesis on the right is invertible and, as $A - z$ is by hypothesis invertible on $D(A)$ with a bounded inverse, we deduce that $A - z'$ is invertible, with inverse

$$(A - z')^{-1} = \frac{1}{z' - z}(A - z)^{-1}\left((z' - z)^{-1} - (A - z)^{-1} \right)^{-1}.$$

This is bounded, as a composition of bounded operators. Conversely, if $z' \in \rho(A)$ we can write

$$(z' - z)^{-1} - (A - z)^{-1} = \frac{1}{(z' - z)}(A - z')(A - z)^{-1}.$$

The operator on the right is invertible on $\mathfrak{H}$ and its inverse is

$$(z' - z)(A - z)(A - z')^{-1} = (z' - z)\left(1 + (z' - z)(A - z')^{-1} \right),$$

which is indeed bounded. We have therefore shown (2.4), and thus completely identified the spectrum of $(A - z)^{-1}$ outside of the point 0, since $z' \mapsto (z' - z)^{-1}$ is bijective from $\mathbb{C} \setminus \{z\}$ into $\mathbb{C} \setminus \{0\}$. The range of the operator $(A - z)^{-1}$ is $D(A)$ and its left inverse is $A - z$. In particular, $(A - z)^{-1}$ is always injective. Therefore, $0 \in \rho((A - z)^{-1})$ if and only if this operator is surjective with a bounded inverse, that is, A is defined on all $D(A) = \mathfrak{H}$ and is continuous. $\square$

2.3 Closure

We discuss here the concept of closure of an operator.

Definition 2.12 (Closed Operator) An operator A with domain $D(A)$ is called **closed** when its graph $G(A)$ is closed in $\mathfrak{H} \times \mathfrak{H}$ or, in other words, if for any sequence $(x_n) \subset D(A)$ such that $x_n \to x$ and $Ax_n \to y$, we then have $x \in D(A)$ and $Ax = y$.

Proposition 2.13 (Spectrum of a Non-closed Operator) *If A, defined on $D(A)$, is not closed, then we have $\sigma(A) = \mathbb{C}$.*

Proof Let us argue by contraposition and show that if $\sigma(A) \neq \mathbb{C}$, then A is necessarily closed. Indeed, suppose that $z \notin \sigma(A)$ and consider a sequence $(x_n) \subset D(A)$ such that $x_n \to x$ and $Ax_n \to y$. Since $(A - z)$ is invertible with a bounded inverse, we have $(A - z)^{-1} A x_n = x_n + z(A - z)^{-1} x_n$. Taking the limit we obtain $(A - z)^{-1} y = x + z(A - z)^{-1} x$, thanks to the continuity of $(A - z)^{-1}$. Thus $x = (A - z)^{-1}(y - zx)$ belongs to $D(A)$ because $(A - z)^{-1}$ takes values in $D(A)$. By composing on the left by $(A - z)$, we find that $Ax = y$.   □

From this result we deduce that defining an operator A on a too small domain is really a bad idea. The operator might not be closed and in this case the spectrum will be equal to the entire complex plane, which would contradict our physical intuition that, for example, the spectrum of a self-adjoint operator should be real.

Example 2.14 (Momentum on $\mathbb{R}$) In $\mathfrak{H} = L^2(\mathbb{R})$, consider the momentum operator $P^{\min}$ defined by $P^{\min} f = -if'$ on $D(P^{\min}) = C_c^\infty(\mathbb{R})$. Then $P^{\min}$ is not closed, so $\sigma(P^{\min}) = \mathbb{C}$. To see that $P^{\min}$ is not closed it is enough to take a function $f \in H^1(\mathbb{R}) \setminus C_c^\infty(\mathbb{R})$, for example $f(x) = e^{-|x|^2}$ and use the density of $C_c^\infty(\mathbb{R})$ in $H^1(\mathbb{R})$ (Theorem A.8 in Appendix A), which provides a sequence $f_n \in C_c^\infty(\mathbb{R})$ such that $f_n \to f$ and $f_n' \to f'$ in $L^2(\mathbb{R})$. The pair $(f, -if')$ is then in the closure of the graph of $P^{\min}$, without belonging to the graph.

If we started by defining an operator A on a very small domain, so that everything is easily defined, it seems natural to close its graph. Unfortunately, the closure of the graph of an non-closed operator is not always a graph. Indeed, if properties *(i)* and *(iii)* of Lemma 2.3 easily pass to the closure, it is not the same for property *(ii)*, which can be lost. This justifies the following definition.

Definition 2.15 (Closure) Let A be an operator defined on its domain $D(A)$. We say that A is **closable** if it admits at least one closed extension. In this case, $\overline{G(A)}$ is the graph of an operator denoted $\overline{A}$, of domain $D(\overline{A})$ and called the **closure of** A. It is the smallest closed extension of $(A, D(A))$.

The definition contains the assertion that it is sufficient to have a closed extension to deduce that the closure of the graph $\overline{G(A)}$ is the graph of an operator. Indeed, if B is a closed extension of A, then $G(A) \subset \overline{G(A)} \subset G(B)$ because $\overline{G(A)}$ is by definition the smallest closed set containing $G(A)$. But then property *(ii)* of Lemma 2.3 is verified, since if $(0, y) \in \overline{G(A)}$ then $(0, y) \in G(B)$, so $y = 0$.

Exercise 2.16 (A Non-closable Operator) In $\mathfrak{H} = L^2(I)$ with $I = (0, 1)$, we call L the operator defined by $(Lf)(x) = f(1/2)$, for example on the domain $D(L) = C^0([0, 1])$. Let $\lambda \in \mathbb{C}$ and $f \in L^2(I)$. Show that there exists a sequence of functions $f_n \in C^0([0, 1])$ such that $f_n(1/2) = \lambda$ for all $n \geq 1$ and $f_n \to f$ strongly in $L^2(I)$. Deduce that the closure of the graph of the operator L is

$$\overline{G(L)} = \Big\{ (f, \lambda\mathbb{1}), \ f \in L^2(I), \ \lambda \in \mathbb{C} \Big\} \subset L^2(I) \times L^2(I)$$

where $\mathbb{1}$ denotes here the constant function equal to 1 on $[0, 1]$. Show that $\overline{G(L)}$ does not satisfy condition *(ii)* of Lemma 2.3. Hence L is not closable.

We now give examples that illustrate the importance of Sobolev spaces, since those appear as the domains of the closures of ordinary differential operators in the Hilbert space $\mathfrak{H} = L^2(\mathbb{R}^d)$.

Theorem 2.17 (Closure of ∂_{x_j} and Δ in $\mathbb{R}^d$) *In $\mathfrak{H} = L^2(\mathbb{R}^d)$, let $P_j^{\min}$ be the operator defined by $P_j^{\min} f = -i\partial_{x_j} f$, on the domain $D(P_j^{\min}) = C_c^\infty(\mathbb{R}^d)$, for $j = 1, \ldots, d$. Then $P_j^{\min}$ is closable and its closure is the operator $\overline{P_j^{\min}} =: P_j$ given by $P_j f = -i\partial_{x_j} f$ on the domain*

$$D(P_j) = \Big\{ f \in L^2(\mathbb{R}^d) \ : \ \partial_{x_j} f \in L^2(\mathbb{R}^d) \Big\}$$

$$= \Big\{ f \in L^2(\mathbb{R}^d) \ : \ k_j \widehat{f}(k) \in L^2(\mathbb{R}^d) \Big\}$$

where $\partial_{x_j} f$ is here understood in the sense of distributions.

Let $A^{\min}$ be the operator defined by $A^{\min} f = -\Delta f$, on the domain $D(A^{\min}) = C_c^\infty(\mathbb{R}^d)$. Then $A^{\min}$ is closable and its closure is the operator $\overline{A^{\min}} =: A$ given by $Af = -\Delta f$ on the domain $D(A) = H^2(\mathbb{R}^d)$.

The i in the definitions of $P_j^{\min}$ and P_j will be useful to make this operator symmetric, by compensating for the minus sign that appears in the integration by parts. The minus sign in front of $-i\partial_{x_j}$ is motivated by the interpretation of this operator as the quantum observable associated with momentum (Chap. 1). The result is exactly the same for the differential operator $f \mapsto \partial_{x_j} f$.

If $d = 1$, we simply have $D(P_1) = H^1(\mathbb{R})$. If $d \geq 2$, then $H^1(\mathbb{R}^d) = \bigcap_{j=1}^d D(P_j)$. This theorem therefore very naturally brings up Sobolev spaces, on which it is essential to define the usual differential operators, when we want them to be closed.

Proof We start by verifying that P_j is indeed closed. Let $(f_n, P_j f_n) = (f_n, -i\partial_{x_j} f_n)$ be a sequence from the graph of P_j that converges in $L^2(\mathbb{R}^d) \times L^2(\mathbb{R}^d)$ to the pair (f, g). We need to show that $f \in D(P_j)$ and that $g = P_j f$.

For this, we integrate by parts against a test function $\varphi \in C_c^\infty(\mathbb{R}^d)$ and, using the convergence in $L^2(\mathbb{R}^d)$, we find

$$i \int_{\mathbb{R}^d} \varphi(x) g(x)\, dx = \lim_{n\to\infty} \int_{\mathbb{R}^d} \varphi(x) \partial_{x_j} f_n(x)\, dx$$

$$= -\lim_{n\to\infty} \int_{\mathbb{R}^d} \partial_{x_j}\varphi(x) f_n(x)\, dx = -\int_{\mathbb{R}^d} \partial_{x_j}\varphi(x) f(x)\, dx.$$

This proves that we have $\partial_{x_j} f = ig$ in the sense of distributions therefore in particular that $\partial_{x_j} f \in L^2(\mathbb{R}^d)$. Thus, we indeed have $f \in D(P_j)$ and $g = P_j f$ so that P_j is closed. As $P_j^{\min} \subset P_j$, we see that $P_j^{\min}$ is closable, with $\overline{G(P_j^{\min})} \subset G(P_j)$. It remains to show that $G(P_j) \subset \overline{G(P_j^{\min})}$. For all $(f, -i\partial_{x_j} f)$ in $L^2(\mathbb{R}^d) \times L^2(\mathbb{R}^d)$, we can find a sequence $f_n \in C_c^\infty(\mathbb{R}^d)$ such that $f_n \to f$ and $\partial_{x_j} f_n \to \partial_{x_j} f$ strongly in $L^2(\mathbb{R}^d)$, by density of $C_c^\infty(\mathbb{R}^d)$ in $D(P_j)$ (the proof is the same as for $H^1(\mathbb{R}^d)$ in Theorem A.8 in Appendix A). Thus, P_j is indeed the closure of $P_j^{\min}$. The proof is similar for A, using in addition the fact that

$$H^2(\mathbb{R}^d) = \{f \in L^2(\mathbb{R}^d) \ : \ \Delta f \in L^2(\mathbb{R}^d)\}$$

which is Lemma A.6 (elliptic regularity) in Appendix A. $\square$

2.4 Adjoint

We can now define the adjoint of an operator. As in finite dimension we want to have

$$\langle v, Au \rangle = \langle A^* v, u \rangle \tag{2.6}$$

for all $u \in D(A)$ and $v \in D(A^*)$. The idea is to define A^* on the largest possible domain $D(A^*)$ so that Equality (2.6) holds. The relation (2.6) can also be written

$$\langle (v, A^* v), (Au, -u) \rangle_{\mathfrak{H} \times \mathfrak{H}} = 0 \tag{2.7}$$

where the scalar product of $\mathfrak{H} \times \mathfrak{H}$ is of course defined by

$$\langle (u_1, u_2), (v_1, v_2) \rangle_{\mathfrak{H} \times \mathfrak{H}} := \langle u_1, v_1 \rangle + \langle u_2, v_2 \rangle.$$

The relation (2.7) suggests defining the graph of the operator A^* as the orthogonal of the rotated graph of A:

$$\boxed{G(A^*) = \{(Au, -u), \ u \in D(A)\}^\perp} \tag{2.8}$$

where the orthogonal is taken in $\mathfrak{H} \times \mathfrak{H}$. Note that $G(A^*)$ is always closed, since it is the orthogonal of a subspace.

First of all, we must show that the set $G(A^*)$ introduced in (2.8) is indeed the graph of an operator A^* on its domain $D(A^*)$. We must therefore verify the conditions *(i)–(iii)* of Lemma 2.3. First, $G(A^*)$ does indeed satisfy property *(i)* since the orthogonal of a set is always a vector space. For *(ii)*, the argument relies on the density of $D(A)$. Indeed, we have $(0, w) \in G(A^*)$ if and only if $\langle w, u \rangle = 0$ for all $u \in D(A)$, according to the definition. Thus

$$D(A)^{\perp} = \{ w \in \mathfrak{H} : (0, w) \in G(A^*) \} \tag{2.9}$$

and our assumption that $D(A)$ is dense, that is $D(A)^{\perp} = \{0\}$, guarantees that *(ii)* is verified. This allows us to define the linear map A^*, possibly on a non-dense domain $D(A^*)$. For the density *(iii)* of the domain $D(A^*)$, we write that $v \in D(A^*)^{\perp}$ if and only if

$$(v, 0) \in G(A^*)^{\perp} = \left(\{(Au, -u), \ u \in D(A)\}^{\perp} \right)^{\perp} = \overline{\{(Au, -u), \ u \in D(A)\}}$$

or, equivalently, $(0, v) \in \overline{G(A)}$. In other words, we have found that

$$D(A^*)^{\perp} = \left\{ v \in \mathfrak{H} : (0, v) \in \overline{G(A)} \right\}. \tag{2.10}$$

Thus, $D(A^*)$ is dense if and only if $\overline{G(A)}$ satisfies property *(ii)* of Lemma 2.3, that is A is closable. The conclusion of this discussion is that

- the assumption that $D(A)$ is dense allows us to define the operator A^* but $D(A^*)$ is not necessarily dense;
- $D(A^*)$ is dense if and only if A is closable.

We will therefore only work with closable operators with a dense domain. The adjoint A^* is then well defined. This allows us to consider $(A^*)^*$, which turns out to be equal to $\overline{A}$.

Lemma 2.18 (Double Adjoint) *Let $(A, D(A))$ be a closable operator. Then we have $(A^*)^* = \overline{A}$ so that $((A^*)^*)^* = A^* = \overline{A}^*$.*

Proof The first equality follows from the fact that the bi-orthogonal coincides with the closure. The graph is also rotated twice in the definition of $(A^*)^*$, which does indeed bring up the graph of $\overline{A}$. Then, it remains to note that $A^* = \overline{A}^*$, since $V^{\perp} = \overline{V}^{\perp}$ for any subspace V.   □

The following lemma will be useful later when we need to study the surjectivity of $A - z$.

Lemma 2.19 *Let $(A, D(A))$ be a closable operator. Then we have*

$$\ker(A^* - \bar{z}) = \operatorname{ran}(A - z)^{\perp} = \operatorname{ran}(\overline{A} - z)^{\perp},$$

for all $z \in \mathbb{C}$.

Proof By definition of A^*, we have $(v, \bar{z}v) \in G(A^*)$ if and only if $\langle v, Au \rangle = \langle \bar{z}v, u \rangle = \langle v, zu \rangle$ for all $u \in D(A)$, which indeed means $v \in \operatorname{ran}(A - z)^{\perp}$. As $A^* = \overline{A}^*$ it follows that $\operatorname{ran}(A - z)^{\perp} = \operatorname{ran}(\overline{A} - z)^{\perp}$.   □

Exercise 2.20 Show that if A is a bounded operator on $\mathfrak{H}$, thus defined on $D(A) = \mathfrak{H}$, then its adjoint A^* is also defined on $D(A^*) = \mathfrak{H}$.

2.5   Symmetry

To define self-adjoint operators, it is important to distinguish the property of symmetry (already encountered for matrices) and the problems related to the domain of definition, which are typical of infinite dimension. We first discuss symmetry.

Definition 2.21 (Symmetry) An operator A defined on the domain $D(A) \subset \mathfrak{H}$ is called **symmetric** when $\langle v, Aw \rangle = \langle Av, w \rangle$ for all $v, w \in D(A)$. Equivalently $A \subset A^*$, that is $G(A) \subset G(A^*)$.

Since A^* is always closed, it follows that a symmetric operator is always closable, with $\overline{A} = (A^*)^*$. A symmetric operator therefore has two notable closed extensions, which are $\overline{A}$ (the smallest closed extension) and A^*. If B is a symmetric extension of A, then we have

$$A \subset B \subset B^* \subset A^*$$

since $A \subset B$ implies $B^* \subset A^*$ by definition of the adjoint. We therefore see that all the symmetric extensions of A are located between A and A^*. If they are also closed, they must be between $\overline{A}$ and A^*.

Exercise 2.22 Show that if $(A, D(A))$ is symmetric, then its closure is also.

Before defining the notion of self-adjointness, we make a small spectral digression and discuss the form of the spectrum of symmetric operators. In infinite dimension, a symmetric operator does not always have a real spectrum, but we will see that there are significant constraints on its spectrum. Only one of the four following cases can occur.

Theorem 2.23 (Spectrum of Symmetric Operators) *Let $(A, D(A))$ be a symmetric operator. Then its spectrum is*

- *either equal to the entire complex plane: $\sigma(A) = \mathbb{C}$;*
- *or equal to the closed upper half-plane: $\sigma(A) = \mathbb{C}_+ = \{z \in \mathbb{C} : \Im z \geq 0\}$;*

- *or equal to the closed lower half-plane:* $\sigma(A) = \mathbb{C}_- = \{z \in \mathbb{C} : \Im z \le 0\};$
- *or non-empty and included in* $\mathbb{R} : \emptyset \ne \sigma(A) \subset \mathbb{R}.$

In all cases, the spectrum never contains any eigenvalue in $\mathbb{C} \setminus \mathbb{R}$. *That is,* $\ker(A - z) = \{0\}$ *if* $\Im(z) \ne 0$. *Furthermore, if* $z \in \mathbb{C} \setminus \mathbb{R}$ *is such that* $A - z$ *is surjective, then* $z \in \rho(A)$ *and* $\sigma(A)$ *is included in the half-plane not containing* z.

We recall that if A is not closed, then $\sigma(A) = \mathbb{C}$ (Proposition 2.13). The theorem states that the spectrum of a symmetric operator is either equal to all of $\mathbb{C}$, equal to a complete half-plane, or included in $\mathbb{R}$. There are no other possibilities. The four cases do happen, as we will see on examples later. Next, the sole information that $A - z$ is surjective for a z with $\Im(z) \ne 0$, is enough to imply that $A = \overline{A}$ and the absence of spectrum on the entire half-plane to which z belongs. We therefore see that the spectrum of symmetric operators is very rigid and cannot be any set of $\mathbb{C}$.

If A is symmetric and bounded on all $D(A) = \mathfrak{H}$, then its spectrum is included in the ball of radius $\|A\|$. Only the fourth assertion is then possible and we conclude that $\sigma(A) \subset \mathbb{R}$.

The proof of the theorem relies on a seemingly innocent equality, which in fact plays a central role in the entire theory. It is the fact that, for all $u \in D(A)$ and $a, b \in \mathbb{R}$,

$$
\begin{aligned}
\|(A - a - ib)u\|^2 &= \langle (A - a - ib)u, (A - a - ib)u \rangle \\
&= \|(A - a)u\|^2 + b^2 \|u\|^2 - 2b\Im\langle (A - a)u, u \rangle \\
&= \|(A - a)u\|^2 + b^2 \|u\|^2,
\end{aligned}
$$

where we have used that $\langle (A - a)u, u \rangle = \langle u, (A - a)u \rangle = \overline{\langle (A - a)u, u \rangle}$ is real for every symmetric operator A. The Pythagoras-type relation

$$\boxed{\|(A - a - ib)u\|^2 = \|(A - a)u\|^2 + b^2 \|u\|^2 \ge b^2 \|u\|^2} \tag{2.11}$$

immediately implies that if $(A - a - ib)u = 0$ with $b \ne 0$, then $u = 0$, so there cannot be any non-real eigenvalue. Hence $\ker(A - z) = \{0\}$ if $\Im(z) \ne 0$. It also implies that if the inverse $(A - a - ib)^{-1}$ exists, then it is automatically bounded by

$$\boxed{\left\|(A - a - ib)^{-1}\right\| \le \frac{1}{|b|}.} \tag{2.12}$$

Thus, the only thing that can happen for a symmetric operator is that $A - a - ib$ is not surjective. But let us now write the proof of Theorem 2.23.

Proof We saw in Lemma 2.8 that the resolvent set $\rho(A)$ was open and we gave an estimate on the radius of the ball included in $\rho(A)$ for all $z \in \rho(A)$, depending on $\|(A - z)^{-1}\|$. The essential fact is that for a symmetric operator, this radius is always at least equal to $|b| = |\Im(z)|$, independently of the operator A, according to (2.12).

Thus, if the resolvent set $\rho(A)$ of the closed symmetric operator A intersects one of the half-planes, say $\{\Im(z) > 0\}$, then we claim that the entire half-plane must be in $\rho(A)$. Suppose by contradiction that $\sigma(A) \cap \{z \; : \; \Im(z) > 0\}$ is not empty and take $z = a + ib \in \mathbb{C} \setminus \mathbb{R}$ on the boundary of this set. Then there exists a sequence $z_n = a_n + ib_n \to z$ with $z_n \in \rho(A)$. As $b_n \to b$, the ball of center z_n and radius $|b|/2$ is then included in $\rho(A)$, but as z belongs to this ball for n large enough we arrive at a contradiction. The existence of a point of $\rho(A)$ in the upper half-plane is therefore sufficient to guarantee the total absence of spectrum in this half-plane. As the argument is the same for the lower half-plane, this shows that the spectrum must satisfy one of the four possibilities of the theorem. Moreover, we have already seen that the eigenvalues had to be real and that, for a symmetric operator, the only thing that could happen was that $A - z$ was not surjective, for $\Im(z) \neq 0$.

In order to conclude the proof of Theorem 2.23, it only remains to show that the spectrum of a symmetric operator is never empty, which we already know if A is bounded by Remark 2.9. Let us argue by contradiction and suppose that A is unbounded with $\sigma(A) = \emptyset$. Then A^{-1} is well defined and bounded, since $0 \in \rho(A)$. It is a symmetric operator because

$$\left\langle w, A^{-1}v \right\rangle = \left\langle AA^{-1}w, A^{-1}v \right\rangle = \left\langle A^{-1}w, AA^{-1}v \right\rangle = \left\langle A^{-1}w, v \right\rangle \qquad (2.13)$$

by symmetry of A on $D(A) = \mathrm{ran}(A^{-1})$. By Lemma 2.11, the assumption $\sigma(A) = \emptyset$ implies $\sigma(A^{-1}) = \{0\}$. However, it is classical that the only bounded symmetric operator with zero spectrum is the null operator. This can be shown using the spectral radius, but a more elementary approach is proposed in Exercise 2.24 below. We therefore deduce that $A^{-1} \equiv 0$, which is obviously absurd and concludes the proof of the theorem.   □

Exercise 2.24 (Bounded Symmetric Operators) Let B be a bounded and symmetric operator on $D(B) = \mathfrak{H}$. Following [Bre11, Prop. 6.9], we will show that

$$m := \inf_{\substack{v \in \mathfrak{H} \\ \|v\|=1}} \langle v, Bv \rangle, \qquad M := \sup_{\substack{v \in \mathfrak{H} \\ \|v\|=1}} \langle v, Bv \rangle$$

both belong to the spectrum of B. We recall that $\langle v, Bv \rangle$ is always real since B is symmetric. Show that for all $v, w \in \mathfrak{H}$ we have

$$|\langle w, (B - m)v \rangle| \leq \langle v, (B - m)v \rangle^{\frac{1}{2}} \langle w, (B - m)w \rangle^{\frac{1}{2}}$$

and deduce that $\|(B - m)v\| \leq \langle v, (B - m)v \rangle^{\frac{1}{2}} \|B - m\|^{\frac{1}{2}}$. Using a minimizing sequence for the minimization problem m, deduce that $m \in \sigma(B)$. Similarly, we have $M \in \sigma(B)$ (we will see later in Theorem 2.33 that $\sigma(B) \subset [m, M]$). Deduce from the fact that $m, M \in \sigma(B)$ that if $\sigma(B) = \{0\}$ then $B \equiv 0$.

2.6 Self-adjointness

2.6.1 Definition

It is now time to introduce the concept of self-adjointness.

Definition 2.25 (Self-adjointness) We say that an operator A, defined on $D(A) \subset \mathfrak{H}$, is **self-adjoint** when we have $A = A^*$, which means that A is symmetric ($A \subset A^*$) and that $D(A^*) = D(A)$. It is called **essentially self-adjoint** if it is symmetric and $\overline{A}$ is self-adjoint.

Recall that for any symmetric operator A we have $A \subset A^*$, which means that we have the inclusion of graphs

$$\{(v, Av) \in D(A) \times \mathfrak{H}\} \subset \{(Aw, -w) \in \mathfrak{H} \times D(A)\}^{\perp} = G(A^*). \tag{2.14}$$

When the ambient Hilbert space $\mathfrak{H}$ is of finite dimension d, the graph and the rotated graph are subspaces of dimension d of $\mathfrak{H} \times \mathfrak{H}$. Since $\dim(\mathfrak{H} \times \mathfrak{H}) = 2d$, the right orthogonal is also of dimension d. Thus, in finite dimension the two sets of (2.14) are necessarily equal for a symmetric matrix. In infinite dimension, the two spaces are not necessarily equal and self-adjointness is expressed in the form

$$\{(v, Av) \in D(A) \times \mathfrak{H}\} = \{(Aw, -w) \in \mathfrak{H} \times D(A)\}^{\perp}. \tag{2.15}$$

If $(A, D(A))$ is a symmetric operator, it is self-adjoint if and only if it verifies the property

$$\boxed{\langle v, Az \rangle = \langle w, z \rangle \ \text{ for all } z \in D(A) \quad \Longrightarrow \quad v \in D(A) \text{ and } Av = w.}$$

We can interpret the left side as a *weak formulation* of the equation $Av = w$ on the right side. A self-adjoint operator is therefore so that any weak solution of $Av = w$ is in fact a strong one.

Note that a self-adjoint operator never has a self-adjoint extension or restriction. Thus, it is never possible to decrease or increase the domain while maintaining self-adjointness. Indeed, recall that if $A \subset B$, then $B^* \subset A^*$, so that the equalities $A = A^*$ and $B = B^*$ immediately imply $A = B$.

If A is a symmetric operator, the self-adjoint extensions of A are all located between $\overline{A}$ and A^*, since they are closed and symmetric, as we explained in the previous section. It turns out that the symmetry of A^* is equivalent to the self-adjointness of $\overline{A}$, that is, to the essential self-adjointness of A.

Exercise 2.26 Let $(A, D(A))$ be a symmetric operator. Show that $\overline{A}$ is self-adjoint if and only if A^* is symmetric.

An easy example is that of a symmetric operator defined on all $\mathfrak{H}$, which is automatically self-adjoint.

Proposition 2.27 (Bounded Self-adjoint Operators) *If A is defined on all* $D(A) = \mathfrak{H}$ *and is symmetric, then A is self-adjoint and bounded.*

Proof As $D(A) \subset D(A^*)$ since A is assumed to be symmetric, we immediately have $D(A) = D(A^*)$ when $D(A) = \mathfrak{H}$. Therefore, A is self-adjoint. In particular, A is closed so, by the closed graph theorem, this means that A is continuous, hence bounded. □

2.6.2 Characterization and Weyl Sequences

For operators defined on a strict domain $D(A)$ of $\mathfrak{H}$, the notion of self-adjointness introduced earlier is fully justified by the following theorem.

Theorem 2.28 (Characterization of Self-adjoint Operators) *Let A be a symmetric operator defined on the domain $D(A) \subset \mathfrak{H}$. The following assertions are equivalent:*

1. *A is self-adjoint, that is, it verifies $D(A^*) = D(A)$;*
2. *the spectrum of A is real: $\sigma(A) \subset \mathbb{R}$;*
3. *there exists $\lambda \in \mathbb{C}$ such that $A - \lambda$ and $A - \bar{\lambda}$ are both surjective, from $D(A)$ to $\mathfrak{H}$.*

If we return to Theorem 2.23 providing the form of the spectrum of closed symmetric operators, we see that only the fourth case where $\sigma(A) \subset \mathbb{R}$ corresponds to that of a self-adjoint operator. The equivalence between the abstract relation $D(A) = D(A^*)$ and the real character of the spectrum shows that the theory developed so far is necessary. In addition to mimicking the case of finite dimension, having a real spectrum is extremely important from a practical point of view as it is one of the foundations of quantum mechanics, as we discussed in Sect. 1.5. To have a real spectrum, one must therefore work with operators verifying $D(A) = D(A^*)$. For differential operators this requires using Sobolev spaces and weak derivatives.

While Assertion 2 of the theorem is very comforting from the point of view of theory, Assertion 3 is very useful from a practical point of view and will be frequently used in the following. It is indeed much quicker to verify than 2 since if $\sigma(A) \subset \mathbb{R}$ then $A - a - ib$ and $A - a + ib$ are surjective for all $a \in \mathbb{R}$ and all $b \in \mathbb{R}^*$.

The proof of Theorem 2.28 relies heavily again on the important relation (2.11) seen in the previous section.

Proof (of Theorem 2.28) Let A be a self-adjoint operator. Let us show Assertion 2, that is, that $A - \lambda : D(A) \to \mathfrak{H}$ is invertible with a bounded inverse, for all $\lambda \in \mathbb{C} \backslash \mathbb{R}$. In fact, according to Theorem 2.23 it is enough to do it for $\pm i$, for example (but the proof is exactly the same for any $z \in \mathbb{C} \setminus \mathbb{R}$). The same theorem tells us that A has no non-real eigenvalues, so that $\ker(A \pm i) = \{0\}$. However, since $A = A^*$ we have $\ker(A \pm i) = \ker(A^* \pm i) = \mathrm{ran}(A \mp i)^{\perp}$ according to Lemma 2.19, which shows that $A \pm i$ has a dense range. To see that $\mathrm{ran}(A \pm i)$ is closed (and finally equal to all $\mathfrak{H}$), we can use the relation (2.11). Indeed, if $(A \pm i)v_n \to w$, we have

$$\|v_n - v_p\| \le \|(A \pm i)(v_n - v_p)\|,$$

which shows that v_n is a Cauchy sequence, and thus converges to a vector v in $\mathfrak{H}$. As $A = A^*$ is closed, we conclude that $v \in D(A)$ and that $(A \pm i)v = w$, that is, $\mathrm{ran}(A \pm i)$ is closed, therefore equal to all $\mathfrak{H}$. Again by Theorem 2.23, we deduce that $\sigma(A) \subset \mathbb{R}$.

As Assertion 2 obviously implies 3, it remains to prove that 3 implies 1. We now assume that $A - \lambda$ and $A - \bar{\lambda}$ are surjective for a $\lambda \in \mathbb{C}$ (real or not) and we want to show that A is self-adjoint, that is, the inclusion $\supset$ in (2.15). Let $(v, w = A^*v) \in G(A^*) = \{(Az, -z), z \in D(A)\}^\perp$, that is, such that $\langle v, Az \rangle = \langle w, z \rangle$ for all $z \in D(A)$. As $A - \bar{\lambda}$ is surjective by hypothesis, there exists $y \in D(A)$ such that $w - \bar{\lambda}v = (A - \bar{\lambda})y$ and we get

$$\langle v, (A - \lambda)z \rangle = \langle w - \bar{\lambda}v, z \rangle = \langle (A - \bar{\lambda})y, z \rangle = \langle y, (A - \lambda)z \rangle,$$

since $z \in D(A)$ and A is symmetric. Moreover, $A - \lambda$ is also surjective, so we can find $z \in D(A)$ such that $(A - \lambda)z = y - v$. We then deduce that $y = v$ and therefore that $v \in D(A)$ and $w = Av$. $\qquad\square$

Exercise 2.29 (Characterization of Essential Self-adjointness) Let A be a symmetric operator defined on the domain $D(A) \subset \mathfrak{H}$. Show that A is essentially self-adjoint (that is, $\overline{A}$ is self-adjoint) if and only if there exists $\lambda \in \mathbb{C} \setminus \mathbb{R}$ such that $A - \lambda$ and $A - \bar{\lambda}$ both have a dense range in $\mathfrak{H}$.

Here is now a result that provides a very useful interpretation of the spectrum of self-adjoint operators in infinite dimension.

Theorem 2.30 (Spectrum of Self-adjoint Operators) *Let A be a self-adjoint operator on $D(A) \subset \mathfrak{H}$, and $\lambda \in \mathbb{R}$. The following assertions are equivalent:*

1. $\lambda \in \sigma(A)$;
2. $\displaystyle\inf_{\substack{v \in D(A) \\ \|v\|=1}} \|(A - \lambda)v\| = 0$;
3. *there exists a sequence* $(v_n) \in D(A)^{\mathbb{N}}$ *such that* $\|v_n\| = 1$ *and* $\|(A - \lambda)v_n\| \to 0$.

The third assertion tells us that the elements of the spectrum are all "quasi-eigenvalues" in the sense that we can solve the equation $Av = \lambda v$ approximately with a sequence v_n, without necessarily being able to take the limit and actually find a solution. In finite dimension, as the unit sphere is compact and A is continuous, we can always pass to the limit and there are only eigenvalues.

Definition 2.31 (Weyl Sequence) A sequence (v_n) satisfying the properties of Assertion 3 of Theorem 2.30 is called a **Weyl sequence**.

Proof (of Theorem 2.30) The equivalence between Assertions 2 and 3 follows from the definition of the infimum. If 2 is true, the inverse of $A - \lambda$, if it exists, cannot be bounded, since this would imply $\|(A - \lambda)^{-1}w\| \leq C\|w\|$ and therefore $1 = \|v\| \leq C\|(A - \lambda)v\|$ by taking $w = (A - \lambda)v$ with $\|v\| = 1$, which would contradict the fact that the infimum in 2 is 0. Therefore, λ is necessarily in the spectrum.

Conversely, if the infimum in 2 is $\varepsilon > 0$, then $\|(A - \lambda)v\| \geq \varepsilon\|v\|$ for all $v \in D(A)$. This inequality then plays a role similar to our relation (2.11), but with λ real. Indeed, this obviously implies that $\ker(A - \lambda) = \{0\}$ and therefore, by Lemma 2.19, that $A - \lambda$ has a dense range. But with the same argument as at the beginning of the proof of Theorem 2.28, we conclude that the image is closed and that the inverse is bounded by $1/\varepsilon$. This shows that $\lambda \notin \sigma(A)$. $\qquad\square$

Remark 2.32 If (v_n) is a Weyl sequence, then it is bounded and therefore has a subsequence (v_{n_k}) that converges weakly to a vector v in $\mathfrak{H}$. It turns out that we can take the limit in the equation $(A - \lambda)v_{n_k} \to 0$ and deduce that $v \in D(A)$ and $(A - \lambda)v = 0$. In particular, if λ belongs to the spectrum of the self-adjoint operator A but is not an eigenvalue, we deduce that $v = 0$. Since this is true for any subsequence, we must therefore have $v_n \rightharpoonup 0$. To justify the limit, we take the scalar product against a fixed vector $w \in D(A)$ and find, thanks to the symmetry of A,

$$0 = \lim_{n_k \to \infty} \langle w, (A - \lambda)v_{n_k} \rangle = \lim_{n_k \to \infty} \langle (A - \lambda)w, v_{n_k} \rangle = \langle (A - \lambda)w, v \rangle.$$

This shows that $\langle Aw, v \rangle = \lambda \langle w, v \rangle$ for all $w \in D(A)$, so that $(v, \lambda v) \in G(A^*)$. Since A is assumed to be self-adjoint, we deduce that $v \in D(A)$ and that $Av = \lambda v$.

We now give an interesting consequence of Theorem 2.30, which is a generalization of Exercise 2.24 to the case of unbounded operators.

Theorem 2.33 (Localization of the Spectrum) *Let A be a self-adjoint operator on the domain $D(A) \subset \mathfrak{H}$. We then have*

$$\inf \sigma(A) = \inf_{\substack{v \in D(A) \\ \|v\|=1}} \langle v, Av \rangle, \qquad \sup \sigma(A) = \sup_{\substack{v \in D(A) \\ \|v\|=1}} \langle v, Av \rangle \qquad (2.16)$$

In particular, $\sigma(A) \subset [a, +\infty)$ with $a > -\infty$ if and only if $\langle v, Av \rangle \geq a\|v\|^2$ for all $v \in D(A)$.

If the spectrum is bounded below, its infimum is actually a minimum since it is a closed set. Otherwise, $\inf \sigma(A) = -\infty$ and the theorem asserts that the quadratic form $v \in D(A) \mapsto \langle v, Av \rangle$ is not bounded below on the unit sphere. The interpretation is the same for the supremum.

Proof We only show the result for the infimum. For the supremum, just replace A by $-A$. Let $m := \inf\{\langle v, Av \rangle : v \in D(A), \|v\| = 1\}$. Let $\lambda \in \sigma(A)$ and (v_n) be an associated Weyl sequence. As $Av_n - \lambda v_n \to 0$ and $\|v_n\| = 1$, it follows from the Cauchy-Schwarz inequality that $\langle v_n, Av_n - \lambda v_n \rangle = \langle v_n, Av_n \rangle - \lambda \to 0$. Since

$\langle v_n, A v_n \rangle \geq m$, this shows after taking the limit that $\lambda \geq m$. This being valid for all $\lambda \in \sigma(A)$, we get $\inf \sigma(A) \geq m$. If the spectrum is not bounded below, then $m = -\infty$.

In Exercise 2.24 we have already shown that $m \in \sigma(A)$ when A is bounded. This implies $m \geq \min \sigma(A)$ and therefore the desired equality in the bounded case. When A is not bounded, we argue by contradiction and assume that $\inf \sigma(A) > m$. Because of the strict inequality, we have in particular $\inf \sigma(A) > -\infty$ and the spectrum of A is therefore bounded below. Let then $m < a < \min \sigma(A)$ so that $a \in \rho(A)$. By definition of m we can find a normalized vector $v \in D(A)$ such that $m < \langle v, Av \rangle < a = a\|v\|^2$, which we rewrite in the form

$$\langle v, (A - a)v \rangle = \left\langle w, (A - a)^{-1} w \right\rangle < 0, \qquad \text{where } w := (A - a)v \neq 0.$$

The operator $(A - a)^{-1}$ is bounded and symmetric according to (2.13). By Exercise 2.24, the strict negativity of $\langle w, (A - a)^{-1} w \rangle$ implies that $(A - a)^{-1}$ has spectrum in $(-\infty, 0)$. By Lemma 2.11 we deduce that A has spectrum in $(-\infty, a)$. This contradicts the assumption that $a < \min \sigma(A)$ and thus concludes the proof of the theorem. $\square$

2.6.3 Diagonal Operators

We conclude this section with a useful result in certain applications, which states that a closed symmetric operator already diagonalized in an orthonormal basis is automatically self-adjoint, and that its spectrum is the closure of the set of its eigenvalues.

Theorem 2.34 (Diagonalized Operators) *Let A be a symmetric operator on a domain $D(A) \subset \mathfrak{H}$, such that there exists a Hilbert basis $(e_n)_{n \geq 1}$ of $\mathfrak{H}$ composed of elements of $D(A)$, which are all eigenvectors: $A e_n = \lambda_n e_n$ with $\lambda_n \in \mathbb{R}$. Then, the closure of A is the operator*

$$\overline{A} v = \sum_{n \geq 1} \lambda_n \langle e_n, v \rangle e_n, \qquad D(\overline{A}) := \left\{ v \in \mathfrak{H} : \sum_{n \geq 1} |\lambda_n|^2 |\langle e_n, v \rangle|^2 < \infty \right\}.$$

$$(2.17)$$

The latter is a self-adjoint operator whose spectrum is

$$\sigma(\overline{A}) = \overline{\{\lambda_n, n \geq 1\}}. \tag{2.18}$$

In other words, $(A, D(A))$ is essentially self-adjoint.

Proof Let us call B the operator defined in (2.17). That is, Bv is the vector whose components in the Hilbert basis $(e_n)_{n \geq 1}$ are $\lambda_n \langle e_n, v \rangle$ on $D(B) = \{v :$

$\sum \lambda_n^2 |\langle e_n, v \rangle|^2 < \infty\}$. This operator is symmetric because for $v, w \in D(B)$ we have by Parseval

$$\langle w, Bv \rangle = \sum_{n \geq 1} \overline{\langle e_n, w \rangle} \langle e_n, Bv \rangle = \sum_{n \geq 1} \lambda_n \underbrace{\overline{\langle e_n, w \rangle}}_{=\overline{\langle e_n, Bw \rangle}} \langle e_n, v \rangle = \langle Bw, v \rangle$$

since the λ_n's are real. To show that B is closed, consider a sequence v_m of $D(B)$ that converges in $\mathfrak{H}$ to v and such that Bv_m converges to a vector w. The norm of Bv_m is

$$\|Bv_m\|^2 = \sum_{n \geq 1} |\langle e_n, Bv_m \rangle|^2 = \sum_{n \geq 1} \lambda_n^2 |\langle e_n, v_m \rangle|^2.$$

Each $\langle e_n, v_m \rangle$ converges to $\langle e_n, v \rangle$ when $m \to \infty$ and it then follows from Fatou's inequality for series that

$$\sum_{n \geq 1} \lambda_n^2 |\langle e_n, v \rangle|^2 \leq \lim_{m \to \infty} \|Bv_m\|^2 = \|w\|^2.$$

The series on the left is therefore convergent, which shows that $v \in D(B)$. By symmetry of B we have

$$\langle e_n, w \rangle = \lim_{m \to \infty} \langle e_n, Bv_m \rangle = \lim_{m \to \infty} \langle Be_n, v_m \rangle = \lambda_n \lim_{m \to \infty} \langle e_n, v_m \rangle = \lambda_n \langle e_n, v \rangle.$$

The components of w on the basis are $\lambda_n \langle e_n, v \rangle$ and therefore we have indeed $w = Bv$. We then remark that the λ_n's are eigenvalues of B because $Be_n = \lambda_n e_n$ by definition, so that $\overline{\{\lambda_n, \ n \geq 1\}} \subset \sigma(B)$ since the spectrum is closed. If z belongs to the complement of $\overline{\{\lambda_n, \ n \geq 1\}}$, there exists $\delta > 0$ such that $|z - \lambda_n| \geq \delta$ for all n. Then $B - z$ is invertible with the explicit bounded inverse given by

$$R_z w := \sum_{n \geq 1} \frac{\langle e_n, w \rangle}{\lambda_n - z} e_n.$$

Indeed, the latter defines a bounded operator with $\|R_z\| \leq \delta^{-1}$ since

$$\sum_{n \geq 1} \frac{|\langle e_n, w \rangle|^2}{|\lambda_n - z|^2} \leq \delta^{-2} \sum_{n \geq 1} |\langle e_n, w \rangle|^2 = \delta^{-2} \|w\|^2.$$

In addition, we have $R_z w \in D(B)$ for all $w \in \mathfrak{H}$, since $\frac{\lambda_n}{\lambda_n - z} = 1 + \frac{z}{\lambda_n - z}$ and hence

$$\sum_{n \geq 1} \frac{\lambda_n^2 |\langle e_n, w \rangle|^2}{|\lambda_n - z|^2} \leq \left(1 + \frac{|z|}{\delta}\right)^2 \|w\|^2.$$

It is now allowed to multiply on the left by $(B-z)$ and we find $(B-z)R_z w = w$, that is, R_z is the inverse of $B - z$. This shows, as we wanted, that $\sigma(B) = \overline{\{\lambda_n, \ n \geq 1\}}$. Its spectrum being real, the symmetric operator B is self-adjoint.

Next we introduce a new operator $A^{\min}$ which is the restriction of B to the space $D(A^{\min})$ composed of *finite* linear combinations of the e_n. This operator is symmetric because B is, and moreover its closure is $\overline{A^{\min}} = B$, which is seen by simply truncating the series. The operator $A^{\min}$ is therefore essentially self-adjoint and we have in particular $(A^{\min})^* = (\overline{A^{\min}})^* = B^* = B$.

Let us finally return to the symmetric operator A of the statement. We have $A^{\min} \subset A$ since the e_n are in $D(A)$ therefore, by symmetry, $A^{\min} \subset A \subset A^* \subset (A^{\min})^* = B$. After closure we immediately obtain that $\overline{A} = B$, as announced. $\qquad \square$

Exercise 2.35 (Reproducing a Given Spectrum in $\mathbb{R}$ or $\mathbb{C}$) Let $(\lambda_n)_{n \geq 1}$ be any sequence in $\mathbb{C}$ and $(e_n)_{n \geq 1}$ be an orthonormal basis of an infinite-dimensional separable Hilbert space $\mathfrak{H}$. Prove that the operator in (2.17) is well defined and closed, and that its spectrum is still given by (2.18). Use this to show that for any closed set $\Sigma \subset \mathbb{C}$, there exists a Hilbert space $\mathfrak{H}$ and a closed operator A such that $\sigma(A) = \Sigma$. If $\Sigma \subset \mathbb{R}$, then A can be chosen self-adjoint.

2.7 Momentum and Laplacian on $\mathbb{R}^d$

In Theorem 2.17, we defined the differential operators

$$P_j^{\min} f = -i \partial_{x_j} f, \qquad D(P_j^{\min}) = C_c^\infty(\mathbb{R}^d),$$

$$A^{\min} f = -\Delta f, \qquad D(A^{\min}) = C_c^\infty(\mathbb{R}^d).$$

We saw that these operators were not closed and that their closures were given by

$$P_j f = -i \partial_{x_j} f, \qquad D(P_j) = \{f \in L^2(\mathbb{R}^d) \ : \ \partial_{x_j} f \in L^2(\mathbb{R}^d)\},$$

$$A f = -\Delta f, \qquad D(A) = H^2(\mathbb{R}^d).$$

We show here that the latter operators are self-adjoint. In other words, the operators $P_j^{\min}$ and $A^{\min}$ are essentially self-adjoint. They admit only one possible self-adjoint extension which is their closure.

Theorem 2.36 (Momentum and Laplacian on $\mathbb{R}^d$) *The operators $P_j = -i\partial_{x_j}$ defined on $D(P_j) = \{f \in L^2(\mathbb{R}^d) \ : \ \partial_{x_j} f \in L^2(\mathbb{R}^d)\} \subset \mathfrak{H} = L^2(\mathbb{R}^d)$ for $j = 1, \ldots, d$ are self-adjoint and their spectrum is*

$$\boxed{\sigma(P_j) = \mathbb{R}.}$$

The operator $A = -\Delta$ defined on $D(A) = H^2(\mathbb{R}^d) \subset \mathfrak{H} = L^2(\mathbb{R}^d)$ is self-adjoint and its spectrum is

$$\sigma(-\Delta) = [0, +\infty).$$

The operators P_j and A have no eigenvalues.

Proof Let us write the proof for the Laplacian, the one for P_j being very similar and left as an exercise. It is classical that the operator $-\Delta$ is symmetric on $H^2(\mathbb{R}^d)$. Indeed, after two integrations by parts we have

$$-\int_{\mathbb{R}^d} \overline{g(x)} \, \Delta f(x) \, dx = -\int_{\mathbb{R}^d} \overline{\Delta g(x)} \, f(x) \, dx,$$

for all $f, g \in C_c^\infty(\mathbb{R}^d)$, a relation that extends to all $H^2(\mathbb{R}^d)$, by density of $C_c^\infty(\mathbb{R}^d)$ in this space. According to Theorem 2.28 with $\lambda = -1 = \overline{\lambda}$, it is then sufficient to show that $A + 1$ is surjective. In other words, for every $g \in L^2(\mathbb{R}^d)$ we want to prove the existence of a function $f \in H^2(\mathbb{R}^d)$ such that $(1 - \Delta) f = g$. By passing to the Fourier transform, we find that $(1 + |k|^2) \widehat{f}(k) = \widehat{g}(k)$, so the function

$$f = \mathcal{F}^{-1} \left(\frac{\widehat{g}(k)}{1 + |k|^2} \right)$$

works. It is indeed in $H^2(\mathbb{R}^d)$ by the characterization (A.9) of this space recalled in Appendix A, since $(1 + |k|^2) \widehat{f}(k) = \widehat{g}(k) \in L^2(\mathbb{R}^d)$.

After one integration by parts we also find

$$\langle f, -\Delta f \rangle = \int_{\mathbb{R}^d} |\nabla f(x)|^2 \, dx \geq 0$$

for all $f \in H^2(\mathbb{R}^d)$ which, by Theorem 2.33, implies $\sigma(A) \subset \mathbb{R}^+$. Let us now show the converse inclusion. For every $k_0 \in \mathbb{R}^d$ and any function $f \in H^2(\mathbb{R}^d)$ normalized in $L^2(\mathbb{R}^d)$, we consider the sequence of functions $f_n(x) = n^{-d/2} f(x/n) e^{ix \cdot k_0}$, whose Fourier transform is $\widehat{f_n}(k) = n^{d/2} \widehat{f}(n(k - k_0))$. We have defined f_n so that $|\widehat{f_n}|^2 \rightharpoonup \delta_{k_0}$ in the sense of measures. We then have

$$\left\| \left(-\Delta - |k_0|^2 \right) f_n \right\|_{L^2(\mathbb{R}^d)}^2 = \int_{\mathbb{R}^d} (|k|^2 - |k_0|^2)^2 |\widehat{f_n}(k)|^2 \, dk$$

$$= \int_{\mathbb{R}^d} \left(\left| k_0 + \frac{p}{n} \right|^2 - |k_0|^2 \right)^2 |\widehat{f}(p)|^2 \, dp$$

$$= \frac{1}{n} \int_{\mathbb{R}^d} \left(2p \cdot k_0 + \frac{|p|^2}{n} \right)^2 |\widehat{f}(p)|^2 \, dp,$$

which tends to 0 when $n \to \infty$. By Theorem 2.30, this shows that $|k_0|^2$ belongs to $\sigma(-\Delta)$ for all $k_0 \in \mathbb{R}^d$. As $|k_0|^2$ can take all possible values in $\mathbb{R}^+$, we have indeed shown that $\sigma(-\Delta) = \mathbb{R}^+$. The spectrum contains no eigenvalues because if we have $(-\Delta - \lambda)f = 0$ for a $f \in H^2(\mathbb{R}^d)$, then we deduce from

$$\|(-\Delta - \lambda)f\|_{L^2(\mathbb{R}^d)}^2 = \int_{\mathbb{R}^d} (|k|^2 - \lambda)^2 |\widehat{f}(k)|^2 \, dk$$

that the Fourier transform $\widehat{f}$ is supported in the sphere of radius $\sqrt{\lambda}$. As the latter has zero measure, it follows that $f \equiv 0$ almost everywhere.   □

2.8   Momentum and Laplacian on an Interval

In this section, we study in detail the operators

$$f \mapsto -if' \quad \text{and} \quad f \mapsto -f'' \quad \text{on } \mathfrak{H} = L^2(I) \text{ with } I = (0, 1) \text{ or } I = (0, \infty).$$

These examples are very instructive and illustrate quite well the notions introduced so far. Unlike the case of the whole space $\mathbb{R}$ of the previous section, we will see that the minimal operators, defined on $C_c^\infty(I)$, can have several self-adjoint extensions, or sometimes even none at all. The boundary of the interval I plays a crucial role in the self-adjointness property.

2.8.1   Momentum on the Finite Interval $I = (0, 1)$

We wish to define the operator $f \mapsto -if'$ on an appropriate domain in the Hilbert space $\mathfrak{H} = L^2(I)$ with $I = (0, 1)$. A first natural idea is to define this operator on the very small domain $C_c^\infty(I)$ because we of course want it to coincide with the usual derivative for very smooth functions. We therefore introduce the operator

$$P^{\min} f = -if', \qquad D(P^{\min}) = C_c^\infty(I)$$

and look for the self-adjoint extensions of $P^{\min}$. Another natural operator is the one defined on the Sobolev space $H^1(I)$

$$P^{\max} f = -if', \qquad D(P^{\max}) = H^1(I).$$

As the space $H^1(I)$ contains exactly the functions of $\mathfrak{H}$ whose derivative in the sense of distributions is still in $\mathfrak{H}$, the operator $P^{\max}$ is the largest possible one we can imagine (when the derivative is interpreted in the sense of distributions), hence the notation $P^{\max}$. Obviously, $P^{\max}$ is by definition an extension of $P^{\min}$. We will see that all the self-adjoint extensions of $P^{\min}$ are located between $P^{\min}$ and $P^{\max}$.

The operator $P^{\min}$ is *symmetric* because the boundary terms vanish when we perform an integration by parts:

$$\left\langle f, P^{\min} g \right\rangle = -i \int_0^1 \overline{f(t)} g'(t) \, dt = i \int_0^1 \overline{f'(t)} g(t) \, dt = \left\langle P^{\min} f, g \right\rangle$$

for all $f, g \in C_c^\infty(I)$. Let us recall that the functions of $H^1(I)$ are all continuous up to the two end points of the interval $I = (0, 1)$, according to Lemma A.2, and that the map

$$f \in H^1(I) \mapsto (f(0), f(1)) \in \mathbb{C}^2 \tag{2.19}$$

is continuous. By density of $C^\infty([0, 1])$ (Exercise A.3), we see that integration by parts remains true in $H^1(I)$, this time with boundary terms:

$$-i \int_0^1 \overline{f(t)} g'(t) \, dt = \int_0^1 \overline{-i f'(t)} g(t) \, dt - i \left(\overline{f(1)} g(1) - \overline{f(0)} g(0) \right),$$

$$\forall f, g \in H^1(I). \tag{2.20}$$

As the boundary terms are generally non zero, the operator $P^{\max}$ is *not symmetric*. For example, $\langle f, P^{\max} g \rangle - \langle P^{\max} f, g \rangle = -i$ for $f(x) = 1$ and $g(x) = x$. This already disqualifies $P^{\max}$ which is therefore not a self-adjoint extension of $P^{\min}$. The operator $P^{\min}$ is not better as it is not closed. As for its closure, it is indeed symmetric, but is not self-adjoint, as stated in the following lemma.

Lemma 2.37 (Closure and Adjoints) *The operator $P^{\max}$ is closed. On the other hand, the operator $P^{\min}$ is not closed and its closure is the operator $P_0 : f \mapsto -if'$ defined on the domain*

$$D(P_0) = H_0^1(I) = \left\{ f \in L^2(I) \ : \ f' \in L^2(I), \ f(0) = f(1) = 0 \right\}.$$

We have $(P^{\min})^* = (P_0)^* = P^{\max}$ and $(P^{\max})^* = P_0$, so that P_0 is not self-adjoint. The spectra are

$$\sigma(P^{\min}) = \sigma(P^{\max}) = \sigma(P_0) = \mathbb{C}. \tag{2.21}$$

The spectrum of $P^{\max}$ is composed only of eigenvalues, while those of $P^{\min}$ and P_0 contain none.

Proof Let us start by showing that $P^{\max}$ is closed. The proof is exactly the same as that of Theorem 2.17. We consider a sequence $f_n \in H^1(I)$ such that

$(f_n, P^{\max} f_n) = (f_n, -if_n') \rightarrow (f, g)$ in $L^2(I) \times L^2(I)$. By integrating by parts against a function $\varphi \in C_c^\infty(I)$ and using the convergence in $L^2(I)$, we find

$$i \int_0^1 \varphi(t) g(t) \, dt = \lim_{n \to \infty} \int_0^1 \varphi(t) f_n'(t) \, dt$$

$$= - \lim_{n \to \infty} \int_0^1 \varphi'(t) f_n(t) \, dt = - \int_0^1 \varphi'(t) f(t) \, dt.$$

This proves that we have $f' = ig$ in the sense of distributions on $I = (0, 1)$. Therefore $f \in H^1(I) = D(P^{\max})$ and $g = -if' = P^{\max} f$. Thus, $P^{\max}$ is closed. As the evaluation at 0^+ and 1^- is continuous on $H^1(I)$, the same proof shows that P_0 is also closed. As P_0 is closed, to have $\overline{P^{\min}} = P_0$ it is enough to show that we can approach any element of the graph of P_0 by a sequence of points from the graph of $P^{\min}$. This follows from the density of $C_c^\infty(I)$ in $H_0^1(I)$ (Exercise A.3 in Appendix A).

The graph of the adjoint $(P^{\min})^*$ is by definition the set of pairs $(f, g) \in L^2(I)$ such that

$$-i \int_0^1 \overline{f(x)} u'(x) \, dx = \int_0^1 \overline{g(x)} u(x) \, dx$$

for all $u \in C_c^\infty(I)$. This precisely means that $-if' = g$ in the sense of distributions, therefore in particular $f \in H^1(I)$ and $(P^{\min})^* f = -if'$. Thus, $(P^{\min})^* \subset P^{\max}$. Conversely, if $f \in D(P^{\max}) = H^1(I)$, the integration by parts (2.20) provides

$$\langle P^{\max} f, u \rangle = i \int_0^1 \overline{f'(t)} u(t) \, dt = -i \int_0^1 \overline{f(t)} u'(t) \, dt = \langle f, P^{\min} u \rangle$$

since the boundary terms vanish when $u \in C_c^\infty(I)$. This shows that $(f, P^{\max} f)$ is orthogonal to all $(P^{\min} u, -u)$ which, according to the definition (2.8) of the adjoint, means that $P^{\max} \subset (P^{\min})^*$. Thus, we have proved that $(P^{\min})^* = P^{\max}$. As $\overline{P^{\min}} = P_0$, we also have $(P_0)^* = P^{\max}$.

By Lemma 2.18 we then know that $(P^{\max})^* = (P^{\min})^{**} = \overline{P^{\min}} = P_0$ but it is useful to verify this by direct calculation. By definition, the graph of the adjoint of $P^{\max}$ is the set of pairs (f, g) such that

$$-i \int_0^1 \overline{f(x)} u'(x) \, dx = \int_0^1 \overline{g(x)} u(x) \, dx,$$

but now for all $u \in H^1(I) = D(P^{\max})$, instead of only $C_c^\infty(I)$ as before. Taking $u \in C_c^\infty(I)$ we find that $-if' = g$ with $f \in H^1(I)$. But we can also consider functions $u \in H^1(I)$ that do not necessarily vanish at the boundary and perform an integration by parts. We find $\overline{f(1)} u(1) - \overline{f(0)} u(0) = 0$ for all $f \in D((P^{\max})^*)$ and all $u \in H^1(I)$. This is equivalent to the condition $f(1) = f(0) = 0$ and we have indeed found P_0.

It remains to show that the spectrum equals the entire complex plane for the considered operators. For the non-closed operator $P^{\min}$, this follows immediately from Lemma 2.8 (but the following proof also works). For $P^{\max}$, we seek to solve the eigenvalue equation $-if' = \lambda f$ whose solutions in the sense of distributions are exactly the $f_\lambda(x) = e^{i\lambda x}$, up to a multiplicative constant. As these functions are in $H^1(I)$ for all $\lambda \in \mathbb{C}$, the spectrum of $P^{\max}$ contains the entire complex plane and is composed only of eigenvalues. None of these functions are in $H_0^1(I)$, however, which shows that $P^{\min}$ and P_0 cannot have any eigenvalues. However, for all $\lambda \in \mathbb{C}$ and all $g \in H_0^1(I)$ we have

$$\left\langle f_{\bar{\lambda}}, (P_0 - \lambda)g \right\rangle = \left\langle (P^{\max} - \bar{\lambda})f_{\bar{\lambda}}, g \right\rangle = 0 \tag{2.22}$$

because $P^{\max} = (P_0)^*$ (or because the boundary terms vanish since $g(0) = g(1) = 0$). This proves that $0 \neq f_{\bar{\lambda}} \in \operatorname{ran}(P_0 - \lambda)^\perp$, so $P_0 - \lambda$ is not surjective, for all $\lambda \in \mathbb{C}$. We therefore have $\sigma(P_0) = \mathbb{C}$, this time without any eigenvalues. □

None of the three operators $P^{\min} \subset P_0 \subset P^{\max}$ is self-adjoint. However, since $P^{\max} = (P^{\min})^*$, we know that all self-adjoint extensions P of $P^{\min}$ satisfy

$$P^{\min} \subsetneqq \overline{P^{\min}} = P_0 \subsetneqq P = P^* \subsetneqq P^{\max}.$$

The following result provides all the symmetric extensions of P_0, which turn out to be all self-adjoint. These are all the possible self-adjoint realizations of the momentum $f \mapsto -if'$ on the interval $I = (0, 1)$.

Theorem 2.38 (Momentum on $I = (0, 1)$: Self-adjointness and Spectrum) *The strict symmetric extensions of P_0 (i.e., whose domain strictly contains $D(P_0)$) are the operators $P_{\mathrm{per},\theta}$ defined by $P_{\mathrm{per},\theta} f = -if'$ on the domain*

$$\boxed{D(P_{\mathrm{per},\theta}) = H_{\mathrm{per},\theta}^1(I) := \left\{ f \in H^1(I) \;:\; f(1) = e^{i\theta} f(0) \right\}}$$

for all $\theta \in [0, 2\pi)$. These operators are all self-adjoint and their spectrum is the lattice $2\pi\mathbb{Z}$ translated by θ:

$$\boxed{\sigma(P_{\mathrm{per},\theta}) = \{k + \theta, \; k \in 2\pi\mathbb{Z}\}.}$$

Each element $k + \theta$ of the spectrum is a simple eigenvalue, with associated eigenfunction $x \mapsto e^{i(k+\theta)x}$.

The index 'per' in the notation of $P_{\mathrm{per},\theta}$ means "periodic". Indeed, for $\theta = 0$ we find the periodicity condition $f(1) = f(0)$, which also describes a particle evolving on a circle. The boundary condition $f(1) = e^{i\theta} f(0)$ is sometimes called the *Born-von Kármán condition* and it is used in the calculation of the spectrum of Schrödinger operators with periodic potential, as we will see later in Chap. 7.

Theorem 2.38 specifies that **the Born-von Kármán boundary conditions are the only possible ones for the operator** $-i\mathrm{d}/\mathrm{d}x$ **to be self-adjoint** on the finite interval $I = (0, 1)$. Note that those contain *only one boundary condition*, which is related to the fact that $f \mapsto -if'$ is a first-order differential operator. A different explanation is that, as $H_0^1(I)$ is of co-dimension 2 in $H^1(I)$, a self-adjoint extension of P_0 must always have a domain of co-dimension 1. This is to ensure that the domain of its adjoint, defined by an orthogonality relation, has the same co-dimension 1.

Proof Let P be a symmetric extension of P_0. As we have $\langle P_0 f, g \rangle = \langle P f, g \rangle = \langle f, Pg \rangle$ for all $u \in H_0^1(I)$, we see that $Pf = -if'$ and that its domain is $D(P) \subset H^1(I)$. This also follows from the fact that P is a restriction of $(P_0)^* = P^{\max}$. The symmetry condition is written, after integration by parts,

$$\overline{f(1)}g(1) = \overline{f(0)}g(0), \qquad \forall f, g \in D(P) \subset H^1(I). \tag{2.23}$$

By taking $f = g$, we first find that $|f(1)|^2 = |f(0)|^2$ for all $f \in D(P)$. As we have assumed that P is a strict extension of P_0 (for which $f(0) = f(1) = 0$), there exists at least one function $f_0 \in D(P)$ such that $|f_0(0)| = |f_0(1)| \neq 0$. By writing $f_0(1)/f_0(0) = e^{i\theta}$, we find that $g(1) = e^{i\theta} g(0)$ for all $g \in D(P)$. This shows that $P \subset P_{\mathrm{per},\theta}$, the operator introduced in the statement. The operators $P_{\mathrm{per},\theta}$ are symmetric (as they satisfy the condition above) and closed. Any function $f \in H_{\mathrm{per},\theta}^1(I)$ can be written in the form

$$f(x) = \frac{f(0)}{f_0(0)} f_0(x) + \underbrace{f(x) - \frac{f(0)}{f_0(0)} f_0(x)}_{\in H_0^1(I)},$$

that is, $H_0^1(I)$ is of co-dimension 1 in $H_{\mathrm{per},\theta}^1(I)$. This shows that any subspace containing f_0 and $H_0^1(I)$ must contain all $H_{\mathrm{per},\theta}^1(I)$. As this is the case for $D(P)$, we have $P_{\mathrm{per},\theta} \subset P$ and we conclude, as desired, that $P = P_{\mathrm{per},\theta}$.

To show the self-adjointness of the operators $P_{\mathrm{per},\theta}$, we could immediately calculate their spectrum and use Theorem 2.28 but it is useful to know how to do it directly from the definition. We have $g \in D((P_{\mathrm{per},\theta})^*)$ if and only if $\langle g, -ih' \rangle = \langle -ig', h \rangle$ for all $h \in D(P_{\mathrm{per},\theta})$. After integration by parts, this provides the condition

$$\overline{g(1)}h(1) = \overline{g(0)}h(0), \qquad \forall h \in H_{\mathrm{per},\theta}^1(I), \quad \forall g \in D((P_{\mathrm{per},\theta})^*).$$

By taking $h = f_0$, we find that $g(1) = e^{i\theta} g(0)$, that is, $(P_{\mathrm{per},\theta})^* \subset P_{\mathrm{per},\theta}$. Thus $P_{\mathrm{per},\theta}$ is self-adjoint.

When $\theta = 0$ we find the periodic problem which has a basis of explicit eigenvectors, given by the Fourier modes. More precisely, by setting $e_k(x) = e^{ikx}$

for $k \in 2\pi\mathbb{Z}$, we see that e_k is an orthonormal basis of $L^2(I)$, composed only of vectors of $H^1_{\text{per}}(I)$ (we omit θ in the index in the case $\theta = 0$). Moreover, we obviously have $P_{\text{per}}e_k = k\,e_k$. By Theorem 2.34 it then immediately follows that

$$\sigma(P_{\text{per}}) = \overline{2\pi\mathbb{Z}} = 2\pi\mathbb{Z}.$$

The case of $\theta \neq 0$ is exactly similar, using the orthonormal basis $\tilde{e}_k(x) = e^{i(k+\theta)x}$ which is the image of the Fourier basis by the isometry $f \mapsto e^{i\theta x} f$. $\square$

Remark 2.39 (The Infinite Matrices of Hilbert) In 1906, Hilbert first proposed studying operators defined by "infinite matrices" in the space $\ell^2(\mathbb{C})$ [Hil06]. It was not until the work of von Neumann [von29b] in the late 1920s that it was finally established that this concept is not a good one. Indeed, if a *bounded* operator is completely characterized by the $\langle v_n, A v_m \rangle$ when $(v_n)_{n \geq 1}$ is a Hilbert basis of a space $\mathfrak{H}$, it is not always the case for unbounded operators. For example, the self-adjoint operators $P_{\text{per},\theta}$ constructed in Theorem 2.38 are all different, with pairwise disjoint spectra. However, as they all coincide with P_0 on $H^1_0(I)$, the numbers $\langle v_n, P_{\text{per},\theta} v_m \rangle$ are simply independent of $\theta \in [0, 2\pi)$ if we choose a basis of $L^2(I)$ composed of elements of $H^1_0(I)$ like $v_n(x) = \sqrt{2}\sin(\pi n x)$. Thus, the $P_{\text{per},\theta}$ all have the same "infinite matrix" in this particular basis.

2.8.2 Momentum on the Half-Line $I = (0, \infty)$

We now discuss the same operator $f \mapsto -if'$ on the half-line $I = (0, \infty)$ instead of the finite interval $(0, 1)$. This is a very different situation from the previous ones, as we will show that this operator admits **no self-adjoint realization!** As before, we introduce the three operators

$$P^{\min} f = -if', \qquad D(P^{\min}) = C^\infty_c(I),$$

$$P_0 f = -if', \qquad D(P_0) = H^1_0(I) = \{f \in H^1(I) \ : \ f(0^+) = 0\},$$

$$P^{\max} f = -if', \qquad D(P^{\max}) = H^1(I).$$

Following step by step the proof of Lemma 2.37, we can prove that

- P_0 and $P^{\max}$ are closed while $P^{\min}$ is not,
- $P^{\min}$ and P_0 are symmetric while $P^{\max}$ is not,
- $\overline{P^{\min}} = P_0$, $\qquad (P^{\min})^* = (P_0)^* = P^{\max}$, $\qquad (P^{\max})^* = P_0$.

It suffices to use that $C_c^\infty(I)$ is dense in $D(P_0)$ (Exercise A.5 in Appendix A) and the integration by parts

$$-i\int_0^\infty \overline{f(t)}g'(t)\,\mathrm{d}t = \int_0^\infty \overline{-if(t)}g'(t)\,\mathrm{d}t + i\overline{f(0)}g(0), \quad \forall f, g \in H^1(I).$$

(2.24)

We leave the proof of these statements as an exercise. Of course, we have $\sigma(P^{\min}) = \mathbb{C}$ because this operator is not closed. The situation changes dramatically for the spectrum of P_0.

Theorem 2.40 (Momentum on the Half-Line) *The operator P_0 admits no strict symmetric extension. Hence the operator $P^{\min}$ admits **no self-adjoint extension**. Moreover, we have*

$$\sigma(P_0) = \mathbb{C}_-,$$

(2.25)

a spectrum that includes no eigenvalues.

Here we encounter the first pathological example of a symmetric operator whose spectrum is a half-plane (read again Theorem 2.23), and which does not admit any self-adjoint extension. These two properties are in fact related, as can be seen through the theory of defect indices (Exercise 2.48).

Physically, the theorem means that there is no way to define the quantum momentum on a half-line. This comes from the impossibility of choosing good boundary conditions, because there is simply no choice between imposing the condition $f(0^+) = 0$ which provides P_0, and imposing no condition at all, which leads to $P^{\max}$. On a bounded interval, we saw in Theorem 2.38 that the self-adjoint extensions were those with a single boundary condition, linking $f(0)$ and $f(1)$. Here the functions of $H^1(I)$ all tend to 0 at infinity. A boundary condition is therefore already imposed at infinity and it is this which prevents the existence of a self-adjoint extension for a first order differential operator.

Proof As every self-adjoint extension of $P^{\min}$ is closed, it is also a self-adjoint extension of P_0. It is therefore sufficient to show that P_0 admits no strict symmetric extension. Let P be such an extension. We of course have $P \subset P^* \subset (P_0)^* = P^{\max}$, so that P is a restriction of $P^{\max}$. The symmetry of P means, after integration by parts on $D(P^{\max}) = H^1(I)$, that $\overline{f(0)}g(0) = 0$ for all $f, g \in D(P)$. As P is by hypothesis a strict extension of P_0, there exists $0 \neq f \in D(P) \setminus D(P_0)$, that is $f \in H^1(I)$ such that $f(0) \neq 0$. But then we deduce that $g(0) = 0$ for all $g \in D(P)$, that is $P \subset P_0$, which is absurd.

For the spectrum, the argument is based on the fact that the solutions of the equation $-if' = \lambda f$ in the sense of distributions on $I = (0, +\infty)$ are exactly the functions $f_\lambda(x) = Ce^{i\lambda x}$. These functions are in $L^2(I)$ only when $\Im(\lambda) > 0$. The function f_λ is of modulus one if $\lambda \in \mathbb{R}$ or explodes exponentially fast at infinity when $\Im(\lambda) < 0$. For $\Im(\lambda) > 0$ these functions are indeed in $H^1(I)$ and are therefore eigenvalues of $P^{\max}$. This shows that $\mathbb{C}_+ \subset \sigma(P^{\max})$, since $\sigma(P^{\max})$ is closed. As none of these functions vanish at the origin, we also see that P_0 has no eigenvalue.

The rest of the reasoning is as seen before on a finite interval. For all $\lambda \in \mathbb{C}$ with $\Im(\lambda) < 0$, we have $f_{\bar{\lambda}} \in \ker(P^{\max} - \bar{\lambda}) = \operatorname{ran}(P_0 - \lambda)^\perp$ which shows that $P_0 - \lambda$ is not surjective and thus $\mathbb{C}_- \subset \sigma(P_0)$. To conclude, according to Theorem 2.23, it is then sufficient to show that $i \notin \sigma(P_0)$. We already know that $\ker(P_0 - i) = \{0\}$ because P_0 is symmetric. In addition, we still have $\operatorname{ran}(P_0 - i)^\perp = \ker(P^{\max} + i) = \{0\}$ because $f_{-i} \notin L^2(I)$. This shows that $\operatorname{ran}(P_0 - i)$ is dense. With the same argument as at the beginning of the proof of Theorem 2.28 we can verify that this space is in fact closed. This implies $i \in \rho(P_0)$ and finally $\sigma(P_0) = \mathbb{C}_-$ by Theorem 2.23. $\square$

Exercise 2.41 Show that the operator $(Bv)(x) = i \int_0^x e^{y-x} v(y)\, dy$ is bounded on $L^2(I)$ with values in $D(P_0) = H_0^1(I)$, and that it is the inverse of $(P_0 - i)$.

Exercise 2.42 Show that the spectrum of $P^{\max}$ is $\sigma(P^{\max}) = \mathbb{C}_+$.

2.8.3 Laplacian on the Finite Interval $I = (0, 1)$

Let us now study the self-adjoint extensions of the Laplacian

$$A^{\min} f = -f'', \qquad D(A^{\min}) = C_c^\infty(I), \qquad I = (0, 1),$$

in the space $\mathfrak{H} = L^2(I)$. As before, the operator $A^{\min}$ is symmetric but not closed. Its closure is the operator $A_0 f = -f''$ defined on the domain

$$D(A_0) = H_0^2(I) = \left\{ f \in H^2(I)\ :\ f(0) = f(1) = f'(0) = f'(1) = 0 \right\},$$

which is the closure of $C_c^\infty(I)$ in $H^2(I)$. It is again natural to introduce the operator $A^{\max} f = -f''$ defined on the domain

$$D(A^{\max}) = H^2(I).$$

This is a closed operator but it is not symmetric because the boundary terms do not cancel in the integration by parts. As in Lemma 2.37

$$(A^{\min})^* = (A_0)^* = A^{\max}, \qquad (A^{\max})^* = A_0,$$

so that A_0 is not self-adjoint. The following result is proved as Lemma 2.37.

Lemma 2.43 (Spectra) *We have*

$$\sigma(A^{\min}) = \sigma(A^{\max}) = \sigma(A_0) = \mathbb{C}. \tag{2.26}$$

The spectrum of $A^{\max}$ is composed only of eigenvalues, while those of $A^{\min}$ and A_0 contain none.

Proof For $\lambda \in \mathbb{C} \setminus \{0\}$ the solutions of the eigenvalue equation $-f'' = \lambda f$ in the sense of distributions on $I = (0, 1)$ are exactly given by $f(x) = \alpha e^{izx} + \beta e^{-izx}$ with $\alpha, \beta \in \mathbb{C}$ where $z \in \mathbb{C} \setminus \{0\}$ is a square root of λ. If $\lambda = 0$, we get $f(x) = \alpha + \beta x$. As all these functions are in $H^2(I)$, this shows that $\sigma(P^{\max})$ equals the whole complex plane $\mathbb{C}$ and is composed only of eigenvalues. However, none of these functions are in $D(A_0)$, except for $\alpha = \beta = 0$. This shows that $A^{\min}$ and A_0 admit no eigenvalues. As we have $\mathrm{ran}(A_0 - \bar{z}^2)^{\perp} = \ker(A^{\max} - z^2) \neq \{0\}$ by Lemma 2.19, this proves that $A_0 - \bar{z}^2$ is not surjective, hence that $\sigma(A_0) = \mathbb{C}$. $\square$

It remains to determine the self-adjoint extensions A of the operator $A^{\min}$, which necessarily satisfy

$$A^{\min} \subsetneq \overline{A^{\min}} = A_0 \subsetneq A = A^* \subsetneq A^{\max}.$$

We will actually find all the extensions of A_0. Unlike the case of the momentum in Theorem 2.38, we will see that some symmetric extensions are self-adjoint while others are not. As in Sect. 2.8.1, the symmetric extensions A of A_0 must satisfy

$$\overline{g'(1)}f(1) - \overline{g(1)}f'(1) + \overline{g(0)}f'(0) - \overline{g'(0)}f(0) = 0, \qquad \forall f, g \in D(A).$$

This condition can also be written in matrix form

$$\left\langle \begin{pmatrix} g(0) \\ g'(0) \\ g(1) \\ g'(1) \end{pmatrix}, \begin{pmatrix} 0 & 1 & 0 & 0 \\ -1 & 0 & 0 & 0 \\ 0 & 0 & 0 & -1 \\ 0 & 0 & 1 & 0 \end{pmatrix} \begin{pmatrix} f(0) \\ f'(0) \\ f(1) \\ f'(1) \end{pmatrix} \right\rangle_{\mathbb{C}^4} = 0$$

and means that $V := \{(f(0), f'(0), f(1), f'(1)) \in \mathbb{C}^4 : f \in D(A)\}$ is an isotropic subspace of the bilinear form associated with the 4×4 matrix

$$M = \begin{pmatrix} 0 & 1 & 0 & 0 \\ -1 & 0 & 0 & 0 \\ 0 & 0 & 0 & -1 \\ 0 & 0 & 1 & 0 \end{pmatrix}. \tag{2.27}$$

Recall that an isotropic space is by definition a subspace of $\mathbb{C}^4$ such that $\langle v, Mv \rangle_{\mathbb{C}^4} = 0$ for all $v \in V$ which, by polarization, is equivalent to $\langle w, Mv \rangle_{\mathbb{C}^4} = 0$ for all $v, w \in V$ or $V \subset (MV)^{\perp}$. Since M is invertible, isotropic subspaces necessarily satisfy

$$\dim(V) \le \dim(MV)^{\perp} = 4 - \dim(MV) = 4 - \dim(V),$$

that is, $\dim(V) \le 2$.

This discussion suggests introducing the operators A_V defined by

$$A_V f = -f'', \quad D(A_V) = \left\{ f \in H^2(I) \; : \; (f(0), f'(0), f(1), f'(1)) \in V \right\} \tag{2.28}$$

for every vector subspace V of $\mathbb{C}^4$. When $V = \{0\}$ we find the boundary conditions $f(0) = f'(0) = f(1) = f'(1) = 0$ and we then have $D(A_V) = H_0^2(I)$, that is, $A_V = A_0$. When $V = \mathbb{C}^4$ there are no conditions and we obtain $A_V = A^{\max}$.

Theorem 2.44 (Laplacian on the Interval $I = (0, 1)$) *For every vector subspace $V \subset \mathbb{C}^4$, the operator A_V defined in (2.28) is closed and its domain can be written*

$$D(A_V) = H_0^2(I) + \text{vect}(f_i), \tag{2.29}$$

where the $f_i \in H^2(I)$ are arbitrary functions of $D(A_V)$ chosen so that the $(f_i(0), f_i'(0), f_i(1), f_i'(1))$ form a basis of V. Its adjoint is

$$\boxed{(A_V)^* = A_{(MV)^\perp}} \tag{2.30}$$

where M is the 4×4 matrix introduced in (2.27). The A_V's provide all the possible extensions of A_0 that are restrictions of $A^{\max}$, as V runs through all vector subspaces of $\mathbb{C}^4$.

The symmetric operators A_V's are those for which V is an isotropic space of M, that is, $M \subset (MV)^\perp$, and these are all possible symmetric extensions of A_0. The self-adjoint extensions of A_0 are the A_V's for which $V = (MV)^\perp$, that is, which are isotropic and of dimension 2. When A_V is symmetric non self-adjoint, we have $\sigma(A_V) = \mathbb{C}$.

The theorem essentially reduces the question of self-adjointness to a finite-dimensional problem involving the space $V \subset \mathbb{C}^4$ of boundary values and the matrix M in (2.27). There is a kind of competition between the symmetry hypothesis, which requires that V is not too large (that is, included in $(MV)^\perp$ so that the boundary terms disappear in the integration by parts), and the self-adjointness which requires that V is large enough. Only isotropic spaces of dimension 2 are then admissible and these are the boundary conditions that we must physically consider. Note that, this time, the space $D(A_0) = H_0^2(I)$ is of co-dimension 4 in $D(A^{\max}) = H^2(I)$ and the domain must be of co-dimension 2. In Table 2.1 we list some famous examples of boundary conditions with V isotropic of dimension two, hence with A_V self-adjoint.

Which boundary condition to use or study? In principle, none is better than the others and they all have their peculiarities. The choice of a boundary condition is always motivated by practical considerations related to the studied model. For example, in order to describe the vibrations of a string that is attached at both of its ends, one should use the Dirichlet condition $f(0) = f(1) = 0$ [CH53, Sec. V.3.1]. As we mentioned for $P = -i d/dx$ in Theorem 2.38, the Born-von

Table 2.1 Some classic boundary conditions, for which the Laplacian A_V defined in (2.28) is self-adjoint. The spaces V provided all satisfy $(MV)^\perp = V$ where M is the matrix of Eq. (2.27)

Dirichlet	$f(0) = f(1) = 0$	$V = \text{Vect}(e_2, e_4)$
Neumann	$f'(0) = f'(1) = 0$	$V = \text{Vect}(e_1, e_3)$
Periodic	$f(0) = f(1)$ and $f'(0) = f'(1)$	$V = \text{Vect}(e_1 + e_3, e_2 + e_4)$
Robin	$af(0) - bf'(0) = af(1) + bf'(1) = 0$ $(a, b) \in \mathbb{R}^2 \setminus \{(0,0)\}$	$V = \text{Vect}(be_1 + ae_2,$ $be_3 - ae_4)$
Born-von Kármán	$e^{i\theta} f(0) - f(1) = e^{i\theta} f'(0) - f'(1) = 0$ $\theta \in [0, 2\pi[$	$V = \text{Vect}(e_1 + e^{i\theta}e_3,$ $e_2 + e^{i\theta}e_4)$

Kármán condition $f(1) - e^{i\theta} f(0) = f'(1) - e^{i\theta} f'(0) = 0$ (which includes the periodic condition) naturally occurs when studying periodic systems (see Chap. 7). The Robin condition often occurs in electromagnetism (where it is sometimes called impedance condition or de Gennes condition) or to describe the cooling of a radiating wire (where it can also be called Fourier condition) [CH53, Sec. V.3.7.2]. It is sometimes simply called the "third boundary condition".

Proof We leave as an exercise the proof of the closure of A_V, which follows from the continuity of the trace map

$$f \in H^2(I) \mapsto (f(0), f'(0), f(1), f'(1)) \in \mathbb{C}^4$$

(see Lemma A.2 in Appendix A). Now consider any extension A of A_0 that is a restriction of $A^{\max}$, i.e., such that $D(A) \subset H^2(I)$ and $Af = -f''$. We introduce the subspace of $\mathbb{C}^4$

$$V = \{(f(0), f'(0), f(1), f'(1)), \quad f \in D(A)\} \subset \mathbb{C}^4,$$

so that $A \subset A_V$. To show that $A = A_V$, we choose a basis v_i of V and f_i functions of $D(A)$ such that $v_i = (f_i(0), f_i'(0), f_i(1), f_i'(1))$. Any function of $D(A_V)$ can be written in the form

$$f(x) = \sum_i \alpha_i f_i(x) + \underbrace{f(x) - \sum_i \alpha_i f_i(x)}_{\in H_0^2(I) = D(A_0)} \tag{2.31}$$

where the α_i are chosen so that $(f(0), f'(0), f(1), f'(1)) = \sum_i \alpha_i v_i$. Since $A_0 \subset A$ we then deduce that $f \in D(A)$, that is, $D(A_V) \subset D(A)$ and thus $A = A_V$ as announced. The A_V are indeed all the extensions of A_0 that are restrictions of $A^{\max}$, their domain being given by (2.29) according to (2.31).

Now let us determine the adjoint of A_V, for V an arbitrary subspace of $\mathbb{C}^4$. As $A_0 \subset A_V$, we have $(A_V)^* \subset (A_0)^* = A^{\max}$, so that $D((A_V)^*) \subset H^2(I)$ and

$(A_V)^* f = -f''$. For all $f, g \in H^2(I)$, the boundary terms of the integration by parts provide

$$\langle g, -f'' \rangle - \langle -g'', f \rangle = \left\langle \begin{pmatrix} g(0) \\ g'(0) \\ g(1) \\ g'(1) \end{pmatrix}, M \begin{pmatrix} f(0) \\ f'(0) \\ f(1) \\ f'(1) \end{pmatrix} \right\rangle_{\mathbb{C}^4}.$$

The adjoint of A_V is characterized by the fact that the term on the right vanishes for all $f \in D(A_V)$. Since $(f(0), f'(0), f(1), f'(1))$ exactly spans V, this is equivalent to saying that $(g(0), g'(0), g(1), g'(1))$ belongs to $(MV)^\perp$, which indeed shows that $(A_V)^* = A_{(MV)^\perp}$. As the symmetry of A_V can be written $A_V \subset (A_V)^* = A_{(MV)^\perp}$, it is indeed equivalent to $V \subset (MV)^\perp$. Self-adjointness is obtained when $V = (MV)^\perp$.

For any $\alpha \in \mathbb{C} \setminus \{0\}$, consider the two functions $f_\alpha(t) := e^{i\alpha t}$ and $f_{-\alpha}(t) := e^{-i\alpha t}$, which are the two eigenfunctions of $A^{\max}$ with eigenvalue α^2. The space generated by the traces of these functions on the boundary of the interval is

$$V_\alpha = \text{vect} \left\{ \begin{pmatrix} 1 \\ i\alpha \\ e^{i\alpha} \\ i\alpha e^{i\alpha} \end{pmatrix}, \begin{pmatrix} 1 \\ -i\alpha \\ e^{-i\alpha} \\ -i\alpha e^{-i\bar{\alpha}} \end{pmatrix} \right\} \subset \mathbb{C}^4$$

and it is of dimension 2 since $\alpha \neq 0$. If $\dim(V) \in \{0, 1\}$ then $\dim((MV)^\perp) \in \{3, 4\}$ so $(MV)^\perp$ has a non-empty intersection with all the V_α. Since $D((A_V)^*) = D(A_{(MV)^\perp})$ contains exactly all the functions whose trace at the boundary belongs to $(MV)^\perp$, it must therefore contain a non-zero function of $\ker(A^{\max} - \alpha^2)$. This shows that

$$\ker((A_V)^* - \alpha^2) = \text{Im}(A_V - \alpha^2)^\perp \neq \{0\}.$$

This means that $A_V - \alpha^2$ is not surjective, and therefore $\alpha^2 \in \sigma(A_V)$. Since the spectrum is closed and $\alpha \in \mathbb{C} \mapsto \alpha^2 \in \mathbb{C}$ is surjective, this shows that $\sigma(A_V) = \mathbb{C}$ when $\dim(V) \in \{0, 1\}$ (which we already knew for $V = \{0\}$ from Lemma 2.43).   □

Example 2.45 (Spectrum of Dirichlet, Neumann and Born-von Kármán Laplacians) The Dirichlet Laplacian obtained for $V = \text{Vect}(e_2, e_4)$ is the operator

$$A_{\text{Dir}} f := -f'', \qquad D(A_{\text{Dir}}) = \left\{ f \in H^2(I) : f(0) = f(1) = 0 \right\}.$$

As $(\sqrt{2}\,\sin(\pi kt))_{k\in\mathbb{N}}$ forms a Hilbert basis[1] of $L^2(I)$ which is composed of eigenfunctions of A_{Dir}, we deduce from Theorem 2.34 that

$$\boxed{\sigma(A_{\mathrm{Dir}}) = \left\{\pi^2 k^2,\ k \in \mathbb{N}\right\}.}$$

Similarly, the spectrum of the Neumann Laplacian

$$A_{\mathrm{Neu}} f := -f'', \qquad D(A_{\mathrm{Neu}}) = \left\{f \in H^2(I)\ :\ f'(0) = f'(1) = 0\right\}$$

is

$$\boxed{\sigma(A_{\mathrm{Neu}}) = \left\{\pi^2 k^2,\ k \in \mathbb{N} \cup \{0\}\right\}}$$

with the eigenfunctions $\sqrt{2}\cos(\pi kt)$ (for $k = 0$ we must remove the $\sqrt{2}$ if we want the function to be normalized in $L^2(I)$). Apart from the origin, the two operators have the same spectrum, but not the same eigenfunctions. This peculiarity will be further studied in Exercise 3.27 in Chap. 3. All eigenvalues are of multiplicity 1.

Finally, the Hilbert basis $(e^{i(k+\theta)x})_{k\in 2\pi\mathbb{Z}}$ already encountered in Sect. 2.8.1 diagonalizes the Born-von Kármán Laplacian

$$A_{\mathrm{per},\theta} f = -f'',$$

$$D(A_{\mathrm{per},\theta}) = \left\{f \in H^2(I)\ :\ f(1) - e^{i\theta} f(0) = f'(1) - e^{i\theta} f'(0) = 0\right\}$$

whose spectrum is

$$\boxed{\sigma(A_{\mathrm{per},\theta}) = \left\{(k + \theta)^2,\ k \in 2\pi\mathbb{Z}\right\}.}$$

Complementary Exercises

Exercise 2.46 (Space $H^1_{\mathrm{per}}(I)$ and Fourier Series)   With $I = (0, 1)$, we denote

$$H^1_{\mathrm{per}}(I) := \left\{f \in H^1(I)\ :\ f(1) = f(0)\right\}$$

which is a closed subspace of $H^1(I)$.

---

[1] To see this, extend a function from $L^2(I)$ to an *odd* function of $(-1, 1)$.

1. We call $c_k(f) = \int_0^1 f(t)e^{-ikt}\,dt = \langle e_k, f \rangle$ the Fourier coefficients of f with $e_k(t) = e^{ikt}$. Show by integrating by parts that for all $f \in H^1_{\mathrm{per}}(I)$, we have $c_k(f') = ikc_k(f)$. Deduce that $\sum_{k\in 2\pi\mathbb{Z}} k^2|c_k(f)|^2 < \infty$.

2. Let $f \in L^2(I)$ be a function such that $\sum_{k\in 2\pi\mathbb{Z}} k^2|c_k(f)|^2 < \infty$. Show that the Fourier series

$$\sum_{k\in 2\pi\mathbb{Z}} c_k(f)e_k(x)$$

converges uniformly on the closed interval $[0, 1]$. Deduce that f coincides almost everywhere with a continuous function, such that $f(0) = f(1)$. Calculate also $\langle f, \varphi' \rangle$ for all $\varphi \in C^\infty_c(I)$ and deduce that $f \in H^1_{\mathrm{per}}(I)$. In conclusion we have shown that

$$H^1_{\mathrm{per}}(I) = \left\{ f \in L^2(I) \,:\, \sum_{k\in 2\pi\mathbb{Z}} |k|^2|c_k(f)|^2 < \infty \right\}.$$

3. Extend the results to $H^1_{\mathrm{per},\theta}(I)$.

Exercise 2.47 (Beware of Commutators) The theory of operators in infinite dimension is more delicate than it seems and can sometimes go against intuition. The excellent article [Gie00], written for physicists, provides several examples of apparent contradictions that result from a misuse of the concept of self-adjointness. Here is one of them.

We work on the unit circle or, equivalently, on $I = (0, 1)$ with the periodic boundary conditions seen in Sect. 2.8.1. We consider the bounded self-adjoint operator

$$X : f(x) \mapsto xf(x), \qquad D(X) = L^2(I)$$

and the operator

$$P_{\mathrm{per}} = -i\frac{\mathrm{d}}{\mathrm{d}x}, \qquad D(P_{\mathrm{per}}) = H^1_{\mathrm{per}}(I)$$

constructed in Theorem 2.38. We have

$$[P_{\mathrm{per}}, X]f = (P_{\mathrm{per}}X - XP_{\mathrm{per}})f = -i\big((xf)' - xf'\big) = -if$$

which is Heisenberg's relation $[P_{\mathrm{per}}, X] = -i$. The functions $e_k(x) = e^{ikx}$ for $k \in 2\pi\mathbb{Z}$ are the normalized eigenvectors of P_{per}, with $P_{\mathrm{per}}e_k = ke_k$. Using the self-adjointness of P_{per}, we find:

$$-i = \langle e_k, -ie_k \rangle = \langle e_k, [P_{\mathrm{per}}, X]e_k \rangle = \langle P_{\mathrm{per}}e_k, Xe_k \rangle - \langle Xe_k, P_{\mathrm{per}}e_k \rangle$$

$$= (k - k)\int_0^1 x|e_k(x)|^2\,\mathrm{d}x = 0.$$

Where is the mistake?

Exercise 2.48 (Deficiency Indices) Let $(A, D(A))$ be a closed symmetric operator.

1. Show that $\|(A + i)v\| = \|(A - i)v\|$ for all $v \in D(A)$. Deduce that the operator $U = (A + i)(A - i)^{-1}$ is an isometry from $\mathrm{ran}(A - i)$ into $\mathrm{ran}(A + i)$.
2. What can we conclude in finite dimension?
3. Let $B = B^*$ be a self-adjoint extension of A. Show that $V = (B + i)(B - i)^{-1}$ is an isometry from $\mathfrak{H}$ into $\mathfrak{H}$, which is an extension of U, that is, such that $Vf = Uf$ for all $f \in D(A)$.
4. Then show that the image of $\mathrm{ran}(A + i)^{\perp} = \ker(A^* - i)$ by V contains $\mathrm{ran}(A - i)^{\perp} = \ker(A^* + i)$.
5. Deduce that a symmetric operator can only have self-adjoint extensions if $\dim \ker(A^* - i) = \dim \ker(A^* + i)$ (which can be finite or infinite).
6. Reinterpret the result of Theorem 2.40 in this perspective.

Exercise 2.49 (Laplacian on the Half-Line) Let $I = (0, \infty)$. Find all the self-adjoint extensions of the Laplacian $A^{\min} f = -f''$ defined on $D(A^{\min}) = C_c^{\infty}(I)$, in the Hilbert space $\mathfrak{H} = L^2(I)$.

Exercise 2.50 (Dirichlet, Neumann and Born-von Kármán Laplacians on a Hypercube) Let $\Omega = (0, 1)^d$ and define the operator

$$A'_{\mathrm{Dir}} f := -\Delta f, \qquad D\left(A'_{\mathrm{Dir}}\right) = \left\{f \in C^2(\overline{\Omega}) : f_{|\partial\Omega} \equiv 0\right\}.$$

The boundary condition simply means that $f(x_1, \ldots, x_d)$ vanishes as soon as one of the x_i equals 0 or 1. Because of the strong regularity C^2, this operator is not closed but we will show that it is essentially self-adjoint. Its closure is the Dirichlet Laplacian on Ω.

1. Using integrations by parts in each direction (or directly Green's formula), show that A'_{Dir} is symmetric.
2. Show that the functions $f_k(x) := 2^{\frac{d}{2}} \prod_{j=1}^{d} \sin(\pi k_j x_j)$ with $k \in \mathbb{N}^d$ form an orthonormal basis of eigenvectors of A'_{Dir}.
3. Deduce from Theorem 2.34 that the closure $(-\Delta)_{\mathrm{Dir}} := \overline{A'_{\mathrm{Dir}}}$ is a self-adjoint operator, with spectrum

$$\sigma\left((-\Delta)_{\mathrm{Dir}}\right) = \pi^2 \left\{\sum_{j=1}^{d} k_j^2, \ k_j \in \mathbb{N}\right\}.$$

This is the Dirichlet Laplacian on the unit cube.

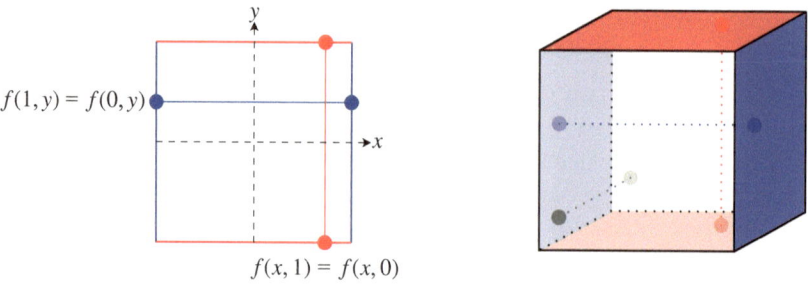

Fig. 2.1 Periodic conditions in 2 and 3 dimensions.

4. Similarly construct the Neumann Laplacian on Ω, with the boundary condition $\partial_n f_{|\partial\Omega} \equiv 0$ where n is the normal to the boundary, which means that $\partial_{x_i} f(x_1, \ldots, x_d)$ vanishes if x_i equals 0 or 1. Show that its spectrum is

$$\sigma\left((-\Delta)_{\text{Neu}}\right) = \pi^2 \left\{\sum_{j=1}^{d} k_j^2, \ k_j \in \mathbb{N} \cup \{0\}\right\}.$$

5. The periodic boundary conditions are obtained by restricting the smooth $\mathbb{Z}^d$-periodic functions to the unit cube Ω. These functions satisfy by definition the relation $f(x + \ell) = f(x)$ for all $x \in \mathbb{R}^d$ and $\ell \in \mathbb{Z}^d$, which implies in particular that $\nabla f(x+\ell) = \nabla f(x)$ for all $x \in \mathbb{R}^d$ and $\ell \in \mathbb{Z}^d$. The only way for x and $x+\ell$ to both belong to $\overline{\Omega}$ is that they are both on the boundary, on two opposite faces (Fig. 2.1). The periodic condition can therefore be expressed in the somewhat complicated form

$$\begin{cases} f(x_1, \ldots, 1, \ldots x_d) = f(x_1, \ldots, 0, \ldots x_d) \\ \partial_{x_j} f(x_1, \ldots, 1, \ldots x_d) = \partial_{x_j} f(x_1, \ldots, 0, \ldots x_d) \end{cases} \quad \forall j = 1, \ldots, d,$$

(2.32)

where it is each time the variable x_j that equals 0 and 1 in the argument of f. Using the Fourier basis $e^{i2\pi k \cdot x}$ with $k \in \mathbb{Z}^d$, construct as previously the periodic Laplacian $(-\Delta)_{\text{per}}$ on the hypercube Ω and show that its spectrum is

$$\sigma\left((-\Delta)_{\text{per}}\right) = 4\pi^2 \left\{\sum_{j=1}^{d} k_j^2, \ k_j \in \mathbb{Z}\right\}.$$

Generalize the above to the Born-von Kármán Laplacian $(-\Delta)_{\text{per},\xi}$ whose boundary conditions take the form

$$\begin{cases} f(x_1, \ldots, 1, \ldots x_d) = e^{i\xi} f(x_1, \ldots, 0, \ldots x_d) \\ \partial_{x_j} f(x_1, \ldots, 1, \ldots x_d) = e^{i\xi} \partial_{x_j} f(x_1, \ldots, 0, \ldots x_d) \end{cases} \quad \forall j = 1, \ldots, d.$$

(2.33)

It is possible to show that

$$D\left((-\Delta)_{\mathrm{Dir}}\right) = \left\{f \in H^2(\Omega) \ : \ f_{|\partial\Omega} \equiv 0\right\} = H^2(\Omega) \cap H_0^1(\Omega),$$

$$D\left((-\Delta)_{\mathrm{Neu}}\right) = \left\{f \in H^2(\Omega) \ : \ \partial_n f_{|\partial\Omega} \equiv 0\right\},$$

and

$$D(-\Delta_{\mathrm{per},\xi}) = H^2_{\mathrm{per},\xi}(\Omega) := \left\{f \in H^2(\Omega) \text{ verifying } (2.33)\right\}$$

(see Sect. A.6 of Appendix A).

Chapter 3
Self-adjointness Criteria: Rellich, Kato and Friedrichs

The aim of this chapter is to provide practical methods for constructing self-adjoint extensions of symmetric operators. A natural approach is to determine conditions on B for an operator in the form $A + B$ to be self-adjoint when A already is. In the first section, we will present a result due to Rellich and Kato that guarantees that $A + B$ is self-adjoint on the same domain as A:

$$D(A + B) = D(A).$$

Next, we will discuss a method due to Friedrichs [Fri34], that allows defining operators in a somewhat indirect way using their quadratic form. This method is so effective that some authors (e.g. [Eva10]) develop the theory of elliptic partial differential equations based almost exclusively on the quadratic form, without ever talking about the associated operator! We will here make the formal link between the question of self-adjointness and methods based on quadratic forms, like the Lax-Milgram theorem. This will allow us to construct self-adjoint realizations of operators in the form $A + B$, by only guaranteeing the equality of the domains of the associated quadratic forms

$$Q(A + B) = Q(A)$$

without the domains of the operators necessarily being related. This second approach only applies to operators whose spectrum is bounded from below or above. Since the energy of quantum systems is most often bounded below, this is a very natural technique when studying the self-adjointness of the quantum Hamiltonian.

Finally, we will apply these techniques to Schrödinger operators, that is, of the form $A = -\Delta + V(x)$. The function V can tend to 0 at infinity, thus describing a quantum particle that is only locally confined, or it can tend to $+\infty$ when $|x| \to +\infty$, which corresponds to a truly confined particle, as is the case for the harmonic oscillator.

© The Editor(s) (if applicable) and The Author(s), under exclusive license
to Springer Nature Switzerland AG 2024
M. Lewin, *Spectral Theory and Quantum Mechanics*, Universitext,
https://doi.org/10.1007/978-3-031-66878-4_3

3.1   Relatively Bounded Perturbations

In this section, we give a simple criterion on a symmetric operator B, that implies that $A + B$ is self-adjoint on the same domain $D(A)$ as the self-adjoint operator A. Then we apply it to the case where $A = -\Delta$ and $B = V(x)$ is the operator of multiplication by a function.

3.1.1   Rellich-Kato Theory

The following result is essentially due to Rellich [Rel37], but it was popularized and extensively used by Kato for Schrödinger operators [Kat51].

Theorem 3.1 (Rellich-Kato) *Let A be a self-adjoint operator on its domain $D(A)$ and B a symmetric operator on the same domain $D(A)$. If there exist $0 \le \alpha < 1$ and $C > 0$ such that*

$$\|Bv\| \le \alpha \|Av\| + C\|v\|, \qquad \forall v \in D(A), \tag{3.1}$$

then the operator $A + B$ is self-adjoint on $D(A)$. Moreover, if $\mathcal{D} \subset D(A)$ is a dense subspace such that $\overline{A_{|\mathcal{D}}} = A$, then we also have $A + B = \overline{(A + B)_{|\mathcal{D}}}$, that is, $A + B$ is essentially self-adjoint on $\mathcal{D}$. Finally, if the spectrum of A is bounded from below, that is $\sigma(A) \subset [a, \infty)$ for some $a \in \mathbb{R}$, then so is the one of $A + B$:

$$\sigma(A + B) \subset \left[a - \frac{C}{1 - \alpha}, +\infty \right). \tag{3.2}$$

Here we have used the notation $A_{|\mathcal{D}}$ for the restriction of A defined on the domain $D(A_{|\mathcal{D}}) = \mathcal{D}$. When we have $\overline{A_{|\mathcal{D}}} = A$ as in the statement, $\mathcal{D}$ is called a **core of the operator** A.

Proof As $A + B$ is symmetric on $D(A)$ by assumption, we show that $A + B \pm i\mu$ is surjective from $D(A)$ to $\mathfrak{H}$, for a well-chosen $\mu \in \mathbb{R}$, in order to then apply Theorem 2.28. We start by writing

$$A + B + i\mu = \left(1 + B(A + i\mu)^{-1} \right)(A + i\mu). \tag{3.3}$$

The operator $A + i\mu$ is invertible from $D(A)$ to $\mathfrak{H}$ for all $\mu \in \mathbb{R}$, since A is self-adjoint. We must therefore find μ such that $1 + B(A + i\mu)^{-1}$ is invertible on $\mathfrak{H}$. By applying inequality (3.1) to the vector $(A + i\mu)^{-1}v \in D(A)$, we deduce that

$$\left\| B(A + i\mu)^{-1}v \right\| \le \alpha \left\| A(A + i\mu)^{-1}v \right\| + C \left\| (A + i\mu)^{-1}v \right\|.$$

As this is true for all $v \in \mathfrak{H}$, we obtain the inequality on the operator norms

$$\left\| B(A + i\mu)^{-1} \right\| \leq \alpha \left\| A(A + i\mu)^{-1} \right\| + C \left\| (A + i\mu)^{-1} \right\|. \tag{3.4}$$

Recall the fundamental relation seen in (2.11)

$$\| (A + i\mu)v \|^2 = \| Av \|^2 + \mu^2 \| v \|^2, \qquad \forall v \in D(A),$$

which implies

$$\| v \|^2 = \left\| A(A + i\mu)^{-1}v \right\|^2 + \mu^2 \left\| (A + i\mu)^{-1}v \right\|^2, \qquad \forall v \in \mathfrak{H},$$

and therefore the estimates

$$\left\| A(A + i\mu)^{-1} \right\| \leq 1, \qquad \left\| (A + i\mu)^{-1} \right\| \leq \frac{1}{|\mu|}. \tag{3.5}$$

After inserting this into (3.4), we obtain

$$\left\| B(A + i\mu)^{-1} \right\| \leq \alpha + \frac{C}{|\mu|}.$$

Since $\alpha < 1$ by hypothesis, the term on the right is strictly less than 1 for $|\mu| > C/(1 - \alpha)$. This proves, as desired, that $1 + B(A + i\mu)^{-1}$ is invertible and therefore, according to (3.3), that $A + B + i\mu$ is invertible on $D(A)$ with a bounded inverse. By changing μ to $-\mu$, we deduce from Theorem 2.28 that $A + B$ is self-adjoint on $D(A)$.

If $\mathcal{D}$ is a core of A, that is, a dense subspace on which A is essentially self-adjoint, then $(A + B)_{|\mathcal{D}}$ is also symmetric. Let $v \in D(A) = D(A + B)$ and (v_n) be a sequence of $\mathcal{D}$ such that $v_n \to v$ and $Av_n \to Av$. Then we also have $Bv_n \to Bv$ according to (3.1). This implies that $(A + B)v_n \to (A + B)v$, that is, $A + B \subset \overline{(A + B)_{|\mathcal{D}}}$. As $A + B$ is self-adjoint, hence closed, we conclude that $A + B = \overline{(A + B)_{|\mathcal{D}}}$.

It remains to show (3.2) when $\sigma(A) \subset [a, +\infty)$. Changing A into $A - a$ we can assume $a = 0$. Then we need to consider the operator $A + B + \mu$ and show that it is invertible with a bounded inverse for all $\mu > C/(1 - \alpha)$, or equivalently $\alpha + C/\mu < 1$. The idea is to follow the above proof with $A + i\mu$ replaced by $A + \mu$ everywhere, but then we need a replacement for (3.5). We claim that

$$\left\| (A + \mu)^{-1} \right\| \leq \frac{1}{\mu}, \qquad \left\| A(A + \mu)^{-1} \right\| \leq 1, \qquad \text{for all } \mu > 0, \text{ when } a = 0. \tag{3.6}$$

This will follow easily from the spectral Theorem 4.4 in Chap. 4 (see also Corollary 4.6) but here we provide a more elementary proof. Lemma 2.11 implies that

$$\sigma\big((A+\mu)^{-1}\big) = \overline{\left\{\frac{1}{\lambda+\mu},\ \lambda \in \sigma(A)\right\}}, \quad \sigma\big(A(A+\mu)^{-1}\big) = \overline{\left\{\frac{\lambda}{\lambda+\mu},\ \lambda \in \sigma(A)\right\}}.$$

The second equality is because $A(A+\mu)^{-1} = 1 - \mu(A+\mu)^{-1}$. Thus we see that

$$\sigma\big((A+\mu)^{-1}\big) \subset \left[0, \frac{1}{\mu}\right], \quad \sigma\big(A(A+\mu)^{-1}\big) \subset [0, 1].$$

It remains to use that for a bounded self-adjoint operator T, $\sigma(T) \subset [0, M]$ implies $\|T\| \leq M$. Indeed, writing $T^2 - z^2 = (T+z)(T-z)$ where $T+z$ is always invertible for $z > 0$, we see that $\sigma(T^2) = \sigma(T)^2 = \{\lambda^2,\ \lambda \in \sigma(T)\}$. Hence we have $\sigma(T^2) \subset [0, M^2]$ which, by Theorem 2.33, implies $\|Tv\|^2 = \langle v, T^2 v\rangle \leq M^2\|v\|^2$ as claimed. $\square$

Definition 3.2 (Relatively Bounded Operators) Let $0 < \alpha < \infty$ and A be a self-adjoint operator. When a symmetric operator B on $D(A)$ satisfies (3.1), it is called A-**bounded**, with relative bound α. If it satisfies (3.1) for all $\alpha > 0$ and some $C = C(\alpha)$, it is called **infinitesimally A-bounded**.

In the proof of the theorem, we showed that $\|B(A + i\mu)^{-1}\| < 1$ for μ large enough. It turns out that this is actually equivalent to the hypothesis (3.1).

Exercise 3.3 Let A be a self-adjoint operator and B a symmetric operator on $D(A)$. Show that B is A-bounded with relative bound < 1 if and only if $B(A + i\mu)^{-1}$ is a bounded operator, with $\limsup_{|\mu| \to +\infty} \|B(A + i\mu)^{-1}\| < 1$. Show that it is infinitesimally A-bounded if and only if $\|B(A + i\mu)^{-1}\|$ tends to zero when $|\mu| \to +\infty$.

3.1.2 Application to Schrödinger Operators

We want to apply the Rellich-Kato theorem to the operators in the form $-\Delta + V(x)$ in $H^2(\mathbb{R}^d)$. We need to control $\|Vf\|$ by the L^2 norms of f and Δf, which is the object of the following theorem.

Theorem 3.4 (Infinitesimally $(-\Delta)$-Bounded Potentials) Let $V \in L^p(\mathbb{R}^d) + L^\infty(\mathbb{R}^d)$ with

$$\begin{cases} p = 2 & \text{if } d \in \{1, 2, 3\}, \\ p > 2 & \text{if } d = 4, \\ p = \frac{d}{2} & \text{if } d \geq 5. \end{cases} \tag{3.7}$$

Then, for every $\varepsilon > 0$, there exists a constant C_ε such that

$$\|Vf\|_{L^2(\mathbb{R}^d)} \le \varepsilon \|\Delta f\|_{L^2(\mathbb{R}^d)} + C_\varepsilon \|f\|_{L^2(\mathbb{R}^d)}, \qquad \forall f \in H^2(\mathbb{R}^d) \tag{3.8}$$

and

$$\left| \int_{\mathbb{R}^d} V(x)|f(x)|^2 \, dx \right| \le \varepsilon \int_{\mathbb{R}^d} |\nabla f(x)|^2 \, dx + C_\varepsilon \int_{\mathbb{R}^d} |f(x)|^2 \, dx, \quad \forall f \in H^1(\mathbb{R}^d). \tag{3.9}$$

Note that condition (3.7) is slightly stronger in dimension $d \le 4$ than that the one we have encountered in (1.37) in Chap. 1. The latter was only used to ensure (3.9). The proof of Theorem 3.4 is exactly the same as that of Lemma 1.10, except that we must bound $\int_{\mathbb{R}^d} V^2 |f|^2$ by the $H^2(\mathbb{R}^d)$ norm of f, and therefore need to use the corresponding Sobolev embedding, recalled in Appendix A in Theorem A.15. We leave it as an exercise.

The following result is a simple consequence of the previous theorem and of Theorem 3.1(Rellich-Kato).

Corollary 3.5 (Self-adjointness of Schrödinger Operators) *Let $V \in L^p(\mathbb{R}^d, \mathbb{R}) + L^\infty(\mathbb{R}^d, \mathbb{R})$ be a real-valued function, with p as in (3.7). Then the operator $f \mapsto -\Delta f + Vf$ is self-adjoint on $D(-\Delta) = H^2(\mathbb{R}^d)$ and its spectrum is bounded below. Furthermore, $f \mapsto -\Delta f + Vf$ is essentially self-adjoint on $C_c^\infty(\mathbb{R}^d)$ or any other dense subspace in $H^2(\mathbb{R}^d)$.*

Proof Since the function V is real, $f \mapsto Vf$ is symmetric. Moreover, this operator is well defined on $H^2(\mathbb{R}^d)$ and it is infinitesimally $(-\Delta)$-bounded, according to (3.8). The result then follows from Theorem 3.1. For the lower bound on the spectrum, we can directly use (3.2). It is also possible to apply Theorem 2.33 and the inequality

$$\langle f, (-\Delta + V)f \rangle_{L^2(\mathbb{R}^d)} = \int_{\mathbb{R}^d} |\nabla f(x)|^2 \, dx + \int_{\mathbb{R}^d} V(x)|f(x)|^2 \, dx$$

$$\ge (1 - \varepsilon) \int_{\mathbb{R}^d} |\nabla f(x)|^2 \, dx - C_\varepsilon \int_{\mathbb{R}^d} |f(x)|^2 \, dx$$

$$\ge -C_\varepsilon \int_{\mathbb{R}^d} |f(x)|^2 \, dx,$$

for $\varepsilon < 1$, according to (3.9). □

The statement means that perturbing the Laplacian by a function in the form $V = V_p + V_\infty$ with $V_p \in L^p(\mathbb{R}^d, \mathbb{R})$ and $V_\infty \in L^\infty(\mathbb{R}^d, \mathbb{R})$ does not change the domain of self-adjointness. Operators in the form $A = -\Delta + V$ are called **Schrödinger operators**.

Example 3.6 (Hydrogen Atom)   The hydrogen atom studied in Chap. 1 is described by the potential $V(x) = -|x|^{-1}$ in dimension $d = 3$. By writing

$$\frac{1}{|x|} = \underbrace{\frac{\mathbb{1}_{B_1}(x)}{|x|}}_{\in L^2(\mathbb{R}^3)} + \underbrace{\frac{\mathbb{1}_{\mathbb{R}^3 \setminus B_1}(x)}{|x|}}_{\in L^\infty(\mathbb{R}^3)}$$

where B_1 is the unit ball, we deduce from Corollary 3.5 that the operator $-\Delta/2 - |x|^{-1}$ is self-adjoint on $H^2(\mathbb{R}^3)$.

3.2   Quadratic Forms and Self-adjointness

In this section we present a completely different technique for finding self-adjoint extensions of operators, which is adapted to the particular case where the associated quadratic form is bounded below. This method is essentially due to Friedrichs [Fri34].

3.2.1   Closure of Quadratic Forms

We consider a Hermitian sesquilinear form $\varphi(\cdot, \cdot)$ on a dense domain Q in a complex Hilbert space $\mathfrak{H}$. This means that $(v, w) \in Q \times Q \mapsto \varphi(v, w) \in \mathbb{C}$ is linear with respect to w and anti-linear with respect to v, with $\varphi(v, w) = \overline{\varphi(w, v)}$. The associated quadratic form is by definition equal to

$$q(v) := \varphi(v, v)$$

and it is real-valued. Recall that we can reconstruct φ from q using the *polarization formula*

$$\varphi(v, w) = \frac{1}{4}\Big(q(v + w) - q(v - w) + iq(v + iw) - iq(v - iw)\Big).$$

It is therefore equivalent to give φ or q. In the following our definitions will apply indifferently to φ or q.

Definition 3.7 (Coercivity, Closure)   Let φ be a sesquilinear form with associated quadratic form $q(v) = \varphi(v, v)$, defined on a dense subspace $Q \subset \mathfrak{H}$. We say that q and φ are **bounded from below** or **semi-bounded** when there exists a constant $\alpha \in \mathbb{R}$ such that

$$q(v) \geq \alpha \|v\|^2, \qquad \forall v \in Q. \tag{3.10}$$

They are called **coercive** when $\alpha > 0$. In this case, φ is a scalar product and $v \mapsto \sqrt{q(v)}$ defines a norm on the space Q. We say that q and φ are **closed** when they are coercive and (Q, φ) is a Hilbert space, that is, is complete: for any sequence (v_n) such that

$$\lim_{n,m \to \infty} q(v_n - v_m) = 0, \tag{3.11}$$

there exists $v \in Q$ such that

$$\lim_{n \to \infty} q(v_n - v) = 0.$$

Finally, we say that q and φ are **closable in** $\mathfrak{H}$ when they admit a closed extension in $\mathfrak{H}$. In this case, there exists a smallest closed extension denoted by $(\overline{Q}, \overline{\varphi})$ and called the closure.

A quadratic form that is bounded-below in the previous sense, is not necessarily bounded-below on the whole of Q! This is due to the term $\|v\|^2$ on the right of (3.10). If $\alpha < 0$ it is perfectly possible that $q(v) < 0$ for some $v \in Q$ and then of course $q(tv) = t^2 q(v) \to -\infty$ when $t \to \infty$. By homogeneity, our assumption (3.10) is rather equivalent to q being bounded from below on Q intersected with the unit sphere of $\mathfrak{H}$.

We have not explicitly written the definition of the extension and restriction of a quadratic form, which is similar to that of operators. In the following, we will only consider quadratic forms that are bounded from below, and will often need them to be in addition coercive. Of course, if q is bounded below by α without being coercive, then $q + a\| \cdot \|^2$ is coercive for $a > -\alpha$, and all the results stated with q being coercive easily extend to the case of a form bounded below by replacing q with $q + a\| \cdot \|^2$.

The following gives a simple criterion for a quadratic form to be closable.

Lemma 3.8 (Closable Quadratic Forms) *Let φ be a coercive sesquilinear form with associated quadratic form $q(v) = \varphi(v, v)$, defined on a dense subspace $Q \subset \mathfrak{H}$. Then φ and q are closable if and only if for any Cauchy sequence $v_n \in Q$ as in (3.11) satisfying in addition $v_n \to 0$ in $\mathfrak{H}$, then we have $q(v_n) \to 0$.*

The statement should be compared with the characterization of a closable operator $(A, D(A))$, in terms of the closure of the graph $\overline{G(A)}$, which must satisfy the property (ii) of Lemma 2.3.

Proof Let us first assume that q admits a closed extension $\tilde{q}$ on a domain $\tilde{Q} \supset Q$. Consider a Cauchy sequence v_n for q such that $v_n \to 0$. Since $\tilde{q}$ coincides with q on Q, v_n is also a Cauchy sequence for $\tilde{q}$. Hence there exists $v \in \tilde{Q}$ such that $\tilde{q}(v_n - v) \to 0$. Since $\tilde{q}$ is coercive by assumption, this gives $v_n \to v$ in $\mathfrak{H}$, hence $v = 0 \in Q$. We thus have $q(v_n) = \tilde{q}(v_n) \to 0$ as desired.

The converse part of the statement is less obvious. Let (v_n) be any Cauchy sequence for q, as in (3.11). From the coercivity (3.10) of q, (v_n) is also a Cauchy sequence in $\mathfrak{H}$, and therefore converges to some $v \in \mathfrak{H}$. Moreover, we have the triangle inequality

$$\lim_{n,m\to\infty} \left| \sqrt{q(v_n)} - \sqrt{q(v_m)} \right| \leq \lim_{n,m\to\infty} \sqrt{q(v_n - v_m)} = 0$$

(since $v \mapsto \sqrt{q(v)}$ is a norm), which shows that $q(v_n)$ converges to a non-negative real number. Intuitively, this limit should be $\bar{q}(v)$ where $\bar{q}$ is the closure of q, to be constructed. This leads us to introduce

$$\tilde{Q} := \left\{ v \in \mathfrak{H} \ : \ \exists (v_n) \in Q^{\mathbb{N}}, \ v_n \to v, \ \lim_{n,m\to\infty} q(v_n - v_m) = 0 \right\}.$$

and $\tilde{q}(v) := \lim_{n\to\infty} q(v_n)$. We need to verify that the latter defines $\tilde{q}$ unambiguously on $\tilde{Q}$, that is, the limit does not depend on the chosen sequence (v_n). Let (v_n') be another Cauchy sequence for q that converges to the same v in $\mathfrak{H}$. Then $w_n := v_n - v_n'$ is also a Cauchy sequence for q and it converges to 0 in $\mathfrak{H}$. The assumption of the lemma tells us that $q(w_n) \to 0$ but then we obtain as desired

$$\left| \sqrt{q(v_n)} - \sqrt{q(v_n')} \right| \leq \sqrt{q(v_n - v_n')} = \sqrt{q(w_n)} \to 0.$$

This proves that $\tilde{q}$ is well defined on $\tilde{Q}$. Note that $Q \subset \tilde{Q}$ and that $\tilde{q} = q$ on $\tilde{Q}$ (take $v_n \equiv v \in Q$). We leave the verification that $\tilde{q}$ is closed as an exercise. In fact, one can show that $\tilde{q} = \bar{q}$ is the closure of q. $\square$

Quadratic forms are not always closable. Here is an example similar to Exercise 2.16 from Chap. 2.

Example 3.9 (Non-closability of δ) Let I be the finite interval $(0, 1)$. Consider the quadratic form

$$q(f) = \int_0^1 |f|^2 + |f(1/2)|^2$$

on the space $Q = C^0([0, 1]) \subset \mathfrak{H} = L^2(I)$, which is coercive as in (3.10) with $\alpha = 1$. By Exercise 2.16 there exists a sequence $f_n \in L^2(I)$ such that $f_n \to 0$ in $L^2(I)$ and $f_n(1/2) = 1$. One can also take, for instance, $f_n(x) = \chi(n(x - 1/2))$ where $\chi \in C_c^\infty((-1/2, 1/2))$ with $\chi(0) = 1$. Then (f_n) is a Cauchy sequence for q, but we have $q(f_n) = \|f_n\|_{L^2}^2 + 1 \to 1 \neq 0$. Hence q is not closable.

3.2.2 Case of Symmetric Operators

Let A be a symmetric operator on its domain $D(A) \subset \mathfrak{H}$. The **quadratic form associated with** A is the one defined by

$$q_A(v) := \langle v, Av \rangle$$

on the domain $Q := D(A)$. The associated sesquilinear form is of course

$$\varphi_A(v, w) := \langle v, Aw \rangle, \qquad \forall v, w \in D(A).$$

Definition 3.10 (Bounded from Below, Coercive) We say that a symmetric operator $(A, D(A))$ is **bounded from below, semi-bounded** or **coercive** when its quadratic form q_A is, as in Definition 3.7. In this case we write $A \geq \alpha$.

When A is self-adjoint, Theorem 2.33 shows that the assumption (3.10) is equivalent to saying that $\sigma(A) \subset [\alpha, \infty)$. But we only assume for now that A is symmetric.

In the same way that every symmetric operator is closable (Sect. 2.5), we can show that every coercive symmetric operator has a quadratic form that is closable.

Theorem 3.11 (Quadratic Form of a Symmetric Operator) *Let $A \geq \alpha > 0$ be a coercive symmetric operator on its domain $D(A)$. Then its quadratic form q_A is always closable. Its closure, denoted $\overline{q_A}$ with the sesquilinear form $\overline{\varphi_A}$, is defined on $Q_A \subset \mathfrak{H}$, which is a complete space for $\overline{q_A}$. Furthermore, we have the following properties:*

- *$D(A)$ is dense in Q_A for the norm $\sqrt{\overline{q_A}}$;*
- *Q_A is dense in $\mathfrak{H}$ for the norm of $\mathfrak{H}$;*
- *for all $u \in D(A)$ and all $v \in Q_A$ we have $\overline{\varphi_A}(v, u) = \langle v, Au \rangle$;*
- *the embedding $Q_A \hookrightarrow \mathfrak{H}$ is continuous, that is $\alpha \|v\|^2 \leq \overline{q_A}(v)$ for all $v \in Q_A$;*
- *if A is closed, the embedding $D(A) \hookrightarrow Q_A$ is also continuous, that is*

$$\overline{q_A}(v) = \langle v, Av \rangle \leq \frac{1}{2}\|v\|_{D(A)}^2 = \frac{\|v\|^2}{2} + \frac{\|Av\|^2}{2}$$

for all $v \in D(A)$.

If A is bounded from below without being coercive, that is, it verifies $\langle v, Av \rangle \geq \alpha \|v\|^2$ for all $v \in D(A)$, we similarly define $Q_A := Q_{A+a}$ and $\overline{q_A} := \overline{q_{A+a}} - a\|\cdot\|^2$ for $a > -\alpha$. It can be verified that this definition is independent of a.

Proof Due to Lemma 3.8, we consider a Cauchy sequence (v_n) for q_A as in (3.11) such that $v_n \to 0$ in $\mathfrak{H}$ and we prove that necessarily $q(v_n) \to 0$. We simply write

$$q_A(v_n - v_m) = q_A(v_n) + q_A(v_m) - 2\Re\langle v_n, Av_m \rangle$$

and pass to the limit $n \to \infty$ first, before we take $m \to \infty$. We obtain

$$\lim_{j \to \infty} q_A(v_j) = \lim_{m \to \infty} \left(\lim_{n \to \infty} \mathfrak{R} \langle v_n, A v_m \rangle \right) = 0$$

since $v_n \to 0$ in $\mathfrak{H}$. Hence q_A is closable. We leave the verification of the other properties as an exercise.   □

Example 3.12 (Laplacian on $\mathbb{R}^d$) Consider the operator $A = -\Delta$ which is self-adjoint on $H^2(\mathbb{R}^d)$, as we have seen in Theorem 2.36. Then we have

$$q_A(f) = - \int_{\mathbb{R}^d} \overline{f(x)} \Delta f(x) \, \mathrm{d}x = \int_{\mathbb{R}^d} |\nabla f(x)|^2 \, \mathrm{d}x, \qquad \forall f \in H^2(\mathbb{R}^d).$$

As q_A is only positive, we make it coercive by looking for example at $A + 1$ so that $q_{A+1}(f) = \|f\|^2_{H^1(\mathbb{R}^d)} \geq \|f\|^2_{L^2(\mathbb{R}^d)}$. The space $H^2(\mathbb{R}^d)$ being dense in $H^1(\mathbb{R}^d)$ for the norm of $H^1(\mathbb{R}^d)$, the completion process provides $Q_{A+1} = H^1(\mathbb{R}^d)$ and $\overline{q_{A+1}}(f) = \|f\|^2_{H^1(\mathbb{R}^d)}$. Thus

$$\overline{q_A}(f) = \overline{q_{A+1}}(f) - \|f\|^2_{L^2(\mathbb{R}^d)} = \int_{\mathbb{R}^d} |\nabla f(x)|^2 \, \mathrm{d}x \quad \text{on} \quad Q_A = H^1(\mathbb{R}^d).$$

3.2.3   Case of Self-adjoint Operators

We will see that the quadratic form is an object that can sometimes be easier to handle than the operator A itself. However, it is legitimate to wonder what relationship there is between A and $\overline{q_A}$. Can we recover A from $\overline{q_A}$? The answer is negative for a general symmetric operator, but it is positive for a self-adjoint operator, which partly justifies the introduction of the notion of quadratic form.

Theorem 3.13 (Characterization of the Domain) *Let A be a bounded-below self-adjoint operator and let $\overline{\varphi_A}$ be the associated closed sesquilinear form, constructed in Theorem 3.11. The following assertions are equivalent:*

(i) $v \in Q_A$ and there exists $z \in \mathfrak{H}$ such that $\overline{\varphi_A}(v, h) = \langle z, h \rangle$ for all $h \in Q_A$;
(ii) $v \in D(A)$ and $Av = z$.

Theorem 3.13 means that we can reconstruct the operator A from the closure of its quadratic form. Namely the domain can be expressed as

$$D(A) = \left\{ v \in Q_A \ : \ \exists z \in \mathfrak{H}, \ \overline{\varphi_A}(v, h) = \langle z, h \rangle, \ \forall h \in Q_A \right\} \tag{3.12}$$

and then $Av = z$ for any $v \in D(A)$. The equation $\overline{\varphi_A}(v, h) = \langle z, h \rangle$ is called the *weak formulation* of the equation $Av = z$ and it is obtained formally by taking the scalar product with h. The "weak" character comes from the fact that we only suppose that v belongs to Q_A.

Proof If $v \in D(A)$ and $Av = z$, then $\overline{\varphi_A}(v, h) = \langle Av, h \rangle = \langle z, h \rangle$ for all $h \in Q_A$. Conversely, if $\overline{\varphi_A}(v, h) = \langle z, h \rangle$ for all $h \in Q_A$, then we can take $h \in D(A)$ and we find $\langle v, Ah \rangle = \langle z, h \rangle$ for all $h \in D(A)$. This means that $(v, z) \in G(A^*) = G(A)$ since A is assumed to be self-adjoint. Therefore $v \in D(A)$ and $Av = z$.   □

Since a self-adjoint operator is fully characterized by the closure $\overline{q_A}$ of its quadratic form, we will always denote, for simplicity,

$$\boxed{q_A := \overline{q_A}, \qquad \varphi_A := \overline{\varphi_A}, \qquad Q(A) := Q_A.}$$

For a symmetric operator, we will keep the notation with the overlining. We can now use the previous result to give a variational characterization of the equation $(A+a)v = z$, a result that takes the same form as the Lax-Milgram theorem [LM54].

Theorem 3.14 (Variational Characterization)   *Let $A \geq \alpha$ be a bounded-below self-adjoint operator and let $a > -\alpha$. Let $z \in \mathfrak{H}$ be arbitrary. Then, the minimization problem*

$$\inf_{w \in Q(A)} \left\{ \frac{1}{2} q_A(w) + \frac{a}{2} \|w\|^2 - \mathfrak{R}\langle w, z \rangle \right\} \tag{3.13}$$

admits the unique minimizer $v = (A+a)^{-1}z \in D(A)$. This latter is also the unique $v \in Q(A)$ satisfying the relation

$$\varphi_A(v, h) + a\langle v, h \rangle = \langle z, h \rangle \tag{3.14}$$

for all $h \in Q(A)$.

The theorem tells us how to find A (or rather $(A+a)^{-1}$) from the quadratic form q_A, by a variational procedure. The vector $v = (A + a)^{-1}z$ is the unique minimizer of the problem (3.13).

Proof We have already seen in Theorem 3.13 that the weak formulation (3.14) combined with the condition that $v \in Q(A)$ was equivalent to the fact that $v \in D(A)$ with $(A+a)v = z$, that is to say $v = (A+a)^{-1}z$. For all $w \in Q(A)$, by completing the square we find

$$\frac{1}{2} q_A(w) + \frac{a}{2} \|w\|^2 - \mathfrak{R}\langle w, z \rangle - \frac{1}{2} q_A(v) - \frac{a}{2} \|v\|^2 + \mathfrak{R}\langle v, z \rangle$$

$$= \frac{1}{2} q_A(w - v) + \frac{a}{2} \|w - v\|^2 \geq \frac{a + \alpha}{2} \|w - v\|^2 \tag{3.15}$$

which is non-negative and vanishes only when $w = v$. This shows that v is the unique minimizer of (3.13).   □

Exercise 3.15 Without using the information that $v = (A + a)^{-1}z$, show that the minimization problem (3.13) admits a minimizer (take a minimizing sequence, show that it is bounded in $Q(A)$ and pass to the weak limit in this space). By a perturbation argument, show that any minimizer satisfies the relation (3.14).

3.2.4 Example of the Laplacian on $I = (0, 1)$

We have seen in Sect. 2.8.3 many possible self-adjoint realizations of the Laplacian on the interval $I = (0, 1)$. These are parameterized by all the isotropic spaces of dimension 2 of the matrix M defined in (2.27). They all have a bounded-below quadratic form (which we will prove later in Theorem 5.28) that we determine here for the particular cases of Table 2.1.

Born-von Kármán Laplacian For the Born-von Kármán condition we have

$$D(A_{\mathrm{per},\theta}) = H^2_{\mathrm{per},\theta}(I)$$

$$= \left\{ u \in H^2(I) \ : \ u(1) = e^{i\theta}u(0), \ u'(1) = e^{i\theta}u'(0) \right\}$$

with $A_{\mathrm{per},\theta}u := -u''$ and $\theta \in [0, 2\pi)$. An integration by parts shows that, for $u \in D(A_{\mathrm{per},\theta})$,

$$q_{A_{\mathrm{per},\theta}}(u) = \int_0^1 |u'(t)|^2 \, \mathrm{d}t.$$

The quadratic form of the Born-von Kármán Laplacian is therefore defined on the closure of $D(A_{\mathrm{per},\theta})$ for the norm $H^1(I)$, that is to say on the domain

$$Q(A_{\mathrm{per},\theta}) = H^1_{\mathrm{per},\theta}(I) = \left\{ u \in H^1(I) \ : \ u(1) = e^{i\theta}u(0) \right\}.$$

The condition on $u'(1)$ and $u'(0)$ is lost because it does not make sense in $H^1(I)$.

Theorem 3.13 tells us how to find back the domain $D(A_{\mathrm{per},\theta})$ from the quadratic form and it is useful to check in practice how the two missing conditions $u'' \in L^2(I)$ and $u'(1) = e^{i\theta}u'(0)$ appear. From (3.12), we have $u \in D(A_{\mathrm{per},\theta})$ if and only if $u \in H^1_{\mathrm{per},\theta}(I)$ and there exists $g \in L^2(I)$ such that

$$\int_0^1 \overline{u'(t)}\varphi'(t) \, \mathrm{d}t + \int_0^1 \overline{u(t)}\varphi(t) \, \mathrm{d}t = \int_0^1 \overline{g(t)}\varphi(t) \, \mathrm{d}t$$

for all $\varphi \in H^1_{\mathrm{per},\theta}(I)$. Taking $\varphi \in C^\infty_c(I)$ we find the equation in the sense of distributions

$$-u'' + u = g$$

which implies $u'' \in L^2(I)$. We can then integrate by parts for all $\varphi \in H^1_{\mathrm{per},\theta}(I)$ and, using the equation, we deduce that the boundary terms must cancel out:

$$\overline{u'(1)}\varphi(1) = \overline{u'(0)}\varphi(0).$$

This implies $u'(1) = e^{i\theta}u'(0)$, as desired.

Robin Laplacian   We now discuss the Robin boundary condition, which we write in the form

$$D(A_{\mathrm{Rob},\theta}) = \Big\{u \in H^2(I) \; : \; \cos(\pi\theta)u(1) + \sin(\pi\theta)u'(1) = 0,$$

$$\cos(\pi\theta)u(0) - \sin(\pi\theta)u'(0) = 0\Big\}$$

with $\theta \in [0, 1)$ and $A_{\mathrm{Rob},\theta}u = -u''$. An integration by parts provides, this time,

$$\langle u, A_{\mathrm{Rob},\theta}u \rangle = \int_0^1 |u'(t)|^2 \, dt$$

$$+ \begin{cases} \dfrac{1}{\tan(\pi\theta)}\left(|u(1)|^2 + |u(0)|^2\right) & \text{for } \theta \in \left(0, \tfrac{1}{2}\right) \cup \left(\tfrac{1}{2}, 1\right), \\ 0 & \text{for } \theta \in \left\{0, \tfrac{1}{2}\right\}. \end{cases}$$

$$(3.16)$$

According to the inequality

$$\sup_{[0,1]} |u|^2 \leq 2 \, \|u\|_{L^2(I)} \, \|u'\|_{L^2(I)} + \|u\|^2_{L^2(I)}, \tag{3.17}$$

proved in (A.5) in Appendix A, the associated norm is still equivalent to that of H^1 and the quadratic form is therefore defined on the domain

$$Q(A_{\mathrm{Rob},\theta}) = \begin{cases} H^1(I) & \text{for } \theta \in (0, 1), \\ H^1_0(I) & \text{for } \theta = 0. \end{cases} \tag{3.18}$$

Indeed, when $\theta \neq 0$ the conditions on $u'(0)$ and $u'(1)$ disappear in $H^1(I)$. We see that:

- Only the Dirichlet boundary condition makes sense in $H^1(I)$, and the quadratic form is then defined on $Q(A_{\mathrm{Rob},0}) = H^1_0(I)$;

- The quadratic forms for the Robin Laplacian with $0 < \theta < 1$ (including the case $\theta = 1/2$ of Neumann) are all defined on the same domain $H^1(I)$, where the boundary condition is invisible;
- The quadratic form associated with $A_{\mathrm{Rob},\theta}$ depends on θ and contains a boundary term only for $\theta \in (0, 1) \setminus \{1/2\}$.

The boundary condition therefore appears much more clearly in the domain of the self-adjoint operators $A_{\mathrm{Rob},\theta}$ than in the associated quadratic forms.

Remark 3.16 When $\theta \to 0^+$ the boundary term in $1/\tan(\pi\theta)$ tends to $+\infty$ and it then plays the role of a *penalization* which leads, in the limit, to the Dirichlet condition $u(0) = u(1) = 0$. We will study the behavior of the eigenvalues of the Robin Laplacian later in Exercise 5.56.

3.2.5 Friedrichs Realization

We have proven that a semi-bounded self-adjoint operator is completely character-ized by its quadratic form. In practice, a converse version of Theorem 3.14 due to Friedrichs is often used. It states that every closed quadratic form q is equal to a q_A for a unique self-adjoint operator A. This is an elegant technique for constructing self-adjoint extensions of semi-bounded symmetric operators. The idea is to start from a bounded-below symmetric operator A defined on a very small domain, then to calculate the associated quadratic form q_A. As this latter is bounded below, it can be closed by Theorem 3.11, which then provides a self-adjoint extension $B = B^* \supset A$. In particular, all bounded-below symmetric operators admit self-adjoint extensions.

Theorem 3.17 (Riesz-Friedrichs) *Let $Q \subset \mathfrak{H}$ be two Hilbert spaces, with norms $\|\cdot\|_Q$ and $\|\cdot\|_{\mathfrak{H}}$ and scalar products $\langle\cdot,\cdot\rangle_Q$ and $\langle\cdot,\cdot\rangle_{\mathfrak{H}}$. We assume that Q is dense and continuously embedded into $\mathfrak{H}$, that is, there exists $\alpha > 0$ such that*

$$\|v\|_Q \geq \alpha\|v\|_{\mathfrak{H}}, \qquad \forall v \in Q. \tag{3.19}$$

Then there exists a unique self-adjoint operator A on its domain $D(A) \subset \mathfrak{H}$, such that $q_A = \|\cdot\|_Q^2$, $\varphi_A = \langle\cdot,\cdot\rangle_Q$ and $Q = Q(A)$. Its domain is given by

$$D(A) := \left\{v \in Q \ : \ \exists z \in \mathfrak{H}, \ \langle v, h\rangle_Q = \langle z, h\rangle_{\mathfrak{H}}, \ \forall h \in Q\right\} \tag{3.20}$$

and then $Av := z$ for $v \in D(A)$.

Moreover, if B is a self-adjoint operator on $D(B) \subset Q$ such that $D(B)$ is dense in Q and $q_B = q$ on $D(B)$, then $B = A$.

The last property is a bit stronger than the uniqueness at the beginning of the theorem, because we do not assume *a priori* that q is the closure q_B of the quadratic form associated with B. But this is what we show in the proof.

Proof The formula (3.20) for the domain suggests to introduce

$$G(A) = \left\{ (v, z) \in Q \times \mathfrak{H} \; : \; \langle v, h \rangle_Q = \langle z, h \rangle_{\mathfrak{H}}, \; \forall h \in Q \right\}.$$

It is clear that $G(A)$ is a subspace of $\mathfrak{H} \times \mathfrak{H}$. If, moreover, $(0, z) \in G(A)$ then we have $\langle z, h \rangle_{\mathfrak{H}} = 0$ for all $h \in Q$, which implies that $z = 0$ since Q is dense in $\mathfrak{H}$. Thus, $G(A)$ is the graph of a linear operator A defined on $D(A)$ in (3.20) by $Av = z$. We will verify a little later that $D(A)$ is a dense subspace of $\mathfrak{H}$. For $v_1, v_2 \in D(A)$, we have

$$\langle Av_1, v_2 \rangle_{\mathfrak{H}} = \langle v_1, v_2 \rangle_Q = \overline{\langle v_2, v_1 \rangle_Q} = \overline{\langle Av_2, v_1 \rangle_{\mathfrak{H}}} = \langle v_1, Av_2 \rangle_{\mathfrak{H}}$$

which shows that the operator A is symmetric on its domain.

Let $z \in \mathfrak{H}$. Since $h \mapsto \langle z, h \rangle_{\mathfrak{H}}$ is a continuous linear form on Q, Riesz's theorem implies the existence of a unique $v \in Q$ such that $\langle v, h \rangle_Q = \langle z, h \rangle_{\mathfrak{H}}$ for all $h \in Q$, that is to say such that $Av = z$. The usual proof consists in minimizing the strictly convex functional $w \in Q \mapsto \|w\|_Q^2 / 2 - \Re\langle z, w \rangle$, a bit like in Theorem 3.14. The existence of a minimizer is guaranteed by the inequality (3.19) and a minimization argument as in Exercise 3.15. Thus, we have shown that the operator A is surjective from $D(A)$ to $\mathfrak{H}$.

Let us now prove that $D(A)$ is dense. Let h be in the orthogonal of $D(A)$ for the scalar product of Q and, since A is surjective, let $v \in D(A)$ be such that $Av = h$. We have $0 = \langle v, h \rangle_Q = \langle h, h \rangle_{\mathfrak{H}}$ which implies $h = 0$, so $D(A)$ is dense in Q for the norm of Q. Since Q is itself supposed to be dense in $\mathfrak{H}$, we conclude that $D(A)$ is dense in $\mathfrak{H}$.

As we have already shown that A is surjective from $D(A)$ to $\mathfrak{H}$, according to Theorem 2.28, it follows that A is self-adjoint. Moreover, we have also proved that $D(A)$ is dense in Q for its norm, and so we indeed have $q_A = \| \cdot \|_Q^2$ as announced.

Finally, we need to show that A is also the only self-adjoint operator whose domain is dense in Q for the associated topology, and whose quadratic form coincides with q on its domain. Let B be such another operator. We have by hypothesis $q_B(v) = q(v)$ for all $v \in D(B)$ which, by polarization, implies $\langle Bv, h \rangle = \langle v, h \rangle_Q$ for all $v, h \in D(B)$. As $D(B)$ is supposed to be dense in Q, this relation remains true for all $h \in Q$. This is exactly the characterization of the fact that $v \in D(A)$ with $Av = Bv$. Thus, $B \subset A$ which implies $A = B$ because both operators are self-adjoint. $\qquad\square$

The previous result allows us to define the Friedrichs extension of a coercive symmetric operator.

Corollary 3.18 (Friedrichs Extension) *Let A be a symmetric operator on its domain $D(A) \subset \mathfrak{H}$, which is bounded below. Let $q = \overline{q_A}$ be the closure of the* *associated quadratic form, constructed in Theorem 3.11, whose domain is denoted Q. Then there exists a unique self-adjoint extension B of A, called the **Friedrichs extension** such that*

$$\boxed{D(A) \subset D(B) \subset Q.}$$

More precisely, we have

$$D(B) = \left\{ v \in Q \ : \ \exists z \in \mathfrak{H}, \ \langle v, h \rangle_Q = \langle z, h \rangle, \ \forall h \in Q \right\}$$

with $Bv := z$, $q = q_B$ and $Q = Q(B)$.

A consequence of this result is that every symmetric operator A that is bounded from below admits at least one self-adjoint extension. The latter is also bounded from below. The other possible extensions do not always have a lower bounded spectrum, however [RS75, pp. 179–180].

Proof We have already seen in Theorem 3.17 that the quadratic form q was associated with a unique self-adjoint operator B. The operator B is an extension of A because we have $D(A) \subset Q$ by construction, so

$$\langle Bv, h \rangle = \langle v, h \rangle_Q = \langle v, Ah \rangle$$

for all $v \in D(B)$ and all $h \in D(A) \subset Q$. This shows that $B \subset A^*$ and therefore $A \subset \overline{A} = (A^*)^* \subset B^* = B$. The uniqueness follows from the last assertion of Theorem 3.17. Indeed, since $D(A)$ is by construction dense in Q, any self-adjoint extension $\tilde{B}$ of A such that $D(A) \subset D(\tilde{B}) \subset Q$ also has its domain $D(\tilde{B})$ dense in Q. Moreover, for all $v \in D(\tilde{B})$ and all $w \in D(A)$

$$\left\langle \tilde{B}v, w \right\rangle = \left\langle v, \tilde{B}w \right\rangle = \langle v, Aw \rangle = \varphi_B(v, w)$$

since $A \subset \tilde{B}$. Since $D(A)$ is by definition dense in $Q = Q(B)$ for the associated norm, the relation remains true in all $D(\tilde{B})$ and we can then use the uniqueness stated in Theorem 3.17.   □

The Friedrichs method allows one to show that $A + B$ is self-adjoint when A is and that B is relatively bounded, but in a sense involving quadratic forms instead of operators, as we saw for the Rellich-Kato theorem. The following result is due to Kato, Lions, Lax, Milgram and Nelson [RS75, p. 323].

Theorem 3.19 (KLMN) *Let A be a self-adjoint operator on its domain $D(A) \subset \mathfrak{H}$, which is coercive, that is, such that $\langle v, Av \rangle \geq \alpha \|v\|^2$ for all $v \in D(A)$ with $\alpha > 0$. Let q_A be the associated closed quadratic form, of domain $Q(A)$. Let b be another quadratic form defined on $Q(A)$, such that*

$$|b(v)| \leq \eta \, q_A(v) + \kappa \|v\|^2, \qquad \forall v \in Q(A),$$

for a real $0 \leq \eta < 1$ and $\kappa > 0$. Then $v \mapsto q_A(v) + b(v)$ is closed and coercive on $Q(A)$. It is therefore associated with a unique self-adjoint operator C, which is such that $Q(C) = Q(A)$.

If b is the quadratic form of a symmetric operator B defined on $D(A)$, then C is the unique self-adjoint extension of $A + B$ defined on $D(A + B) = D(A)$, such that $D(A) \subset D(C) \subset Q(A)$.

Proof As

$$(1 - \eta)q_A(v) - \kappa \, \|v\|^2 \leq q_A(v) + b(v) \leq (1 + \eta)q_A(v) + \kappa \, \|v\|^2 ,$$

the quadratic form $v \mapsto q_A(v) + b(v) + \kappa \, \|v\|^2$ is equivalent to q_A on $Q(A)$. Thus $Q(A)$ is also complete for this new quadratic form and the result follows from Theorem 3.17. □

In general, the operator C can have a very different domain from that of A, even if the form domains are the same. An example is given in Exercise 3.26.

3.3 Quadratic Forms and Schrödinger Operators

In this section we use quadratic forms to construct self-adjoint realizations of Schrödinger operators $H = -\Delta + V(x)$, particularly in the case where the multiplication operator by the function V is not naturally defined on the space $D(-\Delta) = H^2(\mathbb{R}^d)$. This includes situations where V is too singular or diverges at infinity.

3.3.1 Case of Locally Singular Potentials

With the help of the Rellich-Kato theorem, we saw in Corollary 3.5 that Schrödinger operators $H = -\Delta + V(x)$ were self-adjoint on $D(H) = H^2(\mathbb{R}^d)$ when $V \in L^p(\mathbb{R}^d, \mathbb{R}) + L^\infty(\mathbb{R}^d, \mathbb{R})$ with

$$\begin{cases} p = 2 & \text{if } d \in \{1, 2, 3\}, \\ p > 2 & \text{if } d = 4, \\ p = \frac{d}{2} & \text{if } d \geq 5. \end{cases} \tag{3.21}$$

They are even essentially self-adjoint on $C_c^\infty(\mathbb{R}^d)$, which means that there is only one possible self-adjoint realization. On the other hand, we had previously seen in Lemma 1.10 that the corresponding quadratic form

$$\mathcal{E}^V(u) = \int_{\mathbb{R}^d} |\nabla u(x)|^2 \, \mathrm{d}x + \int_{\mathbb{R}^d} V(x)|u(x)|^2 \, \mathrm{d}x$$

was continuous and bounded below under the weaker assumption

$$\begin{cases} p = 1 & \text{if } d = 1, \\ p > 1 & \text{if } d = 2, \\ p = \frac{d}{2} & \text{if } d \geq 3. \end{cases} \tag{3.22}$$

More precisely, Lemma 1.10 implies that

$$c_1 \|u\|_{H^1(\mathbb{R}^d)}^2 \leq \mathcal{E}^V(u) + C \|u\|_{L^2(\mathbb{R}^d)}^2 \leq c_2 \|u\|_{H^1(\mathbb{R}^d)}^2 \tag{3.23}$$

for constants $C, c_1, c_2 > 0$. Thus $\mathcal{E}^V + C\| \cdot \|_{L^2(\mathbb{R}^d)}^2$ defines a quadratic form equivalent to the square of the norm of $H^1(\mathbb{R}^d)$. The two assumptions (3.21) and (3.22) differ in dimension $d \leq 4$. The energy is bounded-below for potentials that can be more locally singular than those for which we can show that the operator is self-adjoint on $H^2(\mathbb{R}^d)$.

Let us for instance work in dimension $d = 1$. Assumption (3.22) covers potentials behaving as

$$V(x) \underset{x \to 0}{\sim} \frac{c}{|x|^\alpha}, \qquad c \neq 0 \tag{3.24}$$

with $\alpha < 1$ while the corresponding operator is self-adjoint on $H^2(\mathbb{R})$ only under the assumption that $\alpha < 1/2$. In fact, when $1/2 \leq \alpha < 1$, the operator $-\Delta + V(x)$ is not even defined on $C_c^\infty(\mathbb{R})$ because for any continuous function u such that $u(0) = 1$, we will have $Vu \notin L^2(\mathbb{R})$. Thus the operator $-\Delta + V(x)$ is not well defined on $H^2(\mathbb{R})$ either. When V is singular only at the origin, where it behaves as (3.24), it is possible to start by defining the operator $-\Delta + V(x)$ on $C_c^\infty(\mathbb{R} \setminus \{0\})$. None of its self-adjoint extensions will contain $H^2(\mathbb{R})$. We see from this example that the minimal domain that can be used as a starting point for the operator $-\Delta + V(x)$ may depend on the local singularities of the function V. This greatly complicates the construction of possible self-adjoint extensions. However, these subtleties disappear when using the quadratic form, which is always defined on all $H^1(\mathbb{R}^d)$.

Corollary 3.20 (Friedrichs Realization for Singular Potentials) *Let* $V \in$ $L^p(\mathbb{R}^d, \mathbb{R}) + L^\infty(\mathbb{R}^d, \mathbb{R})$ *with p satisfying (3.22). Then there exists a unique self-adjoint operator H such that $D(H)$ is dense in $H^1(\mathbb{R}^d)$ and whose quadratic form is*

$$\langle u, Hu \rangle = \int_{\mathbb{R}^d} |\nabla u(x)|^2 \, dx + \int_{\mathbb{R}^d} V(x)|u(x)|^2 \, dx$$

for all $u \in D(H)$. This operator is defined on the domain

$$\boxed{D(H) = \left\{ u \in H^1(\mathbb{R}^d) \; : \; -\Delta u + Vu \in L^2(\mathbb{R}^d) \right\}}$$

by $Hu := -\Delta u + Vu$ *where each term is understood in the sense of distributions or in* $H^{-1}(\mathbb{R}^d)$. *We then have* $Q(H) = H^1(\mathbb{R}^d)$.

Proof The quadratic form is coercive and closed on $H^1(\mathbb{R}^d)$ by (3.23). The Riesz-Friedrichs Theorem 3.17 or the KLMN Theorem 3.19 then provide a unique self-adjoint operator H whose quadratic form is $\mathcal{E}^V$. The domain $D(H)$ is the set of functions $u \in H^1(\mathbb{R}^d)$ for which there exists $h \in L^2(\mathbb{R}^d)$ such that

$$\int_{\mathbb{R}^d} \nabla\overline{v(x)} \cdot \nabla u(x)\,\mathrm{d}x + \int_{\mathbb{R}^d} V(x)\overline{v(x)}u(x)\,\mathrm{d}x = \int_{\mathbb{R}^d} \overline{v(x)}h(x)\,\mathrm{d}x$$

for all $v \in H^1(\mathbb{R}^d)$. This is exactly the definition of the fact that $-\Delta u + Vu = h \in L^2(\mathbb{R}^d)$, with each term interpreted in $H^{-1}(\mathbb{R}^d)$. As $C_c^\infty(\mathbb{R}^d)$ is dense in $H^1(\mathbb{R}^d)$, it is equivalent to take $v \in C_c^\infty(\mathbb{R}^d)$, in which case each term is now seen as a distribution. □

The uniqueness of H under the assumption that $D(H)$ is dense in $H^1(\mathbb{R}^d)$ is important from a physical point of view. It means that the constraint that the energy is finite on the domain of H is sufficient to completely determine the operator H.

It is important to note that, on $D(H)$, only the sum $-\Delta u + Vu$ belongs to $L^2(\mathbb{R}^d)$. The two terms may not belong to $L^2(\mathbb{R}^d)$ separately (otherwise the domain would be included in $H^2(\mathbb{R}^d)$). It is even possible that we have $H^2(\mathbb{R}^d) \cap D(H) = \{0\}$, that is, the two terms $-\Delta u$ and Vu are *never* in $L^2(\mathbb{R}^d)$ (Exercise 3.26 below).

Remark 3.21 If V satisfies the stronger condition (3.21), the Friedrichs realization has the domain $D(H) = H^2(\mathbb{R}^d)$. This is because $H^2(\mathbb{R}^d) \subset H^1(\mathbb{R}^d)$ and is dense in this space. In particular, Δu and Vu are both in $L^2(\mathbb{R}^d)$ for $u \in D(H) = H^2(\mathbb{R}^d)$.

3.3.2   Case of an Arbitrary Positive Potential

So far, we have mainly studied Schrödinger operators $-\Delta + V(x)$ describing a particle in an external potential that remains bounded at infinity. It is also customary to study *confined* particles, which means that V tends to $+\infty$ at infinity. The most classical example is the harmonic oscillator $V(x) = \omega^2|x|^2$. With the technique of quadratic forms, we can define $H = -\Delta + V(x)$ for a potential whose positive part is more or less arbitrary (locally integrable), bounded or not.

Let us consider a real-valued potential that we write in the form $V(x) = V_+(x) - V_-(x)$ where $V_+ = \max(V, 0)$ and $V_- = \max(-V, 0)$. We assume that $V_- \in L^p(\mathbb{R}^d, \mathbb{R}) + L^\infty(\mathbb{R}^d, \mathbb{R})$ with p satisfying Assumption (3.22), so that the associated potential energy can be controlled by the kinetic energy. For V_+ we have more freedom, as it is enough that the quadratic form is well defined, the term $\int_{\mathbb{R}^d} V_+|u|^2$ being always non-negative. We then just ask that $V_+ \in L^1_{\text{loc}}(\mathbb{R}^d, \mathbb{R})$. The energy

$$\mathcal{E}^V(u) = \int_{\mathbb{R}^d} |\nabla u(x)|^2 \, dx + \int_{\mathbb{R}^d} V(x)|u(x)|^2 \, dx \tag{3.25}$$

is well defined on $C_c^\infty(\mathbb{R}^d)$ (because V_+ is locally integrable). It is also well defined on the larger space

$$\mathcal{V} := \left\{ u \in H^1(\mathbb{R}^d) \; : \; \sqrt{V_+}\, u \in L^2(\mathbb{R}^d) \right\}, \tag{3.26}$$

which is complete when equipped with the norm

$$\|u\|_{\mathcal{V}} := \sqrt{\|u\|^2_{H^1(\mathbb{R}^d)} + \int_{\mathbb{R}^d} V_+|u|^2}.$$

When C is large enough, $\sqrt{\mathcal{E}^V + C\|\cdot\|^2_{L^2(\mathbb{R}^d)}}$ provides an equivalent norm thanks to our assumptions on V_-, as we saw in Exercise 1.28. The space (3.26) is the largest subspace of $L^2(\mathbb{R}^d)$ on which the energy $\mathcal{E}^V$ is well defined. Moreover, $C_c^\infty(\mathbb{R}^d)$ is dense in $\mathcal{V}$ for the associated norm (we leave this as an exercise). The following result is proved exactly like Theorem 3.22.

Theorem 3.22 (Friedrichs Realization for $V_+ \in L^1_{\text{loc}}$) *Let V be a real-valued measurable function with $V_- = \max(-V, 0) \in L^p(\mathbb{R}^d, \mathbb{R}_+) + L^\infty(\mathbb{R}^d, \mathbb{R}_+)$ where p satisfies (3.22) and $V_+ = \max(V, 0) \in L^1_{\text{loc}}(\mathbb{R}^d, \mathbb{R}_+)$. Then there exists a unique self-adjoint operator H such that $D(H)$ is dense in $\mathcal{V}$ and whose quadratic form is*

$$\langle u, Hu \rangle = \int_{\mathbb{R}^d} |\nabla u(x)|^2 \, dx + \int_{\mathbb{R}^d} V(x)|u(x)|^2 \, dx$$

for all $u \in D(H)$. This operator is defined on the domain

$$\boxed{D(H) = \left\{ u \in \mathcal{V} \; : \; -\Delta u + Vu \in L^2(\mathbb{R}^d) \right\}}$$

by $Hu := -\Delta u + Vu$ where each term is understood in the sense of distributions. We then have $Q(H) = \mathcal{V}$.

If $V \in L^2_{\text{loc}}(\mathbb{R}^d)$, we can introduce the operator $H^{\min} = -\Delta + V$ defined on $D(H^{\min}) = C_c^\infty(\mathbb{R}^d)$. The Friedrichs extension H constructed in the previous theorem is then the unique self-adjoint extension satisfying the condition $D(H^{\min}) \subset D(H) \subset \mathcal{V}$. Are there other extensions not included in $\mathcal{V}$ or is the Friedrichs extension H the only possible one? In other words, when is $H^{\min}$ essentially self-adjoint? A famous result due to Simon and Kato [RS75, Thm. X.29] asserts that $\overline{H^{\min}} = H$ with the only assumptions that $V_+ \in L^2_{\mathrm{loc}}(\mathbb{R}^d)$ and $V_- \in L^p(\mathbb{R}^d) + L^\infty(\mathbb{R}^d)$ where p satisfies the stronger condition (3.21). In this case we also have

$$D(H) = D(H^{\max}) = \left\{ u \in L^2(\mathbb{R}^d) : -\Delta u + V u \in L^2(\mathbb{R}^d) \right\},$$

that is, we do not need to assume that $u \in \mathcal{V}$.

3.3.3 Separability, Harmonic Oscillator

A disadvantage of the Friedrichs method is that it can be difficult to determine more precisely the domain of the constructed operator. For $u \in D(-\Delta + V)$ we always have $-\Delta u + V u \in L^2(\mathbb{R}^d)$ but the two terms of this sum are distributions that are not necessarily individually in $L^2(\mathbb{R}^d)$. We say that the operator is **separable** when the domain of the Friedrichs extension simply equals

$$D(-\Delta + V) = D(-\Delta) \cap D(V) = \left\{ u \in H^2(\mathbb{R}^d) : V u \in L^2(\mathbb{R}^d) \right\}, \qquad (3.27)$$

that is, when $-\Delta u$ and $V u$ are always both in $L^2(\mathbb{R}^d)$.

In this section we examine the case of a quite regular potential V (differentiable once) but diverging at infinity. The typical example we have in mind is the harmonic oscillator for which $V(x) = \omega^2 |x|^2$ is even C^∞. We will see that Equality (3.27) is true when the gradient of V does not grow faster than $V^{3/2}$ up to a multiplicative constant. This assumption is verified for any potential V that behaves polynomially at infinity like the harmonic oscillator.

As it is the identification of the domain that interests us, we will assume for simplicity that $V \geq 0$. By the Rellich-Kato Theorem 3.1, the following result will also be true for $-\Delta + V + W$ with $W \in L^\infty(\mathbb{R}^d)$ or even $W \in L^p(\mathbb{R}^d) + L^\infty(\mathbb{R}^d)$ with a p satisfying (3.21). We will also assume that V is *locally bounded*. This allows us to define $H^{\min} = -\Delta + V$ on

$$D(H^{\min}) = C_c^\infty(\mathbb{R}^d).$$

By Corollary 3.18, $H^{\min}$ admits a *unique self-adjoint extension* H whose domain $D(H)$ is dense in the energy space

$$\mathcal{V} = \left\{ u \in H^1(\mathbb{R}^d), \quad \sqrt{V} u \in L^2(\mathbb{R}^d) \right\}.$$

More precisely

$$D(H) = \left\{ u \in H^1(\mathbb{R}^d), \quad \sqrt{V} u \in L^2(\mathbb{R}^d), \quad (-\Delta + V) u \in L^2(\mathbb{R}^d) \right\}.$$

We want to deduce from the above information that necessarily Δu and Vu both belong to $L^2(\mathbb{R}^d)$. The assumption that V is locally bounded and that $u \in H^1(\mathbb{R}^d)$ immediately imply that the functions of the domain are all in $H^2_{\text{loc}}(\mathbb{R}^d)$ and the main question is whether Δu is square integrable over the entire space $\mathbb{R}^d$.

Theorem 3.23 (Separability) *Let the space dimension be $d \geq 1$. Let V be a continuous and non-negative function on $\mathbb{R}^d$, whose gradient ∇V (in the sense of distributions) is a function that satisfies almost everywhere*

$$|\nabla V(x)| \leq \alpha V(x)^{\frac{3}{2}} + \kappa \tag{3.28}$$

with $0 \leq \alpha < \sqrt{2}$ and $\kappa \geq 0$. Then the domain of the Friedrichs extension H of the operator $H^{\min}$ is equal to

$$\boxed{D(H) = \left\{ u \in H^2(\mathbb{R}^d), \quad Vu \in L^2(\mathbb{R}^d) \right\}} \tag{3.29}$$

and we have the series of inequalities

$$\frac{1}{2} \|\Delta u\|^2_{L^2(\mathbb{R}^d)} + \frac{\sqrt{2} - \alpha}{\sqrt{2}} \|Vu\|^2_{L^2(\mathbb{R}^d)}$$

$$\leq \|(-\Delta + V)u\|^2_{L^2(\mathbb{R}^d)} + \frac{3\kappa^{\frac{4}{3}}}{2} \|u\|^2_{L^2(\mathbb{R}^d)}$$

$$\leq 2\|\Delta u\|^2_{L^2(\mathbb{R}^d)} + 2\|Vu\|^2_{L^2(\mathbb{R}^d)} + \frac{3\kappa^{\frac{4}{3}}}{2} \|u\|^2_{L^2(\mathbb{R}^d)} \tag{3.30}$$

for all $u \in D(H)$. The operator $H^{\min} = -\Delta + V$ is essentially self-adjoint on $D(H^{\min}) = C^\infty_c(\mathbb{R}^d)$ with, of course, $\overline{H^{\min}} = H$.

The inequality (3.30) means that we have the equivalence of norms

$$\|u\|_{L^2(\mathbb{R}^d)} + \|(-\Delta + V)u\|_{L^2(\mathbb{R}^d)} \approx \|\Delta u\|_{L^2(\mathbb{R}^d)} + \|Vu\|_{L^2(\mathbb{R}^d)} + \|u\|_{L^2(\mathbb{R}^d)}$$

on $D(H)$. As the norms $\|u\|_{H^1(\mathbb{R}^d)}$ and $\|\sqrt{V}u\|_{L^2(\mathbb{R}^d)}$ do not appear in (3.30), it is also possible to show that $H = H^{\max}$, the maximal operator defined on

$$D(H^{\max}) = \left\{ u \in L^2(\mathbb{R}^d) \; : \; (-\Delta + V)u \in L^2(\mathbb{R}^d) \right\}.$$

However, the proof of (3.30) is simplified if we use the additional information that $u \in H^1(\mathbb{R}^d)$ and $\sqrt{V}u \in L^2(\mathbb{R}^d)$ as provided by the Friedrichs construction.

Our regularity assumptions on V are far too strong and we encourage the reader to determine those for which the proof below works with minor changes.

In Exercise 3.28, we construct a potential $V \geq 0$ satisfying

$$|\nabla V(x)| \leq 2V(x)^{\frac{3}{2}} + \kappa,$$

such as $-\Delta + V$ is *not separable*. This proves that the power $3/2$ is optimal and cannot be increased in (3.28). This also implies that the theorem cannot be true if we allow α to be greater than or equal to 2. In dimension $d \geq 2$, Theorem 3.23 remains true if we replace the assumption $\alpha < \sqrt{2}$ with the optimal constraint $\alpha < 2$, as shown by Everitt and Giertz [EG74, EG78] (see also [EZ78, Oka82]). In dimension $d = 1$ we expect the optimal constant to be rather $4/\sqrt{3}$ instead of 2 [Atk73]. The proof with the slightly more restrictive constraint $\alpha < \sqrt{2}$ is simpler and works in any dimension. The one we now provide is inspired by [Dav83, Lem. 4].

Proof The last inequality of (3.30) follows from the triangle inequality and $(a + b)^2 \leq 2a^2 + 2b^2$. It is valid for all $u \in H^2(\mathbb{R}^d)$ such that $Vu \in L^2(\mathbb{R}^d)$. We now show the first inequality starting with the case of $u \in C_c^\infty(\mathbb{R}^d)$. We write

$$\|(-\Delta + V)u\|_{L^2(\mathbb{R}^d)}^2 = \int_{\mathbb{R}^d} |\Delta u(x)|^2 \, dx + \int_{\mathbb{R}^d} V(x)^2 |u(x)|^2 \, dx$$

$$- 2\Re \int_{\mathbb{R}^d} V(x)\overline{u(x)}\Delta u(x) \, dx.$$

Using

$$2\Re \overline{u(x)}\Delta u(x) = 2\Re \, \mathrm{div}\big(\overline{u(x)}\nabla u(x)\big) - 2|\nabla u(x)|^2$$

and integrating by parts, we get

$$\|(-\Delta + V)u\|_{L^2(\mathbb{R}^d)}^2 = \int_{\mathbb{R}^d} |\Delta u(x)|^2 \, dx + \int_{\mathbb{R}^d} V(x)^2 |u(x)|^2 \, dx$$

$$+ 2\int_{\mathbb{R}^d} V(x)|\nabla u(x)|^2 \, dx + 2\Re \int_{\mathbb{R}^d} \overline{u(x)}\nabla V(x) \cdot \nabla u(x) \, dx. \qquad (3.31)$$

The last term is the only one that can be negative. If it can be controlled by the first three terms, we can then show that Δu, Vu and $\sqrt{V}\nabla u$ must all belong to $L^2(\mathbb{R}^d)$ when $(-\Delta + V)u \in L^2(\mathbb{R}^d)$. The assumption (3.28) and the Cauchy-Schwarz inequality give

$$2\left|\overline{u(x)}\nabla V(x) \cdot \nabla u(x)\right| \leq 2\alpha |u(x)| V(x)^{\frac{3}{2}} |\nabla u(x)| + 2\kappa |u(x)| \, |\nabla u(x)|$$

$$\leq \frac{\alpha}{\sqrt{2}} \left(V(x)^2 |u(x)|^2 + 2V(x)|\nabla u(x)|^2 \right)$$

$$+ 2\kappa |u(x)| \, |\nabla u(x)|. \qquad (3.32)$$

We have thus shown the lower bound

$$\|(-\Delta + V)u\|^2_{L^2(\mathbb{R}^d)} \geq \int_{\mathbb{R}^d} |\Delta u(x)|^2 \, dx + \frac{\sqrt{2} - \alpha}{\sqrt{2}} \int_{\mathbb{R}^d} V(x)^2 |u(x)|^2 \, dx$$

$$+ (2 - \sqrt{2}\alpha) \int_{\mathbb{R}^d} V(x) |\nabla u(x)|^2 \, dx - 2\kappa \, \|u\|_{L^2(\mathbb{R}^d)} \, \|\nabla u\|_{L^2(\mathbb{R}^d)} \,. \qquad (3.33)$$

To estimate the last term, we use the Cauchy-Schwarz inequality in Fourier

$$\|\nabla u\|^2_{L^2(\mathbb{R}^d)} = \int_{\mathbb{R}^d} |k|^2 |\widehat{u}(k)|^2 \, dk$$

$$\leq \left(\int_{\mathbb{R}^d} |\widehat{u}(k)|^2 \, dk \right)^{\frac{1}{2}} \left(\int_{\mathbb{R}^d} |k|^4 |\widehat{u}(k)|^2 \, dk \right)^{\frac{1}{2}}$$

$$= \|u\|^{\frac{1}{2}}_{L^2(\mathbb{R}^d)} \|\Delta u\|^{\frac{1}{2}}_{L^2(\mathbb{R}^d)} \,.$$

With the inequality $2ab \leq a^4/2 + 3b^{4/3}/2$, this provides

$$2\kappa \, \|u\|_{L^2(\mathbb{R}^d)} \, \|\nabla u\|_{L^2(\mathbb{R}^d)} \leq \frac{1}{2} \|\Delta u\|^2_{L^2(\mathbb{R}^d)} + \frac{3\kappa^{\frac{4}{3}}}{2} \|u\|^2_{L^2(\mathbb{R}^d)} \,.$$

By inserting all of this into (3.33), we have shown the first inequality in (3.30) for $u \in C^\infty_c(\mathbb{R}^d)$.

We now need to show that this inequality continues to hold for $u \in D(H)$. Let $u \in H^1(\mathbb{R}^d)$ such that $\sqrt{V} \, u \in L^2(\mathbb{R}^d)$ and $-\Delta u + Vu \in L^2(\mathbb{R}^d)$. As V is locally bounded, we have $\Delta u \in L^2_{\text{loc}}(\mathbb{R}^d)$ and as $\nabla u \in L^2(\mathbb{R}^d)$ by hypothesis, we deduce that $u \in H^2_{\text{loc}}(\mathbb{R}^d)$. Let $\chi \in C^\infty_c(\mathbb{R}^d)$ be a positive radial function with support in the ball of radius 2, which equals $\chi \equiv 1$ on the ball of radius 1 and satisfies $0 \leq \chi \leq 1$ everywhere. We then set $\chi_R(x) = \chi(x/R)$. Thanks to the fact that $u \in H^2_{\text{loc}}(\mathbb{R}^d)$, we can calculate as before

$$\int_{\mathbb{R}^d} \chi_R(x)^2 \big| -\Delta u(x) + V(x)u(x) \big|^2 \, dx$$

$$= \int_{\mathbb{R}^d} \chi_R(x)^2 |\Delta u(x)|^2 \, dx + \int_{\mathbb{R}^d} \chi_R(x)^2 V(x)^2 |u(x)|^2 \, dx$$

$$+ 2 \int_{\mathbb{R}^d} \chi_R(x)^2 V(x) |\nabla u(x)|^2 \, dx + 2\Re \int_{\mathbb{R}^d} \chi_R(x)^2 \overline{u(x)} \nabla V(x) \cdot \nabla u(x) \, dx$$

$$+ 4\Re \int_{\mathbb{R}^d} \overline{u(x)} V(x) \chi_R(x) \nabla \chi_R(x) \cdot \nabla u(x) \, dx. \qquad (3.34)$$

The fourth term is estimated exactly as before using Inequality (3.32). The fifth term is controlled using $|\nabla \chi_R(x)| \le \|\nabla \chi\|_{L^\infty(\mathbb{R}^d)}/R$ and the Cauchy-Schwarz inequality in the form

$$\left| \int_{\mathbb{R}^d} \overline{u(x)} V(x) \chi_R(x) \nabla \chi_R(x) \cdot \nabla u(x)\, dx \right|$$

$$\le \frac{\|\nabla \chi\|_{L^\infty(\mathbb{R}^d)}}{R} \|\chi_R u V\|_{L^2(\mathbb{R}^d)} \|\nabla u\|_{L^2(\mathbb{R}^d)}$$

$$\le \frac{1}{2R} \int_{\mathbb{R}^d} \chi_R(x)^2 V(x)^2 |u(x)|^2 \, dx + \frac{\|\nabla \chi\|_{L^\infty(\mathbb{R}^d)}^2}{2R} \|\nabla u\|_{L^2(\mathbb{R}^d)}^2.$$

We have therefore proved the localized inequality

$$\int_{\mathbb{R}^d} \chi_R(x)^2 \big| - \Delta u(x) + V(x)u(x)\big|^2 \, dx$$

$$\ge \int_{\mathbb{R}^d} \chi_R(x)^2 |\Delta u(x)|^2 \, dx + \left(\frac{\sqrt{2}-\alpha}{\sqrt{2}} - \frac{2}{R} \right) \int_{\mathbb{R}^d} \chi_R(x)^2 V(x)^2 |u(x)|^2 \, dx$$

$$+ (2 - \sqrt{2}\alpha) \int_{\mathbb{R}^d} \chi_R(x)^2 V(x)|\nabla u(x)|^2 \, dx - 2\kappa \|u\|_{L^2(\mathbb{R}^d)} \|\nabla u\|_{L^2(\mathbb{R}^d)}$$

$$- \frac{2\|\nabla\chi\|_{L^\infty(\mathbb{R}^d)}^2}{R} \|\nabla u\|_{L^2(\mathbb{R}^d)}^2.$$

Since $\alpha < \sqrt{2}$ and $\nabla u, (-\Delta + V)u \in L^2(\mathbb{R}^d)$, this shows that $\int_{\mathbb{R}^d} \chi_R^2 |\Delta u|^2$, $\int_{\mathbb{R}^d} \chi_R^2 V^2 |u|^2$ and $\int_{\mathbb{R}^d} \chi_R^2 V |\nabla u|^2$ are uniformly bounded with respect to R. The corresponding functions are therefore in $L^2(\mathbb{R}^d)$ and by letting $R \to \infty$ we obtain the claimed inequality (3.30).

To show that $H = H^{\min}$ we must prove that for any function $u \in D(H)$ there exists a sequence $u_n \in C_c^\infty(\mathbb{R}^d)$ such that $u_n \to u$ and $H u_n \to H u$ in $L^2(\mathbb{R}^d)$. From (3.30), it suffices to show that $u_n \to u$ in $H^2(\mathbb{R}^d)$ and $V u_n \to V u$ in $L^2(\mathbb{R}^d)$. We can start by approximating u by a function v_n in $D(H)$ with compact support (but not necessarily C^∞). Using the fact that V is locally bounded, the result will then follow from the usual density of $C_c^\infty(\mathbb{R}^d)$ in $H^2(\mathbb{R}^d)$. We can simply take, $v_n(x) = \chi(x/n)u(x)$. Then

$$\lim_{n\to\infty} \int_{\mathbb{R}^d} V(x)^2 |v_n(x) - u(x)|^2 \, dx = \lim_{n\to\infty} \int_{\mathbb{R}^d} V(x)^2 \big(1 - \chi(x/n)\big)^2 |u(x)|^2 \, dx = 0$$

by dominated convergence, since $Vu \in L^2(\mathbb{R}^d)$. The argument is the same to conclude that $v_n \to u$ in $L^2(\mathbb{R}^d)$. Moreover,

$$\Delta(\chi(x/n)u(x)) = u(x)\frac{(\Delta\chi)(x/n)}{n^2} + 2\frac{\nabla u(x) \cdot (\nabla\chi)(x/n)}{n} + \chi(x/n)\Delta u(x)$$

which strongly converges to Δu in $L^2(\mathbb{R}^d)$. This implies the convergence of v_n to u in $H^2(\mathbb{R}^d)$ by elliptic regularity. $\qquad\square$

Remark 3.24 For $u \in C_c^\infty(\mathbb{R}^d)$, the last term of (3.31) is also equal to

$$2\Re \int_{\mathbb{R}^d} \overline{u(x)}\nabla V(x) \cdot \nabla u(x)\, dx = -\int_{\mathbb{R}^d} |u(x)|^2 \Delta V(x)\, dx$$

and the argument is simpler if we add assumptions on ΔV. For the harmonic oscillator $V(x) = \omega^2|x|^2$, this quantity is just equal to $-2d\omega^2 \int_{\mathbb{R}^d} |u|^2$.

Example 3.25 (Domain of the Harmonic Oscillator) The harmonic oscillator corresponds to $V(x) = \omega^2|x|^2$ where $\omega > 0$ is related to the stiffness of the spring to which the quantum particle is attached. Theorem 3.23 implies that the Friedrichs realization of the operator $H = -\Delta + \omega^2|x|^2$ is separable:

$$D\left(-\Delta + \omega^2|x|^2\right) = \left\{u \in H^2(\mathbb{R}^d) \,:\, |x|^2u \in L^2(\mathbb{R}^d)\right\}. \qquad (3.35)$$

3.3.4 Laplacian on a Bounded Domain $\Omega \subset \mathbb{R}^d$ *

We have already defined the Dirichlet and Neumann Laplacians on a hypercube in Exercise 2.50. In this section we briefly present the definition on any bounded domain $\Omega \subset \mathbb{R}^d$, using the theory of quadratic forms. We will use some properties of Sobolev spaces on a domain Ω recalled in Appendix A.

Let Ω be a bounded open set of $\mathbb{R}^d$. We call **Dirichlet Laplacian** the unique self-adjoint extension $(-\Delta)_{\mathrm{Dir}}$ of the operator $(-\Delta)^{\min}$ defined on $D((-\Delta)^{\min}) = C_c^\infty(\Omega)$, such that

$$q_{(-\Delta)_{\mathrm{Dir}}}(u) = \int_\Omega |\nabla u(x)|^2\, dx, \qquad \text{on} \quad Q((-\Delta)_{\mathrm{Dir}}) = H_0^1(\Omega).$$

This is the Friedrichs extension of $(-\Delta)^{\min}$ obtained by Corollary 3.18, since $H_0^1(\Omega)$ is by definition the closure of $C_c^\infty(\Omega)$ in $H^1(\Omega)$. When the boundary of Ω is piecewise Lipschitz, $H_0^1(\Omega)$ is also the space of functions of $H^1(\Omega)$ that vanish at the boundary (Theorem A.9 in Appendix A), where the restriction is well defined in $L^2(\partial\Omega)$. This is the Dirichlet boundary condition. The domain is by definition given by

$$D\big((-\Delta)_{\text{Dir}}\big)$$
$$= \left\{ u \in H_0^1(\Omega) \ : \ \exists v \in L^2(\Omega), \ \int_\Omega \overline{\nabla h} \cdot \nabla u = \int_\Omega \overline{h} v, \quad \forall h \in H_0^1(\Omega) \right\}.$$

By taking $h \in C_c^\infty(\Omega)$ we find at least that $(-\Delta)_{\text{Dir}} u = -\Delta u$ in the sense of distributions on Ω.

We call **Neumann Laplacian** the unique self-adjoint extension $(-\Delta)_{\text{Neu}}$ of $(-\Delta)^{\min}$ whose quadratic form is equal to

$$q_{(-\Delta)_{\text{Neu}}}(u) = \int_\Omega |\nabla u(x)|^2 \, dx \qquad \text{on} \quad Q((-\Delta)_{\text{Neu}}) = H^1(\Omega). \tag{3.36}$$

Its domain is therefore

$$D\big((-\Delta)_{\text{Neu}}\big)$$
$$= \left\{ u \in H^1(\Omega) \ : \ \exists v \in L^2(\Omega), \ \int_\Omega \overline{\nabla h} \cdot \nabla u = \int_\Omega \overline{h} v, \quad \forall h \in H^1(\Omega) \right\}$$

and similarly $(-\Delta)_{\text{Neu}} u = -\Delta u$ in the sense of distributions. Unfortunately, the Neumann boundary condition $\partial_n u_{|\partial\Omega} = 0$ (where n is the outward normal at the boundary $\partial\Omega$) does not make sense on the domain of the quadratic form and justifying it on the domain of the operator can be difficult (read about this in Sect. A.6 of Appendix A).

When the boundary of Ω is piecewise Lipschitz, $f_{|\partial\Omega}$ is well defined in $L^2(\partial\Omega)$ for $f \in H^1(\Omega)$, according to Theorem A.9 in Appendix A. We call **Robin Laplacian of parameter** $\theta \in (0,1)$ the unique self-adjoint extension $(-\Delta)_{\text{Rob},\theta}$ of $(-\Delta)^{\min}$ whose quadratic form is equal to

$$q_{(-\Delta)_{\text{Rob},\theta}}(u) = \int_\Omega |\nabla u(x)|^2 \, dx + \frac{1}{\tan(\pi\theta)} \int_{\partial\Omega} |u(x)|^2 \, dx,$$
$$\text{on} \quad Q((-\Delta)_{\text{Rob},\theta}) = H^1(\Omega). \tag{3.37}$$

This quadratic form is equivalent to the $H^1(\Omega)$ norm thanks to the trace inequality (A.15) of Theorem A.9. For $\theta = 1/2$, we use the convention $1/\tan(\pi/2) = 0$ and the second term is just absent. We thus obtain the Neumann Laplacian: $(-\Delta)_{\text{Rob},1/2} = (-\Delta)_{\text{Neu}}$. When $\theta \to 0^+$ we have $1/\tan(\pi\theta) \to +\infty$. The boundary term in the energy plays the role of a penalization. For $\theta = 0$ we therefore agree that $1/\tan(\pi 0^+) = +\infty$, so that the quadratic form is finite only when $u_{|\partial\Omega} \equiv 0$, that is $u \in H_0^1(\Omega)$. This is the Dirichlet Laplacian: $(-\Delta)_{\text{Rob},0} := (-\Delta)_{\text{Dir}}$. The behavior is less obvious when $\theta \to 1^-$ because the boundary term tends to $-\infty$. A detailed study of this limit will be carried out in dimension $d = 1$ in Exercise 5.56.

Finding the explicit domain of the previous operators can turn out to be quite difficult. When Ω is piecewise C^2 and convex in the vicinity of the singularities of its boundary, it is possible to show that

$$D((-\Delta)_{\text{Rob},\theta}) = \left\{ u \in H^2(\Omega) \; : \; \cos(\pi\theta)u_{|\partial\Omega} + \sin(\pi\theta)\partial_n u_{|\partial\Omega} \equiv 0 \right\} \subset H^2(\Omega),$$
(3.38)

as one might expect. For this, one must use the elliptic regularity Theorem A.20. However, it can really happen that the domain is not included in $H^2(\Omega)$, for example when Ω is not regular and not convex in the vicinity of the singularities of its boundary. This is for example the case for a domain in the plane $\mathbb{R}^2$ which has corners with a reentrant angle [Gri85, Dau88].

The operators $(-\Delta)_{\text{Rob},\theta}$ play a central role in physical applications. For example, in dimension $d = 2$ the Dirichlet Laplacian $(-\Delta)_{\text{Dir}}$ describes the small vertical oscillations of a drum attached to its edge. Its eigenvalues are related to the vibration frequencies of the drum. According to Newton's law, the Robin Laplacian describes the temperature of a radiating solid body immersed in a thermal bath, see [CH53, Sec. V.3.7.2] and [HÖ12, Sec. 1-5]. The rate of temperature change in the direction normal to the edge is assumed to be proportional to the temperature jump between the interior and exterior of the domain Ω. In this case $\tan(\pi\theta) = h/k$ where k is the thermal conductivity of the considered material and h its surface heat transfer coefficient.

Complementary Exercises

Exercise 3.26 (We Can Have $H^2(\mathbb{R}^d) \cap D(H) = \{0\}$) Let us work in dimension $d \leq 3$, so that all functions of $H^2(\mathbb{R}^d)$ are continuous. Let (R_n) be any dense sequence of $\mathbb{R}^d$ and consider the potential

$$V(x) = 1 + \sum_{n \geq 1} \frac{1}{n^2 |x - R_n|^\alpha}.$$

Show that $\mathcal{E}^V$ in Formula (3.25) is well defined and equivalent to the norm of $H^1(\mathbb{R}^d)$ for all $0 < \alpha < \min(2, d)$, but that the operator $H = -\Delta + V$ defined in Corollary 3.20 satisfies $H^2(\mathbb{R}^d) \cap D(H) = \{0\}$ as soon as $d/2 \leq \alpha < \min(2, d)$. What is happening for $2 \leq \alpha < 3$ in dimension $d = 3$?

Exercise 3.27 (Operators in the Form A^*A) Let $(A, D(A))$ be any **closed** operator (not necessarily self-adjoint) on a separable Hilbert space $\mathfrak{H}$. The goal is to show that the operator A^*A defined on the domain

$$D(A^*A) = \{v \in D(A) \; : \; Av \in D(A^*)\}$$
(3.39)

by $A^*Av = A^*(Av)$ is self-adjoint.

1. Prove that the quadratic form $q(v) := \|Av\|^2 + \|v\|^2$ is closed on the domain $Q = D(A)$. Deduce the existence of a unique associated self-adjoint operator, which we write in the form $B + 1$ and give its domain.
2. Show that B is precisely the operator A^*A defined in (3.39). In particular $D(A^*A)$ is dense and A^*A is self-adjoint.
3. We now assume that $B = A^*A$ is diagonal in an orthonormal basis $(e_j)_{j\geq 1}$, with $A^*Ae_j = b_j e_j$ and $\sigma(A^*A) = \overline{\{b_j, \ j \geq 1\}}$, as in Theorem 2.34.

   a. Show that the Ae_j's are pairwise orthogonal and calculate their norm.
   b. Show that $D(A) = \{v \in \mathfrak{H} : \sum_j b_j |\langle e_j, v\rangle|^2 < \infty\}$ and that $Av = \sum_j \langle e_j, v\rangle Ae_j$, for all $v \in D(A)$, where the sum is convergent in $\mathfrak{H}$.
   c. Calculate A^* as well as the operator defined in the other direction AA^*, in terms of the e_j's and Ae_j's.
   d. Deduce that A^*A and AA^* have the same spectrum, except possibly the point 0:

   $$\sigma(A^*A) \setminus \{0\} = \sigma(AA^*) \setminus \{0\}. \tag{3.40}$$

   This result is true without assuming that the operators are diagonalizable in a basis. In this case the proof relies on the polar decomposition for unbounded operators.

4. Show that the Dirichlet and Neumann Laplacians on $I = (0, 1)$ satisfy

   $$L_D = (P_0)^* P_0, \qquad L_N = P_0(P_0)^*$$

   where the operators are defined in the previous sense and $P_0 f = -if'$ on $D(P_0) = H_0^1(I)$ (refer back to Sect. 2.8.1). Comment on this result in light of (3.40).

Exercise 3.28 (A Non-separable Potential) Here we construct an example proving that Theorem 3.23 is false when $\alpha \geq 2$ in (3.28). We consider the space dimension $d = 2$. Let $\chi \in C_c^\infty(B_2)$ be a function that equals 1 on the unit ball B_1 and satisfies $0 \leq \chi \leq 1$ everywhere.

1. Following [EG78], we set $f(x) = |x|\chi(x)$. Show that $f \in H^1(\mathbb{R}^2)$, $f/|x| \in L^2(\mathbb{R}^2)$, $(-\Delta + 1/|x|^2)f \in L^2(\mathbb{R}^2)$, but $\Delta f \notin L^2(\mathbb{R}^2)$.
2. Construct a sequence $f_n \in C_c^\infty(\mathbb{R}^2 \setminus \{0\})$ such that $f_n \to f$ in $H^1(\mathbb{R}^2)$, $(-\Delta + 1/|x|^2)f_n \to (-\Delta + 1/|x|^2)f$ in $L^2(\mathbb{R}^2)$ and $\|\Delta f_n\|_{L^2} \to +\infty$. For example, take $f_n(x) = \left(1 - \chi(x/\varepsilon_n)\right)|x|^{\frac{1}{n}} f(x)$ with ε_n tending to 0 fast enough.

Now we "blow up the singularity" to obtain a continuous potential on $\mathbb{R}^2$ but which is very large in some places. We first set

$$V_n(x) = \frac{\chi(x/R)}{\max(\varepsilon_n^2/4, |x|^2)}$$

where $R \geq 2$ and the truncation of the denominator were chosen so that $V_n(x) = 1/|x|^2$ on the support of f_n. Then we take

$$u(x) = \sum_n \beta_n f_n(x - X_n), \qquad V(x) = \sum_n V_n(x - X_n)$$

where the translations $X_n \in \mathbb{R}^2$ are chosen so that the functions have disjoint supports in both sums, for example $X_n = (10n, 0)$.

3. Show that $|\nabla V| \leq 2V^{3/2} + C/R^3$ for a constant C that will be determined.
4. Show that $\|u\|_{H^1}^2 = \sum_n \beta_n^2 \|f_n\|_{H^1}^2$, that $\|\Delta u\|_{L^2}^2 = \sum_n \beta_n^2 \|\Delta f_n\|_{H^1}^2$ and finally that $\|(-\Delta + V)u\|_{L^2}^2 = \sum_n \beta_n^2 \|(-\Delta + 1/|x|^2) f_n\|_{L^2}^2$.
5. Deduce a choice of β_n so that $u \in H^1(\mathbb{R}^2)$ and $(-\Delta + V)u \in L^2(\mathbb{R}^2)$ but $\Delta u \notin L^2(\mathbb{R}^2)$.
6. We now consider a space dimension $d \geq 3$. Verify that the same properties hold for the potential $V(x_1, x_2)$ and the function $u(x_1, x_2)w(x_3, \ldots, x_d)$ with separated variables, with $w \in C_c^\infty(\mathbb{R}^{d-2})$.

Exercise 3.29 (Chladni Figures) In 1787, the German physicist and musician Ernst Chladni discovered the figures that now bear his name. By placing sand on a metal plate fixed at its center and vibrating the plate with a bow, the grains of sand form very characteristic figures (see Fig. 3.1 for an example). The corresponding mathematical problem is an eigenvalue equation for a self-adjoint realization of the square Laplacian Δ^2 on the plate, the latter being represented by a bounded open set $\Omega \subset \mathbb{R}^2$. The grains of sand are placed on the nodal set $u^{-1}(\{0\})$ of the corresponding eigenfunction u. That we have four derivatives is due to the high rigidity of the plate compared, for example, to the skin of a drum which is more flexible and described by the Laplacian. Sophie Germain and Gustav Kirchhoff determined that the energy of the vertical deformations $u : \Omega \to \mathbb{R}$ could be written in the form

$$\mathcal{E}(u) = \int_\Omega |\Delta u|^2 - 2(1 - \mu) \left(\partial_x^2 u\, \partial_y^2 u - |\partial_x \partial_y u|^2 \right),$$

where $0 < \mu < 1$ is a constant associated with the material [CH53, GK12]. If the plate is fixed at a point $x_0 \in \Omega$ we add the constraints that $u(x_0) = 0$ and $\nabla u(x_0) = 0$. Determine the self-adjoint operator associated with the quadratic form $\mathcal{E}$, and in particular the boundary conditions of Ω.

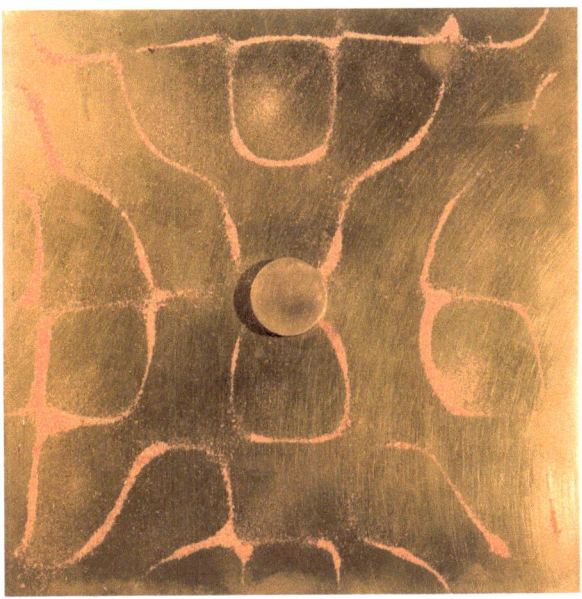

Fig. 3.1 Chladni figure obtained at the Institute of Mathematics and Statistics of the University of São Paulo in Brazil. © CC-BY-SA 4.0 Matemateca IME-USP/Rodrigo Tetsuo Argenton. Wikimedia Community User Group Brasil

Chapter 4
Spectral Theorem and Functional Calculus

In this chapter we explain how to diagonalize self-adjoint operators in infinite dimensions. Our goal is to show a result similar to the famous theorem due to Cauchy in 1826, which states that "every Hermitian matrix is diagonalizable in an orthonormal basis". As the spectrum of self-adjoint operators in infinite dimensions does not always contain only eigenvalues, as we have already seen in several examples, the situation is of course more subtle. Actually, in infinite dimensions there does not exist "one" spectral theorem, but several results equivalent to each other, each author preferring to highlight one or the other. As in [RS72, Dav95] we give priority here to the theorem which states that *every self-adjoint operator is unitarily equivalent to a multiplication operator*, that is, in the form $f(x) \mapsto a(x)f(x)$ on a space $L^2(\mathbb{R}^d, d\mu)$. When the measure μ is a sum of k deltas at distinct points $x_1, \ldots, x_k$, then $L^2(\mathbb{R}^d, d\mu) \simeq \mathbb{C}^k$ and a multiplication operator is nothing else than a diagonal matrix, with coefficients $a(x_i)$. We mention other statements, in particular the construction of functional calculus which is an important tool in the proof of the spectral theorem, as well as a representation formula based on spectral projections.

The spectral theorem, in any of its equivalent formulations, is without doubt the main mathematical result of the spectral theory of unbounded self-adjoint operators.

4.1 Multiplication Operators

Let B be a Borel set of $\mathbb{R}^d$ and μ a positive Borel measure on B, which we assume to be **locally finite**, which means that $\mu(B \cap B_R) < \infty$ for all $R > 0$, where B_R is the open ball of radius R centered at the origin. Here we will study some particular operators on the Hilbert space $\mathfrak{H} = L^2(B, d\mu)$ (all functions are complex-valued). The assumption that μ is locally finite is enough to ensure that $L^2(B, d\mu)$ contains

© The Editor(s) (if applicable) and The Author(s), under exclusive license
to Springer Nature Switzerland AG 2024
M. Lewin, *Spectral Theory and Quantum Mechanics*, Universitext,
https://doi.org/10.1007/978-3-031-66878-4_4

all functions of $L_c^\infty(B, d\mu)$, that is, μ-essentially bounded with compact support, which then form a dense subspace of $\mathfrak{H}$. Indeed, for all $v \in L^2(B, d\mu)$ we can write

$$v = v\mathbb{1}(|v| > R) + v\mathbb{1}(|v| \le R)\mathbb{1}(|x| > R') + v\mathbb{1}(|v| \le R)\mathbb{1}(|x| \le R') \quad (4.1)$$

where the first two functions tend to 0 in $L^2(B, d\mu)$ by dominated convergence, when $R, R' \to \infty$. The choice of the space $L^2(B, d\mu)$ allows us to cover many examples in a unified way, by varying the measure μ.

Example 4.1 (Hermitian Matrices) Let $B = \{x_1, \ldots, x_k\}$ be a set of k distinct points in $\mathbb{R}^d$ and $\mu = \sum_{j=1}^k \delta_{x_j}$. Then $\mathfrak{H} = L^2(B, d\mu) \simeq \mathbb{C}^k$, so that every self-adjoint operator on $\mathfrak{H}$ identifies with a Hermitian matrix M of size $k \times k$. The image of a function $f \in L^2(B, d\mu)$ is by definition the function $g \in L^2(B, d\mu)$ such that $g(x_i) = \sum_{j=1}^k M_{ij} f(x_j)$. A diagonal matrix M corresponds to the operator $f \mapsto af$ where a is the function defined by $a(x_i) = M_{ii}$.

The previous example naturally leads to the following definition.

Definition 4.2 (Multiplication Operators) Let a be a function of $L_{loc}^2(B, d\mu)$ (with complex values), that is, such that $\int_{B \cap B_R} |a|^2 d\mu < \infty$ for every ball $B_R \subset \mathbb{R}^d$ of radius R, centered at the origin. We call **multiplication operator** and denote by M_a, the operator defined by

$$(M_a v)(x) = a(x)v(x), \qquad D(M_a) = \left\{v \in L^2(B, d\mu) \; : \; av \in L^2(B, d\mu)\right\}.$$

The main result of this section is the following (see Fig. 4.1).

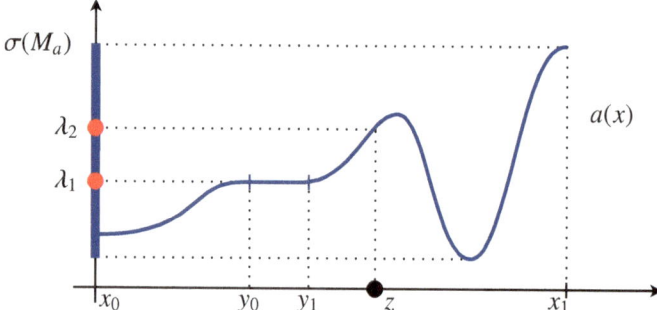

Fig. 4.1 Example of a function a defined on the interval $B = [x_0, x_1]$, which we equip with the measure $\mu = \text{Leb}_{[x_0, x_1]} + \delta_z$ (the Lebesgue measure to which we add an atom at z). The spectrum of the corresponding multiplication operator M_a is the image $a([x_0, x_1])$, with only two eigenvalues $\lambda_1 = a(y_0)$ and $\lambda_2 = a(z)$. The first is of infinite multiplicity and $\ker(M_a - \lambda_1)$ is the space of square integrable functions that are supported in the interval $[y_0, y_1]$. The second is of multiplicity 1 and $\ker(M_a - \lambda_2) \simeq \mathbb{C}$ is the space of functions that vanish μ-almost everywhere on $[x_0, x_1] \setminus \{z\}$.

Theorem 4.3 (Multiplication Operators) *Let $a \in L^2_{loc}(B, d\mu)$.*

(i) *The operator $(M_a, D(M_a))$ is closed.*
(ii) *Its spectrum is*

$$\sigma(M_a) = Ess\,Ran(a), \tag{4.2}$$

the essential range of the function a defined by

$$Ess\,Ran(a) = \Big\{ y \in \mathbb{C} \; : \; \mu(\{x \; : \; |a(x) - y| \leq \varepsilon\}) > 0, \; \forall \varepsilon > 0 \Big\}.$$

(iii) *The eigenvalues of M_a are the $\lambda \in Ess\,Ran(a)$ such that $\mu(\{a = \lambda\}) > 0$ and the corresponding eigenspace is $L^2(\{a = \lambda\}, d\mu)$, the space of all square-integrable functions with support in the set $\{a = \lambda\}$, defined μ-almost everywhere.*
(iv) *$(M_a, D(M_a))$ is a bounded operator if and only if the function a is a μ-essentially bounded function. In this case, we have*

$$\|M_a\| = \|a\|_{L^\infty(B, d\mu)}. \tag{4.3}$$

(v) *$(M_a, D(M_a))$ is self-adjoint if and only if the function a is real-valued (bounded or not).*

Proof We have already studied operators in this form, for example the Laplacian or the momentum after applying the Fourier transform in Sect. 2.7 and operators diagonalized in an orthonormal basis in Theorem 2.34 and Exercise 2.35. The proof in the general case is very similar. Consider a sequence $(v_n) \subset D(M_a)$ such that $v_n \to v$ and $av_n \to w$ strongly in $L^2(B, d\mu)$. As $a \in L^2_{loc}(B, d\mu)$ and v_n converges strongly in $L^2(B, d\mu)$, we have $av_n \to av$ strongly in $L^1(B \cap B_R, d\mu)$ where we recall that B_R is the open ball of radius R, centered at the origin. This proves that $w = av$. Thus, $av \in L^2(B, d\mu)$ and therefore $v \in D(M_a)$. In other words, M_a is closed.

Let then $\lambda \in \mathbb{C} \setminus Ess\,Ran(a)$, which means according to the definition that there exists $\varepsilon > 0$ such that $|a - \lambda| \geq \varepsilon$ μ-almost everywhere. In particular, we have $|(a - \lambda)^{-1}| \leq \varepsilon^{-1}$ μ-almost everywhere. For all $v \in L^2(B, d\mu)$, the function $v(a - \lambda)^{-1}$ is in $D(M_a)$ since

$$\int_B \frac{|v(x)|^2}{|a(x) - \lambda|^2}\, d\mu(x) \leq \varepsilon^{-2} \int_B |v(x)|^2\, d\mu(x) < \infty$$

and

$$\int_B \frac{|a(x)|^2|v(x)|^2}{|a(x) - \lambda|^2}\, d\mu(x) \leq \left(1 + |\lambda|\varepsilon^{-1}\right)^2 \int_B |v(x)|^2\, d\mu(x) < \infty,$$

where we used that $\frac{a}{a-\lambda} = 1 + \frac{\lambda}{a-\lambda}$. We then see that $v \mapsto v(a-\lambda)^{-1}$ is the inverse of $(M_a - \lambda)$ and that it is bounded by ε^{-1}. Thus, $\lambda \notin \sigma(M_a)$, which shows that $\sigma(M_a) \subset \text{Ess Ran}(a)$.

Conversely, if $\lambda \in \text{Ess Ran}(a)$, then we have $\mu(\{|a-\lambda| \le 1/n\}) > 0$ for all n so there exists a radius R_n such that $0 < \mu(\{|a-\lambda| \le 1/n\} \cap B_{R_n}) < \infty$. We prefer to work with a set of finite measure, since in principle we can have $\mu(\{|a-\lambda| \le 1/n\}) = +\infty$. Then, the function

$$v_n = \frac{\mathbb{1}_{\{|a-\lambda| \le 1/n\} \cap B_{R_n}}}{\mu(\{|a-\lambda| \le 1/n\} \cap B_{R_n})^{1/2}}$$

is normalized in $L^2(B, \mathrm{d}\mu)$. As we have

$$\|(M_a - \lambda)v_n\|^2 = \frac{\int_B \mathbb{1}_{\{|a-\lambda| \le 1/n\} \cap B_{R_n}}(x)|a(x) - \lambda|^2 \, \mathrm{d}\mu(x)}{\mu(\{|a-\lambda| \le 1/n\} \cap B_{R_n})} \le \frac{1}{n^2}$$

we see that $M_a - \lambda$ cannot have a bounded inverse, therefore $\lambda \in \sigma(M_a)$, which remained to be shown to have the equality $\sigma(M_a) = \text{Ess Ran}(a)$.

The operator M_a is symmetric if and only if

$$\int_B \left(a(x) - \overline{a(x)}\right)\overline{v(x)}u(x) \, \mathrm{d}\mu(x) = 0$$

for all $u, v \in D(M_a)$. By taking $u = |\Im(a)|\mathbb{1}_{B_R}$ and $v = \text{sgn}(\Im(a))\mathbb{1}_{B_R}$ for any ball B_R, we deduce that $\Im(a) = 0$ μ-almost everywhere, that is, a is real. Then $\sigma(M_a) = \text{Ess Ran}(a) \subset \mathbb{R}$ and M_a is self-adjoint according to Theorem 2.28.

On the other hand, v is an eigenvalue of M_a if and only if $(a(x) - \lambda)v(x) = 0$ μ-almost everywhere, which is equivalent to saying that v has its support in the set where $a = \lambda$. Such functions are non-zero if and only if this set has non-zero measure.

Finally, if a is bounded, we obviously have $\|M_a\| \le \|a\|_{L^\infty(B, \mathrm{d}\mu)}$. The opposite inequality is obtained by constructing a sequence v_n in the same way as before, on the set $\{|a| \ge \|a\|_{L^\infty(B, \mathrm{d}\mu)} - 1/n\}$. Similarly, we can show that M_a is not bounded if a is not. □

4.2   Spectral Theorem

We are now ready to diagonalize all self-adjoint operators. The following statement means that every self-adjoint operator is in fact a multiplication operator on a good space.

Theorem 4.4 (Spectral Theorem) *Let $(A, D(A))$ be a self-adjoint operator on a separable Hilbert space $\mathfrak{H}$. Then there exists $d \geq 1$, a Borel set $B \subset \mathbb{R}^d$, a locally finite Borel measure μ on B, a real-valued locally bounded function $a \in L^\infty_{\text{loc}}(B, d\mu)$, and an isomorphism $U : \mathfrak{H} \to L^2(B, d\mu)$ such that*

$$UAU^{-1} = M_a, \qquad UD(A) = D(M_a).$$

More precisely, it is possible to take $d = 2$, $B = \sigma(A) \times \mathbb{N} \subset \mathbb{R}^2$, $a(s, n) = s$ and μ a finite measure on B.

The fact that one can take $B = \sigma(A) \times \mathbb{N} \subset \mathbb{R}^2$ and $a(s, n) = s$ means that all the structure of the operator A can be contained in the measure μ. In most cases it is not necessary to know that one can make this particular choice, but sometimes this simplifies some arguments. The preimage of any y by the function $a(s, n) = s$ is given by

$$\{a = y\} = \bigcup_{n \in \mathbb{N}} \{(y, n)\}.$$

In particular, according to Theorem 4.3, we have

$$\ker(A - \lambda) = U^{-1} L^2(\{\lambda\} \times \mathbb{N}, d\mu)$$

and the multiplicity of a possible eigenvalue is given by

$$\dim \ker(A - \lambda) = \#\{n \in \mathbb{N} : \mu(\{(\lambda, n)\}) > 0\}.$$

Thus, in the Cartesian product $\sigma(A) \times \mathbb{N}$, the set $\mathbb{N}$ is used to account for the multiplicity of eigenvalues.

Remark 4.5 The representation as a multiplication operator is not at all unique. For example, we have already seen that the Laplacian on $\mathbb{R}^d$ was unitarily equivalent to the multiplication operator by the function $a(k) = |k|^2$ on $L^2(\mathbb{R}^d)$, so with $B = \mathbb{R}^d$, μ the Lebesgue measure, and U the Fourier transformation. In dimension $d = 1$, we will show in Exercise 4.46 that it can also be represented by the function $\tilde{a}(s, n) = s$ on $\tilde{B} = \mathbb{R}_+ \times \{0, 1\}$. Two copies of the spectrum are necessary and they correspond to the decomposition of $L^2(\mathbb{R})$ into even and odd functions. In a way, the continuous spectrum is of multiplicity two. In higher dimension $d \geq 2$, it is possible to construct an explicit unitary based on hyperspherical coordinates for which $\tilde{B} = \mathbb{R}^+ \times \mathbb{N} \subset \mathbb{R}^2$, but the spectrum must be repeated an infinite number of times.

The proof of Theorem 4.4 is provided later in Sect. 4.4. We will now give a selection of some results that immediately follow from the spectral theorem. We will see multiple other consequences later. The first concerns the norm of the resolvent $(A - z)^{-1}$ for all $z \in \rho(A)$. We have already seen and widely used that

$$\left\| (A - z)^{-1} \right\| \leq \frac{1}{|\Im(z)|}$$

when $z \in \mathbb{C} \backslash \mathbb{R}$ and A is self-adjoint. With the spectral theorem we can now calculate the exact value of the norm in question.

Corollary 4.6 (Norm of the Resolvent) *Let A be a self-adjoint operator on the domain $D(A) \subset \mathfrak{H}$. For all $z \in \mathbb{C} \backslash \sigma(A)$, we have*

$$\left\| (A - z)^{-1} \right\| = \frac{1}{\mathrm{d}(z, \sigma(A))} \tag{4.4}$$

(the distance from z to the closed set $\sigma(A)$ in the complex plane).

Proof Up to an isometry, we can assume that $A = M_a$ is the operator of multiplication by $a(s, n) = s$ on $\mathfrak{H} = L^2(\sigma(A) \times \mathbb{N}, \mathrm{d}\mu)$. Then, by Theorem 4.3, we have

$$\left\| (A - z)^{-1} \right\| = \left\| (a - z)^{-1} \right\|_{L^\infty(\sigma(A) \times \mathbb{N}, \mathrm{d}\mu)} = \frac{1}{\mathrm{Ess\,Inf}|s - z|}$$

$$= \frac{1}{\min_{s \in \sigma(A)} |s - z|} = \frac{1}{\mathrm{d}(z, \sigma(A))}.$$

$\square$

Here is a second corollary of the spectral theorem.

Corollary 4.7 (Every Isolated Point of the Spectrum Is an Eigenvalue) *Let A be a self-adjoint operator on the domain $D(A) \subset \mathfrak{H}$. If $\lambda \in \sigma(A)$ is an isolated point of the spectrum (that is, $[\lambda - \varepsilon, \lambda + \varepsilon] \cap \sigma(A) = \{\lambda\}$ for $\varepsilon > 0$ small enough), then λ is an eigenvalue of A.*

Proof In the representation where A is the operator of multiplication by the function $(s, n) \mapsto s$, we have

$$\mu\left([\lambda - \varepsilon, \lambda + \varepsilon] \times \mathbb{N}\right) = \sum_{n \in \mathbb{N}} \mu\left([\lambda - \varepsilon, \lambda + \varepsilon] \times \{n\}\right) \neq 0$$

for all $\lambda \in \sigma(A)$ and all $\varepsilon > 0$, since λ must belong to the essential range of $a(s, n) = s$, by Theorem 4.3. If λ is isolated, since $B = \sigma(A) \times \mathbb{N}$ we therefore have

$$\mu(\{\lambda\} \times \mathbb{N}) = \mu\left([\lambda - \varepsilon, \lambda + \varepsilon] \times \mathbb{N}\right) > 0$$

for $\varepsilon > 0$ small enough. This shows that λ is an eigenvalue because $\{\lambda\} \times \mathbb{N} = a^{-1}(\{\lambda\})$.

$\square$

4.3   Functional Calculus

We now use the spectral theorem to define $f(A)$ for a large class of functions f.

Consider any representation where the operator A becomes a multiplication operator M_a by a function $a \in L^\infty_{\text{loc}}(B, d\mu)$. Let $f : \mathbb{R} \to \mathbb{C}$ be a **locally bounded Borel function**. We can define the operator $f(A)$ by requiring it to be the operator of multiplication by the function $f(a)$ in the representation where A has been diagonalized:

$$f(A) := U^{-1} M_{f(a)} U, \qquad D\big(f(A)\big) = U^{-1} D\big(M_{f(a)}\big). \tag{4.5}$$

Recall that M_a and its domain were defined in Sect. 4.1. The assumption that f is locally bounded serves to ensure that $f(a)$ is also locally bounded μ-almost everywhere (since a is), so that $M_{f(a)}$ is well defined. From Theorem 4.3, the spectrum $\sigma(f(A))$ is the μ-essential range of the function $f(a)$. In general it cannot simply be expressed in terms of $\sigma(A)$, more information on the nature of the spectrum is needed (in particular the location of the eigenvalues). However, if f is a continuous function, we find

$$\sigma(f(A)) = \overline{f(\sigma(A))} = \overline{\{f(\lambda), \ \lambda \in \sigma(A)\}}. \tag{4.6}$$

Note that if we have $f(a) = g(a)$ μ-almost everywhere on B, then $f(A) = g(A)$. Thus, we should rather work with the equivalence classes associated with the relation where $f \sim g$ if and only if $f(a) = g(a)$ μ-almost everywhere on B. However, these classes are not known *a priori* because the measure μ and the function a are abstract objects provided by the spectral theorem, about which we have little information. For this reason, it is more convenient to work with Borel functions without making any "almost everywhere" identification. Another observation is that $f(A) = 0$ if f vanishes on the spectrum $\sigma(A) = \text{Ess Ran}(a)$, since in this case $f(a) = 0$ μ-almost everywhere. Finally, if f is real, then the operator $f(A)$ thus defined is self-adjoint on its domain $D(f(A))$, by Theorem 4.3. If f is a bounded function on $\sigma(A)$ (real or not), then $f(A)$ is a bounded operator and moreover $\|f(A)\| = \|f(a)\|_{L^\infty(B, d\mu)} \leq \sup |f|$.

Since there is not a unique way to represent A as a multiplication operator, it is necessary to verify that our definition (4.5) of $f(A)$ does not depend on the chosen representation. We begin by addressing this question in the particular case of **bounded Borel functions**. This is a simpler situation, since the corresponding operators $f(A)$ are bounded, hence defined on the whole space $\mathfrak{H}$. Moreover, bounded Borel functions form a commutative C^*-algebra, which will be useful for the proof.

Theorem 4.8 (Functional Calculus for Bounded Borel Functions) *Let* $(A, D(A))$ *be a self-adjoint operator. There exists a unique map*

$$f \in \mathscr{L}^\infty(\mathbb{R}, \mathbb{C}) \mapsto f(A) \in \mathcal{B}(\mathfrak{H})$$

defined on the algebra $\mathscr{L}^\infty(\mathbb{R}, \mathbb{C})$ *of bounded Borel functions on* $\mathbb{R}$, *with values in the algebra* $\mathcal{B}(\mathfrak{H})$ *of bounded operators on* $\mathfrak{H}$, *such that*

(i) *it is a morphism of* C^*-*algebras, that is* $(f + \alpha g)(A) = f(A) + \alpha g(A)$, $(fg)(A) = f(A)g(A)$, $\mathbb{1}(A) = \mathbb{1}_{\mathfrak{H}}$ *and* $\overline{f}(A) = f(A)^*$;

(ii) *It is continuous, that is* $\|f(A)\| \leq \sup_{x \in \mathbb{R}} |f(x)|$;

(iii) *if* $f(x) = (x - z)^{-1}$ *with* $z \in \mathbb{C} \setminus \mathbb{R}$, *then* $f(A) = (A - z)^{-1}$;

(iv) *if* f *vanishes on* $\sigma(A)$, *then* $f(A) = 0$;

(v) *if* $|f_n(x)| \leq C$ *and* $f_n(x) \to f(x)$ *for all* $x \in \mathbb{R}$, *then* $f_n(A)v \to f(A)v$ *for all* $v \in \mathfrak{H}$.

Our definition (4.5) satisfies all the required assumptions, regardless of the chosen representation, which provides the existence part of the theorem. It is the uniqueness that interests us here, and it implies that the definition (4.5) does not depend on the chosen representation for A as a multiplication operator, at least when f is bounded.

The statement of Theorem 4.8 may seem a bit oversized to the reader, if we only care about uniqueness. In fact, it turns out that Theorem 4.8 is equivalent to the spectral Theorem 4.4! Since the works of F. Riesz [Rie13], it has even become classical to first construct the functional calculus in order to deduce that A is unitarily equivalent to a multiplication operator. With this point of view, the existence part becomes more difficult to show. We will return to this in Sect. 4.4 which contains the coupled proof of the two Theorems 4.4 and 4.8.

Remark 4.9 (Spectral Measure)   Let v be any normalized vector of $\mathfrak{H}$. By functional calculus, the map

$$f \in C_b^0(\mathbb{R}, \mathbb{R}) \mapsto \varphi_v(f) := \langle v, f(A)v \rangle$$

is a continuous linear form. For any non-negative function $0 \leq f \in C_b^0(\mathbb{R}, \mathbb{R})$, we can write $f(A) = (\sqrt{\overline{f}^2})(A) = \sqrt{f}(A)^2$ and obtain

$$\varphi_v(f) = \langle v, f(A)v \rangle = \left\| \sqrt{f}(A)v \right\|^2 \geq 0.$$

In other words, our linear form φ_v is non-negative. It is also normalized, $\varphi_v(1) = 1$, since $1(A) = \mathrm{Id}_{\mathfrak{H}}$. By the Riesz-Markov theorem, there exists a unique *Borel probability measure* $\mu_{A,v}$ on $\mathbb{R}$ such that

$$\langle v, f(A)v \rangle = \int_{\mathbb{R}} f(s) \, d\mu_{A,v}(s)$$

for all $f \in C_b^0(\mathbb{R})$. This is by definition the **spectral measure associated with the self-adjoint operator** A **and the vector** v, which we had mentioned during the presentation of the abstract formalism of quantum mechanics in Sect. 1.5. By the spectral Theorem 4.4, with $B = \sigma(A) \times \mathbb{N}$ and $a(s, n) = s$, we can also express $\mu_{A,v}$ in the form

$$d\mu_{A,v}(s) = \sum_{n \in \mathbb{N}} |Uv(s, n)|^2 d\mu(s, n)$$

which means more precisely that

$$\int_{\mathbb{R}} f(s) \, d\mu_{A,v}(s) = \int_{\sigma(A) \times \mathbb{N}} f(s) |Uv(s, n)|^2 d\mu(s, n).$$

In other words, $\mu_{A,v}$ is the cylindrical projection on $\sigma(A)$ of the probability measure $|Uv(s, n)|^2 d\mu(s, n)$ on $\sigma(A) \times \mathbb{N}$. We then have $v \in D(A)$ if and only if $\mu_{A,v}$ has a moment of order two, $\int_{\mathbb{R}} s^2 \, d\mu_{A,v}(s) < \infty$, and in this case $\int_{\mathbb{R}} s^2 \, d\mu_{A,v}(s) = \|Av\|^2$. We also have $\int_{\mathbb{R}} s \, d\mu_{A,v}(s) = \langle v, Av \rangle$, as mentioned in Sect. 1.5.

We will now discuss the extension of functional calculus to **locally bounded Borel functions**, which requires working with a domain that depends on f as in (4.5). The following result is a consequence of Theorem 4.8 in the bounded case.

Corollary 4.10 (Functional Calculus for Locally Bounded Borel Functions) *Let* $(A, D(A))$ *be a self-adjoint operator and* $f : \mathbb{R} \mapsto \mathbb{C}$ *a locally bounded Borel function. The operator* $f(A)$ *defined in (4.5) is independent of the isomorphism* U *used to represent* A *as a multiplication operator.*

Proof Recall that the domain of the operator $M_{f(a)}$ of multiplication by the function $f(a)$ on $L^2(B, d\mu)$ is given by

$$D(M_{f(a)}) = \left\{ v \in L^2(B, d\mu) : \int_B |f(a(x))|^2 |v(x)|^2 \, d\mu(x) < \infty \right\}.$$

By introducing the truncated function $f_n = f\mathbb{1}(|f| \leq n)$, which is a bounded Borel function, we have the equivalence

$$v \in D(M_{f(a)}) \Longleftrightarrow \limsup_{n \to \infty} \|f_n(a)v\|^2 = \limsup_{n \to \infty} \int_B |f_n(a(x))|^2 |v(x)|^2 d\mu(x) < \infty$$
(4.7)

and in this case $f(a)v = \lim_{n \to \infty} f_n(a)v$ in $L^2(B, d\mu)$. Indeed, if $v \in D(M_{f(a)})$ the integral on the right of (4.7) converges to $\int_B |f(a)|^2 |v|^2 d\mu$, by monotone convergence, and $f_n(a)v \to f(a)v$ strongly in $L^2(B, d\mu)$. Conversely, by Fatou's theorem we have

$$\int_B |f(a(x))|^2 |v(x)|^2 d\mu(x) \leq \liminf_{n \to \infty} \int_B |f_n(a(x))|^2 |v(x)|^2 d\mu(x)$$

which is therefore finite when the term on the right remains bounded. This allows us to characterize $f(A)$ and its domain by properties involving only $f_n(A)$, as follows:

$$
\begin{cases}
v \in D\big(f(A)\big) \Longleftrightarrow \limsup_{n \to \infty} \| f_n(A)v \| < \infty, \\
f(A)v = \lim_{n \to \infty} f_n(A)v.
\end{cases}
\tag{4.8}
$$

This shows that the definition (4.5) is independent of the choice of the representation of A as a multiplication operator, since this is the case for $f_n(A)$ according to Theorem 4.8. $\square$

In addition to the resolvent $(A - z)^{-1}$ obtained for $f(x) = (x - z)^{-1}$, we will encounter in the rest of this chapter several operators in the form of $f(A)$ that play an important role:

- The **spectral projectors** are the $\mathbb{1}_F(A)$ where F is any Borel set of $\mathbb{R}$, and they play the same role in infinite dimension as the projectors on the eigenspaces of Hermitian matrices. They will be studied in Sect. 4.5.
- The **powers** A^k with $k \in \mathbb{N}$ are obtained for $f(x) = x^k$. If $A \geq 0$ (that is, $\sigma(A) \subset \mathbb{R}_+$) we can also consider the operators A^α where α is any positive real number. If A is coercive then we can even take $\alpha \in \mathbb{R}$. The powers and the link with the quadratic form of A will be discussed in Sect. 4.6.
- The **Schrödinger propagator** is the unitary operator defined by e^{-itA} for all $t \in \mathbb{R}$, corresponding to $f(x) = e^{-itx}$. It is used to solve the time-dependent Schrödinger equation

$$
i \frac{\mathrm{d}}{\mathrm{d}t} v(t) = A v(t),
$$

as we will see in Sect. 4.7.1.
- The **heat kernel** is the operator e^{-tA} corresponding to $f(x) = e^{-tx}$. It is used to solve the heat equation

$$
\frac{\mathrm{d}}{\mathrm{d}t} v(t) = -A v(t),
$$

as will be discussed in Sect. 4.7.2.

4.4 Proof of Theorems 4.4 and 4.8

In this section, we present the coupled proof of the spectral Theorem 4.4 and of Theorem 4.8 on bounded Borel functional calculus. We argue in three steps:

1. We construct the continuous functional calculus, that is, a weaker version of Theorem 4.8 where we assume that $f \in \mathbb{C} + C_0^0(\mathbb{R}, \mathbb{C})$ is continuous and has equal limits at $\pm\infty$.

2. We deduce from this the spectral Theorem 4.4.
3. An argument from measure theory will finally provide us with the uniqueness of the extension of functional calculus to all bounded Borel functions, which will conclude the proof of Theorem 4.8.

Step 1: Functional Calculus for Continuous Functions

Here we will provide a partial proof of Theorem 4.8, in the sense that we will not show the properties for the entire class of bounded measurable functions but only for the sub-class of continuous functions that have equal limits at $\pm\infty$:

$$C_{\text{lim}}^0(\mathbb{R}) := \mathbb{C} + C_0^0(\mathbb{R}, \mathbb{C})$$

$$= \left\{ f \in C^0(\mathbb{R}) \ : \ \lim_{x \to +\infty} f(x) \text{ and } \lim_{x \to -\infty} f(x) \text{ exist and are equal} \right\}.$$

This is a particular unital sub-algebra of the algebra of bounded Borel functions $\mathscr{L}^\infty(\mathbb{R}, \mathbb{C})$, which will be sufficient to prove the spectral Theorem 4.4. While it is more traditional to work in $C_0^0(\mathbb{R})$ (the algebra of continuous functions that tend to 0 at infinity), we prefer to add the constant functions to have a unital algebra. We therefore show the following theorem.

Theorem 4.11 (Functional Calculus in $C_{\text{lim}}^0(\mathbb{R})$) *Let A be a self-adjoint operator on $D(A) \subset \mathfrak{H}$. There exists a unique map*

$$f \in C_{\text{lim}}^0(\mathbb{R}) \mapsto f(A) \in \mathcal{B}(\mathfrak{H})$$

which is a continuous morphism of C^*-Banach algebras, that is to say, such that

(i) $(f + \alpha g)(A) = f(A) + \alpha g(A)$, $(fg)(A) = f(A)g(A)$, $\mathbb{1}(A) = \mathbb{1}_{\mathfrak{H}}$ *and* $\overline{f}(A) = f(A)^*$;

(ii) $\|f(A)\| \leq \max_{\mathbb{R}} |f|$;

(iii) *if $f(x) = (x - z)^{-1}$ with $z \notin \mathbb{R}$, then $f(A) = (A - z)^{-1}$.*

To simplify the argument, we have removed several of the assertions of Theorem 4.8 which are not necessary to show the spectral theorem, and will then follow immediately from the latter.

Our proof of Theorem 4.11 is based on the Stone-Weierstrass theorem which specifies that the algebra $\mathcal{A}$ generated by the functions $x \mapsto (x - z)^{-1}$ with $z \in \mathbb{C} \backslash \mathbb{R}$ and the constant functions is dense in $C_{\text{lim}}^0(\mathbb{R}, \mathbb{C})$. The argument is quite standard, except perhaps for the estimate in (ii). We will see that the latter follows from a purely algebraic property, namely the fact that the algebra $\mathcal{A}$ is **stable by the square root**.

Proof (of Theorem 4.11) Consider the algebra $\mathcal{A}$ generated by the constant function equal to 1 and the $x \mapsto (x-z)^{-1}$ with $z \in \mathbb{C} \backslash \mathbb{R}$, that is to say the one composed of finite sums of finite products of such functions. The Stone-Weierstrass theorem in its local version (see also Exercise 4.44) implies that $\mathcal{A}$ is dense in $C_{\lim}^{0}(\mathbb{R}, \mathbb{C})$ for the uniform norm. It is therefore sufficient to construct the functional calculus on $\mathcal{A}$, with the properties (i)–(iii). The result on $C_{\lim}^{0}(\mathbb{R})$ then immediately follows from (iii) and the density of $\mathcal{A}$.

The difficult part of the proof is the continuity property (ii), because it is quite clear how $f(A)$ should be defined for f a finite linear combination of products of $(x-z)^{-1}$, so that (i) and (iii) are true. Note first that for all $z, z' \in \mathbb{C} \backslash \mathbb{R}$, the operators $(A-z)^{-1}$ and $(A-z')^{-1}$ commute:

$$(A-z)^{-1}(A-z')^{-1} = (A-z')^{-1}(A-z)^{-1}.$$

To see this we can use the resolvent formula

$$(A-z)^{-1} = (A-z')^{-1} + (z-z')(A-z)^{-1}(A-z')^{-1} \tag{4.9}$$

which is proved by multiplying on the left by $A - z = A - z' + z - z'$ and which implies

$$(A-z)^{-1}(A-z')^{-1} = (z-z')^{-1}\left((A-z)^{-1} - (A-z')^{-1}\right)$$

where the term on the right is invariant when z and z' are exchanged. We will also need the fact that the adjoint of $(A-z)^{-1}$ is

$$\left[(A-z)^{-1}\right]^{*} = (A-\bar{z})^{-1} \tag{4.10}$$

which follows from the relation

$$\left\langle f, (A-z)^{-1}g \right\rangle = \left\langle (A-\bar{z})(A-\bar{z})^{-1}f, (A-z)^{-1}g \right\rangle = \left\langle (A-\bar{z})^{-1}f, g \right\rangle$$

since A is symmetric and $A - \bar{z}$ is invertible, for all $z \in \mathbb{C} \backslash \mathbb{R}$. Now we can define for all

$$f(x) = c + \sum_{j=1}^{J} \alpha_j \prod_{k=1}^{k_j} (x - z_{j,k})^{-1} \in \mathcal{A}$$

the operator

$$f(A) := c + \sum_{j=1}^{J} \alpha_j \prod_{k=1}^{k_j} (A - z_{j,k})^{-1}.$$

The order of the operators does not matter since they all commute. This definition is of course forced by properties (i) and (iii). The set of all these operators forms an abelian C^*-algebra of bounded operators

$$\mathcal{R} = \{f(A) \ : \ f \in \mathcal{A}\} \subset \mathcal{B}(\mathfrak{H})$$

which is called *the resolvent algebra* and which satisfies the assumptions (i) and (iii) of Theorem 4.11. The difficulty is now to show the continuity property (ii), which will automatically follow from the following lemma.   □

Lemma 4.12 (Stability of $\mathcal{A}$ By the Square Root) *Let $f \in \mathcal{A}$ be a function that is non-negative on $\mathbb{R}$. Then there exists $g \in \mathcal{A}$ such that $f = |g|^2$. This implies $f(A) = |g|^2(A) = g(A)^*g(A) \geq 0$, that is,*

$$\langle v, f(A)v \rangle = \left\| g(A)v \right\|^2 \geq 0 \tag{4.11}$$

for every $v \in \mathfrak{H}$.

Proof (of the Lemma) By reducing to the same denominator, we see that the functions $f \in \mathcal{A}$ can all be written in the form of a reduced rational fraction $f = P/Q$ where Q has all its roots in $\mathbb{C} \setminus \mathbb{R}$ and where $\mathrm{d}°(P) \leq \mathrm{d}°(Q)$. Conversely, by the decomposition into simple elements, any rational fraction satisfying these two properties belongs to $\mathcal{A}$. Let us write

$$P(x) = c \prod_{\ell=1}^{J} (x - \alpha_\ell)^{p_\ell} \prod_{j=1}^{J'} (x - z_j)^{p'_j}, \qquad Q(x) = \prod_{j=1}^{K} (x - \xi_j)^{q_j}$$

where the $\alpha_\ell \in \mathbb{R}$ are the real roots of P and the $z_j, \xi_j \in \mathbb{C} \setminus \mathbb{R}$ are the non-real roots of P and Q, respectively. We see that f is real-valued on $\mathbb{R}$ if and only if $P/Q = \overline{P/Q}$ or

$$c \prod_{j=1}^{J'} (x - z_j)^{p'_j} \prod_{j=1}^{K} (x - \overline{\xi_j})^{q_j} = \overline{c} \prod_{j=1}^{J'} (x - \overline{z_j})^{p'_j} \prod_{j=1}^{K} (x - \xi_j)^{q_j}.$$

The equality of the highest degree terms provides $c = \overline{c}$, that is, the constant c is real. As $(x - \overline{z_j})^{p'_j}$ divides the term on the right but not $\overline{Q}$ because the fraction is reduced, we deduce that $\overline{z_j}$ is a root of P with multiplicity at least equal to p'_j. By reversing the argument, and then using the same method for Q, we finally find that f is real-valued if and only if it takes the form

$$f(x) = c \prod_{\ell=1}^{J} (x - \alpha_\ell)^{p_\ell} \frac{\prod_{j=1}^{J'} |x - z_j|^{2p'_j}}{\prod_{j=1}^{K} |x - \xi_j|^{2q_j}}.$$

Finally, f is positive if and only if $c \geq 0$ and all p_ℓ are even, which indeed means that $f = |g|^2$ with

$$g(x) = \sqrt{c} \prod_{\ell=1}^{J} (x - \alpha_\ell)^{\frac{p_\ell}{2}} \frac{\prod_{j=1}^{J'} (x - z_j)^{p'_j}}{\prod_{j=1}^{K} (x - \xi_j)^{q_j}}$$

and concludes the proof of the lemma. $\square$

The lemma allows us to immediately show the continuity property (ii). Indeed, for all $f \in \mathcal{A}$, we have $0 \leq \|f\|^2_{L^\infty(\mathbb{R})} - |f|^2 \in \mathcal{A}$, so by (4.11) in Lemma 4.12 we deduce that

$$0 \leq \left\langle v, \left(\|f\|^2_{L^\infty(\mathbb{R})} - f(A)^* f(A) \right) v \right\rangle = \|f\|^2_{L^\infty(\mathbb{R})} \|v\|^2 - \|f(A)v\|^2$$

for all $v \in \mathfrak{H}$. Thus, $\|f(A)\| \leq \|f\|_{L^\infty(\mathbb{R})}$, as needed to be shown.

At this point we have shown that (i)–(iii) hold on the sub-algebra $\mathcal{A}$. The corresponding map $f \mapsto f(A)$ is continuous by (ii). By density of $\mathcal{A}$ in $C^0_{\lim}(\mathbb{R})$ (Stone-Weierstrass), it admits a unique extension to the whole of $C^0_{\lim}(\mathbb{R})$ and it satisfies all the same properties. This concludes the proof of Theorem 4.11.

Step 2: Proof of the Spectral Theorem 4.4

We will now prove the spectral theorem using Theorem 4.11, that is, the existence of functional calculus on $C^0_{\lim}(\mathbb{R})$. Let $v \neq 0$ be any vector of $\mathfrak{H}$ and consider the linear form

$$\varphi_v(f) := \langle v, f(A)v \rangle$$

which, by Theorem 4.11, is continuous on $C^0_0(\mathbb{R})$, with $|\varphi_v(f)| \leq \|v\|^2 \|f\|_{L^\infty(\mathbb{R})}$. Moreover, for any non-negative function $f \in C^0_0(\mathbb{R})$, we have $f = g^2$ with $g = \sqrt{f} \in C^0_0(\mathbb{R})$, so that

$$\varphi_v(f) = \langle v, g(A)^* g(A)v \rangle = \|g(A)v\|^2 \geq 0.$$

By the Riesz-Markov theorem, this means that there exists a unique Borel measure $\mu_{A,v}$, positive and bounded on $\mathbb{R}$, such that

$$\varphi_v(f) = \langle v, f(A)v \rangle = \int_{\mathbb{R}} f(s) \, d\mu_{A,v}(s)$$

for all $f \in C^0_0(\mathbb{R})$, and of total mass $\mu_{A,v}(\mathbb{R}) \leq \|v\|^2$. The measure $\mu_{A,v}$ is called the **spectral measure associated with the vector** v and it has already been discussed in Remark 4.9. We can immediately show that $\mu_{A,v}$ vanishes outside of $\sigma(A)$.

Lemma 4.13 *Let* $v \in \mathfrak{H}$ *be a normalized vector and* $\mu_{A,v}$ *the associated spectral measure. We have* $\mu_{A,v}\big(\mathbb{R} \setminus \sigma(A)\big) = 0$.

Proof (of the Lemma) Assume that $\mathbb{R} \setminus \sigma(A)$ is non-empty and take λ_0 in this open set. Lemma 2.8 implies that $(A - z)^{-1}$ is uniformly bounded on a small ball in the complex plane. For example $\|(A - z)^{-1}\| \leq 2$ for all $z \in B_r(\lambda_0)$, the ball of radius

$$r = \frac{1}{2\|(A - \lambda_0)^{-1}\|}.$$

Let $z_n = \lambda + i/n$ with $|\lambda - \lambda_0| \leq r/2$ and $n \geq 2/r$ so that z_n is in $B_r(\lambda_0)$. We then have

$$\left\|(A - z_n)^{-1}v\right\|^2 = \left\langle v, (A - \overline{z_n})^{-1}(A - z_n)^{-1}v\right\rangle = \int_{\mathbb{R}} \frac{d\mu_{A,v}(s)}{(s - \lambda)^2 + n^{-2}}.$$

By integrating over $[\lambda_0 - r/2, \lambda_0 + r/2]$, we find by Fubini

$$\int_{\lambda_0 - \frac{r}{2}}^{\lambda_0 + \frac{r}{2}} \left\|(A - \lambda - i/n)^{-1}v\right\|^2 d\lambda = \int_{\lambda_0 - \frac{r}{2}}^{\lambda_0 + \frac{r}{2}} \left(\int_{\mathbb{R}} \frac{d\mu_{A,v}(s)}{(s - \lambda)^2 + n^{-2}}\right) d\lambda$$

$$\geq \int_{\lambda_0 - \frac{r}{4}}^{\lambda_0 + \frac{r}{4}} \left(\int_{s - \frac{r}{4}}^{s + \frac{r}{4}} \frac{d\lambda}{(s - \lambda)^2 + n^{-2}}\right) d\mu_{A,v}(s)$$

$$= \mu_{A,v}\left(\left[\lambda_0 - \frac{r}{4}, \lambda_0 + \frac{r}{4}\right]\right) \int_{-\frac{r}{4}}^{\frac{r}{4}} \frac{d\tau}{\tau^2 + n^{-2}}$$

$$= 2n \arctan\left(\frac{rn}{4}\right) \mu_{A,v}\left(\left[\lambda_0 - \frac{r}{4}, \lambda_0 + \frac{r}{4}\right]\right).$$

On the second line, we first restricted the integral over s to the interval $[\lambda_0 - r/4, \lambda_0 + r/4]$ and then restricted the one on λ using that for all such s we have $[s - r/4, s + r/4] \subset [\lambda_0 - r/2, \lambda_0 + r/2]$. With the uniform bound $\|(A - z_n)^{-1}\| \leq 2$ on the left term, we obtain

$$\mu_{A,v}\left(\left[\lambda_0 - \frac{r}{4}, \lambda_0 + \frac{r}{4}\right]\right) \leq \frac{2r}{n \arctan\left(\frac{rn}{4}\right)}.$$

The right term tends to 0 when $n \to \infty$ and we get

$$\mu_{A,v}\left(\left[\lambda_0 - \frac{r}{4}, \lambda_0 + \frac{r}{4}\right]\right) = 0.$$

As the open set $\mathbb{R} \setminus \sigma(A)$ is a countable union of such intervals, this concludes the proof of the lemma.   □

Let us continue the proof of the spectral theorem. We have $C_0^0(\mathbb{R}) \subset L^2(\mathbb{R}, d\mu_{A,v}) = L^2(\sigma(A), d\mu_{A,v})$ and $C_0^0(\mathbb{R})$ is even dense in this Hilbert space. We can make this space appear very naturally by studying the sesquilinear form

$$\langle g(A)v, f(A)v \rangle = \langle v, (\overline{g}f)(A)v \rangle = \int_{\sigma(A)} \overline{g(s)} f(s) d\mu_{A,v}(s). \tag{4.12}$$

This equality shows that the linear map

$$f \mapsto f(A)v \tag{4.13}$$

is an isometry from $C_0^0(\mathbb{R}) \subset L^2(\sigma(A), d\mu_{A,v})$ into $\mathfrak{H}$. Thus, after closure we find that the space

$$X_v := \overline{\{f(A)v, \ f \in C_0^0(\mathbb{R})\}} \subset \mathfrak{H} \tag{4.14}$$

is isometric to $L^2(\sigma(A), d\mu_{A,v})$. As $(x - z)^{-1} f \in C_0^0(\mathbb{R})$ for all $z \in \mathbb{C} \setminus \mathbb{R}$ and all $f \in C_0^0(\mathbb{R})$, we have by continuity of $(A - z)^{-1}$ that $(A - z)^{-1} X_v \subset X_v$ for all $z \in \mathbb{C} \setminus \mathbb{R}$. Such a space is called *invariant* (Exercise 4.45). Another important remark for the following is that, even if $(A - z)^{-1}$ is not self-adjoint since z is complex, $(A - z)^{-1}$ also leaves invariant $(X_v)^\perp$, since if $w \in (X_v)^\perp$

$$\left\langle f(A)v, (A - z)^{-1} w \right\rangle = \left\langle (A - \bar{z})^{-1} f(A)v, w \right\rangle = 0$$

for all $f \in C_0^0(\mathbb{R})$. Finally, we have, with $r_z(s) = (s - z)^{-1}$,

$$\left\langle g(A)v, (A - z)^{-1} f(A)v \right\rangle = \langle v, (\overline{g}r_z f)(A)v \rangle = \int_{\sigma(A)} \frac{\overline{g(s)} f(s)}{s - z} d\mu_{A,v}(s), \tag{4.15}$$

a relationship that extends to all X_v by continuity and means that, after applying the inverse of (4.13), the operator $(A - z)^{-1}$ restricted to the invariant subspace X_v is nothing else than the multiplication operator by the function $s \mapsto (s - z)^{-1}$ in $L^2(\sigma(A), d\mu_{A,v})$.

The proof of the theorem is complete if we can find a vector $v \neq 0$ in $\mathfrak{H}$ such that $X_v = \mathfrak{H}$ (such a vector v is called a *cyclic vector*). Indeed, in this case we have shown that $(A - z)^{-1}$ is unitarily equivalent to the multiplication operator by $s \mapsto (s - z)^{-1}$ in $L^2(\mathbb{R}, d\mu_{A,v})$, which implies that A is the multiplication operator by s in this space. We can therefore take $d = 1$, $B = \sigma(A)$ and $\mu = \mu_{A,v}$. Otherwise, we must iterate the argument.

Lemma 4.14 (Decomposition into Invariant Cyclic Subspaces)   *Let A be a self-adjoint operator on $D(A) \subset \mathfrak{H}$. Then there exists a family of vectors $v_1, v_2, \ldots$ (finite or infinite) such that*

$$\mathfrak{H} = \bigoplus_n \mathcal{X}_{v_n} \tag{4.16}$$

where the terms of the direct sum are pairwise orthogonal.

Proof (of the Lemma)   Let us take an orthonormal basis (e_j) of $\mathfrak{H}$ (always assumed separable) and simply set $v_1 = e_1$. If $\mathcal{X}_{v_1} \neq \mathfrak{H}$, consider the smallest $j \geq 2$ such that $e_j \notin \mathcal{X}_{v_1}$. We then take $v_2 = P^{\perp}_{\mathcal{X}_{v_1}} e_j$ (the orthogonal projection of e_j onto the orthogonal of $\mathcal{X}_{v_1}$. As $v_2 \in (\mathcal{X}_{v_1})^{\perp}$, the entire space $\mathcal{X}_{v_2}$ is orthogonal to $\mathcal{X}_{v_1}$. To see this, it is enough to notice that $\langle f(A)v_1, g(A)v_2 \rangle = \langle (\bar{g}f)(A)v_1, v_2 \rangle = 0$, for all $f, g \in C_0^0(\mathbb{R})$, a relationship that persists after closure. By iterating the argument we will have either written $\mathfrak{H}$ as a finite direct sum after a finite number of steps, or constructed a sequence v_n such that $\mathrm{vect}(e_1, \ldots, e_n) \subset \bigoplus_{k=1}^n \mathcal{X}_{v_k}$ for all n. This indeed shows (4.16).   □

For all $z \notin \mathbb{R}$, each of the spaces $\mathcal{X}_{v_n}$ is invariant by $(A - z)^{-1}$, as well as its orthogonal. The bounded operator $(A - z)^{-1}$ is therefore block diagonal. Moreover, $(A - z)^{-1}$ restricted to each of these spaces is isometric to the multiplication operator by the function $s \mapsto (s - z)^{-1}$ on the space $L^2(\sigma(A), \mathrm{d}\mu_{A,v_n})$. We can put all this information together by defining the measure μ on $\sigma(A) \times \mathbb{N} \subset \mathbb{R}^2$ which is μ_{A,v_n} on each of the $\sigma(A) \times \{n\}$ and the appropriate isometry on all $\mathfrak{H}$. If there are only a finite number of v_n, we simply take μ zero on the other copies of $\sigma(A)$. As $\mathcal{X}_{\lambda v} = \mathcal{X}_v$ for all $\lambda \neq 0$, we can choose the norms of each of the v_n so that the sum $\sum_n \|v_n\|^2$ converges, which implies that the measure μ has a finite total mass, since $\mu_{A,v_n}(\mathbb{R}) \leq \|v_n\|^2$ by construction.

The operator $(A - z)^{-1}$ is in this representation the multiplication operator by the function $(s, n) \mapsto (s - z)^{-1}$. This shows as we wanted that A is unitarily equivalent to the multiplication operator by the function $(s, n) \mapsto s$, and completes the proof of the spectral Theorem 4.4.   □

Step 3: Proof of Theorem 4.8 (Functional Calculus for Bounded Measurable Functions)

We have already explained in (4.5) how to construct $f(A)$ for any measurable function using the spectral theorem. Our definition (4.5) satisfies all the properties (i)–(v) of Theorem 4.8 (assertion (v) follows from the dominated convergence theorem). It remains to show uniqueness, which will in particular prove that our definition (4.5) is indeed independent of the chosen representation.

Consider therefore another map $f \mapsto f(A)'$ satisfying all the assumptions of the theorem. Because of the fact that for $f(x) = (x - z)^{-1}$ we have $f(A)' = (A - z)^{-1}$ according to (iii) and as it is a morphism of $*$-algebras according to (i), we see that $f(A)' = f(A)$ for all $f \in \mathcal{A}$ (the resolvent algebra). By density of the latter and by continuity (ii), this implies that $f(A) = f(A)'$ for all $f \in C_0^0(\mathbb{R})$.

For a vector $v \in \mathfrak{H}$, consider now the two linear forms $\ell(f) := \langle v, f(A)v \rangle = \int_{\sigma(A)} f(s) \, d\mu_{A,v}(s)$ and $\ell'(f) := \langle v, f(A)'v \rangle$, which coincide on $C_0^0(\mathbb{R})$. By linearity we can restrict ourselves to real-valued f's. By (v) we know that ℓ' is continuous for dominated convergence. However, by the monotone class theorem, a continuous and positive linear form on $C_0^0(\mathbb{R}, \mathbb{R})$ admits a unique extension to $\mathscr{L}^\infty(\mathbb{R}, \mathbb{R})$ which is continuous for the dominated convergence. This extension is the one given by the corresponding unique bounded Borel measure $\mu_{A,v}$. This proves that $\langle v, f(A)v \rangle = \langle v, f(A)'v \rangle$ for all $v \in \mathfrak{H}$ and all $f \in \mathscr{L}^\infty(\mathbb{R}, \mathbb{R})$ therefore, by polarization, that $f(A) = f(A)'$ as announced.   □

Note that for the uniqueness proof we do not need the assertion (iv) of Theorem 4.8, that $f(A) = 0$ for f supported outside of $\sigma(A)$.

4.5 Spectral Projections

Spectral projections generalize the projections onto the eigenspaces of Hermitian matrices. They play a central role in the spectral analysis of self-adjoint operators, as we will see in the next chapter. We will also learn in Sect. 7.4 of Chap. 7 that they are useful to describe certain infinite quantum systems.

Definition 4.15 (Spectral Projections) Let A be a self-adjoint operator on $D(A) \subset \mathfrak{H}$ and F a Borel set in $\mathbb{R}$. The associated **spectral projection** is the operator $\mathbb{1}_F(A)$, defined using functional calculus (Theorem 4.8). This is an orthogonal projection, that is, $\mathbb{1}_F(A)^2 = \mathbb{1}_F(A)$ and $\mathbb{1}_F(A)^* = \mathbb{1}_F(A)$. The corresponding **spectral subspace** is the closed space $\mathbb{1}_F(A)\mathfrak{H} = \mathrm{ran}(\mathbb{1}_F(A)) = \ker(\mathbb{1}_{\mathbb{R}\setminus F}(A))$.

Using the spectral Theorem 4.4 in the form $A = U^{-1}M_a U$ with $a(s, n) = s$ in $L^2(\sigma(A) \times \mathbb{N}, d\mu)$ and functional calculus of Theorem 4.8, we see that

$$\mathbb{1}_F(A) = U^{-1}M_{\mathbb{1}_F(a)}U = U^{-1}M_{\mathbb{1}_{F \times \mathbb{N}}}U.$$

In other words, after diagonalization the spectral projection $\mathbb{1}_F(A)$ becomes the multiplication by the characteristic function of the set $F \times \mathbb{N}$. Its range is therefore the space of functions supported in this set,

$$\mathrm{ran}(\mathbb{1}_F(A)) = U^{-1}L^2(F \times \mathbb{N}, d\mu),$$

and its rank is the corresponding dimension

$$\mathrm{rank}\big(\mathbb{1}_F(A)\big) = \dim L^2(F \times \mathbb{N}, \mathrm{d}\mu). \tag{4.17}$$

The following result is an immediate consequence of these properties.

Theorem 4.16 (Spectral Projections) *We have the properties*

(*i*) $\mathbb{1}_F(A) = \mathbb{1}_F(A)^* = \big(\mathbb{1}_F(A)\big)^2$;

(*ii*) $\mathbb{1}_\emptyset(A) = 0$, $\mathbb{1}_\mathbb{R}(A) = 1 = \mathrm{Id}_{\mathfrak{H}}$;

(*iii*) *if* $F = \cup_{n\geq 1} F_n$, *then* $\mathbb{1}_F(A)v = \lim\limits_{N\to\infty} \mathbb{1}_{\cup_{n=1}^N F_n} v$ *for all* $v \in \mathfrak{H}$;

(*iv*) $\mathbb{1}_{F_1}(A)\mathbb{1}_{F_2}(A) = \mathbb{1}_{F_1 \cap F_2}(A)$.

Moreover, the spectral subspace $\mathbb{1}_F(A)\mathfrak{H} = \ker(\mathbb{1}_{\mathbb{R}\setminus F}(A))$ *is an invariant subspace of A, on which A is bounded when F is bounded.*

One can show that for any family of orthogonal projections $F \mapsto P_F$ satisfying the properties (*i*)–(*iv*) of the previous theorem, there exists a unique self-adjoint operator $(A, D(A))$ such that $P_F = \mathbb{1}_F(A)$. Hence the spectral projections completely characterize a self-adjoint operator.

Let us now discuss the link between spectral projections and the spectrum.

Lemma 4.17 (Spectrum via Spectral Projections) *Let A be a self-adjoint operator on its domain* $D(A) \subset \mathfrak{H}$ *and* $\lambda \in \mathbb{R}$. *Then*

(*i*) $\lambda \in \sigma(A)$ *if and only if* $\mathbb{1}_{(\lambda-\varepsilon, \lambda+\varepsilon)}(A) \neq 0$ *for all* $\varepsilon > 0$;

(*ii*) λ *is an eigenvalue of A, that is,* $\ker(A - \lambda) \neq \{0\}$, *if and only if* $\mathbb{1}_{\{\lambda\}}(A) \neq 0$. *In this case,* $\mathbb{1}_{\{\lambda\}}(A)$ *is the orthogonal projection onto the corresponding eigenspace* $\ker(A - \lambda)$.

Proof As explained above, after diagonalization the operator $\mathbb{1}_{(\lambda-\varepsilon, \lambda+\varepsilon)}(A)$ becomes the orthogonal projections onto $L^2((\lambda - \varepsilon, \lambda + \varepsilon) \times \mathbb{N}, \mathrm{d}\mu)$. This space is non trivial if and only if $\mu((\lambda - \varepsilon, \lambda + \varepsilon) \times \mathbb{N}) > 0$. This is the case for every $\varepsilon > 0$ if and only is λ belongs to $\sigma(A)$, because $(\lambda - \varepsilon, \lambda + \varepsilon) \times \mathbb{N} = \{|a - \lambda| < \varepsilon\}$ for $a(s, n) = s$ and $\sigma(A) = \sigma(M_a)$ is the essential range of a, by Theorem 4.3. Similarly, λ is an eigenvalue if and only if $\mu(\{a = \lambda\}) = \mu(\{\lambda\} \times \mathbb{N}) > 0$. □

The interpretation of $\mathbb{1}_F(A)$ is therefore simple when the set F only contains eigenvalues and it coincides with what we know in finite dimension. Namely, $\mathbb{1}_F(A)$ is the orthogonal projection onto the direct sum of the corresponding eigenspaces.

When $\lambda \in \sigma(A)$ is not an eigenvalue, we have $\mathbb{1}_{\{\lambda\}}(A) = 0$ and it is needed to enlarge the interval to obtain a non trivial operator. Lemma 4.17 stipulates that the spectral projection $\mathbb{1}_{(\lambda-\varepsilon, \lambda+\varepsilon)}(A)$ does not vanish for any $\varepsilon > 0$. Because of the spectral theorem, we should think of $\mathbb{1}_{(\lambda-\varepsilon, \lambda+\varepsilon)}(A)$ as the orthogonal projection onto the space of vectors for which, loosely speaking, "$|A - \lambda| < \varepsilon$". The mathematically correct statement of course requires diagonalizing A and then considering functions supported on the set $\{|a-\lambda| < \varepsilon\}$. In this direction of thinking, we mention for instance that

$$\|(A - \lambda)v\| \leq \varepsilon \|v\|, \qquad \forall v \in \text{ran}\big(\mathbb{1}_{(\lambda-\varepsilon,\lambda+\varepsilon)}(A)\big) \qquad (4.18)$$

and

$$\|(A - \lambda)v\| \geq \varepsilon \|v\|, \qquad \forall v \in D(A) \cap \text{ran}\big(\mathbb{1}_{\mathbb{R}\setminus(\lambda-\varepsilon,\lambda+\varepsilon)}(A)\big). \qquad (4.19)$$

For the first inequality we only need $v \in \mathfrak{H}$ since A is bounded on the range of the projection, but for the second we have to assume $v \in D(A)$. Of course, these inequalities are obtained by writing

$$\|(A - \lambda)v\|^2 = \int_{\sigma(A)\times\mathbb{N}} (s - \lambda)^2 |Uv(s, n)|^2 \, d\mu(s, n)$$

and using that $Uv(s, n)$ is supported on the set where $(a - \lambda)^2 = (s - \lambda)^2$ is either $< \varepsilon^2$ or $\geq \varepsilon^2$.

Remark 4.18 (Spectral Projections and Weyl Sequences) Recall Theorem 2.30 which stipulates that one can characterize the spectrum in terms of Weyl sequences $v_n \in D(A)$ with $\|v_n\| = 1$ and $(A - \lambda)v_n \to 0$. For any $\lambda \in \sigma(A)$, one can easily construct such a sequence using Lemma 4.17. It suffices to take any normalized v_n in the range of the projection $\mathbb{1}_{(\lambda-1/n,\lambda+1/n)}(A)$, which is never empty. We then get the desired convergence $(A - \lambda)v_n \to 0$ by (4.18).

Conversely, any Weyl sequence has to live close to the range of the spectral projections of a neighborhood of λ. Indeed, let us consider an arbitrary Weyl sequence (v_n) and fix an $\varepsilon > 0$. Using this time (4.19) and the fact that $P(A - \lambda)v_n \to 0$ since $P = \mathbb{1}_{\mathbb{R}\setminus(\lambda-\varepsilon,\lambda+\varepsilon)}(A)$ is bounded, we deduce that

$$v_n - \mathbb{1}_{(\lambda-\varepsilon,\lambda+\varepsilon)}(A)v_n = \mathbb{1}_{\mathbb{R}\setminus(\lambda-\varepsilon,\lambda+\varepsilon)}(A)v_n \xrightarrow[n\to\infty]{} 0$$

$$\text{for any } \lambda\text{-Weyl sequence and any } \varepsilon > 0. \qquad (4.20)$$

This proves the claim that Weyl any sequence has to get arbitrarily close to the spectral subspace associated with a neighborhood of λ. In fact,

$$w_n = \frac{\mathbb{1}_{(\lambda-\varepsilon,\lambda+\varepsilon)}(A)v_n}{\|\mathbb{1}_{(\lambda-\varepsilon,\lambda+\varepsilon)}(A)v_n\|} = (1 + o(1))\mathbb{1}_{(\lambda-\varepsilon,\lambda+\varepsilon)}(A)v_n$$

is a new Weyl sequence that lives exactly in that subspace.

In the same way that it is possible to characterize any Borel measure ν by its distribution function $\lambda \mapsto \nu((-\infty, \lambda])$, it is possible to characterize a self-adjoint operator using the family of spectral projectors $P(\lambda) = \mathbb{1}_{(-\infty, \lambda]}(A)$.

Let us first illustrate this claim in finite dimension. For a $d \times d$ Hermitian matrix M, with ordered eigenvalues $\lambda_1 < \cdots < \lambda_{d'}$ (possibly with multiplicity, so $d' \leq d$), the operator $P(\lambda) := \mathbb{1}_{]-\infty, \lambda]}(M)$ is the orthogonal projector onto the direct sum of the eigenspaces corresponding to the eigenvalues $\lambda_j \leq \lambda$. It is a $d \times d$ matrix whose derivative can be calculated in the sense of distributions. As the function has jumps at the eigenvalues and is constant between two distinct λ_j, we see that

$$\frac{\mathrm{d}P}{\mathrm{d}\lambda}(\lambda) = \sum_{j=1}^{d} \delta_{\lambda_j}(\lambda) \mathbb{1}_{\{\lambda_j\}}(M).$$

In particular, we have the representation

$$\boxed{M = \int_{\mathbb{R}} \lambda \, \mathrm{d}P(\lambda)} \tag{4.21}$$

because $M = \sum_{j=1}^{d} \lambda_j \mathbb{1}_{\{\lambda_j\}}(M)$. This is called the **spectral resolution of the matrix** M. For every continuous function f, we also have

$$f(M) = \int_{\mathbb{R}} f(\lambda) \, \mathrm{d}P(\lambda). \tag{4.22}$$

It seems natural to wonder if it is possible to write a similar formula in infinite dimension for any self-adjoint operator. Let A be a self-adjoint operator and fix a vector $v \in \mathfrak{H}$. Then the function

$$P_v(\lambda) := \langle v, \mathbb{1}_{(-\infty, \lambda]}(A)v \rangle$$

is bounded and non-decreasing. Its derivative in the sense of distributions is therefore a measure, which turns out to be the spectral measure $\mu_{A,v}$ that we have already encountered before (Remark 4.9) and which is such that

$$\langle v, f(A)v \rangle = \int_{\mathbb{R}} f(\lambda) \, \mathrm{d}\mu_{A,v}(\lambda) = \int_{\mathbb{R}} f(\lambda) \, \mathrm{d}\langle v, \mathbb{1}_{(-\infty, \lambda]}(A)v \rangle \tag{4.23}$$

for any bounded Borel function f. This corresponds to taking the scalar product with v in (4.22). We also have

$$\langle v, Av \rangle = \int_{\mathbb{R}} \lambda \, \mathrm{d}\mu_{A,v}(\lambda) = \int_{\mathbb{R}} \lambda \, \mathrm{d}\langle v, \mathbb{1}_{]-\infty, \lambda]}(A)v \rangle \tag{4.24}$$

if in addition $v \in D(A)$, which is the infinite-dimensional version of the spectral resolution (4.21). This spectral resolution of an operator A is often encountered in the literature. In fact, several works (for example [Tes09]) start by constructing the operators $\mathbb{1}_{(-\infty,\lambda]}(A)$, then deduce the functional calculus via Formula (4.23) and finally the spectral theorem. This is another equivalent version of the theory.

To conclude this section, we mention several very useful formulas for spectral projections, based on complex analysis.

Theorem 4.19 (Cauchy's Formula) *Let $(A, D(A))$ be a self-adjoint operator and $a, b \in \rho(A) \cap \mathbb{R}$ with $a < b$. Then we have the formula*

$$\mathbb{1}_{(a,b)}(A) = \mathbb{1}_{[a,b]}(A) = -\frac{1}{2i\pi} \oint_{\mathscr{C}} (A - z)^{-1} dz \qquad (4.25)$$

for every closed loop $\mathscr{C}$ in the complex plane, oriented anticlockwise, which encloses the interval $[a, b]$ and crosses the real axis at a and b (Fig. 4.2).

Proof As $a, b \in \rho(A)$ which is open, these points are at strictly positive distance from the spectrum. The integral on the right of (4.25) converges, since the resolvent is a smooth function of z, with norm bounded by the inverse of the distance to the spectrum (according to (4.4)) and the curve $\mathscr{C}$ also remains at a positive distance from the spectrum. The result then follows from the spectral theorem and Cauchy's formula which stipulates that $(2i\pi)^{-1} \oint_{\mathscr{C}} (x - z)^{-1} dz$ is equal to -1 if the residue x is in the region delimited by the curve $\mathscr{C}$, and 0 otherwise. $\square$

Formula (4.25) does not allow obtaining $\mathbb{1}_{[a,b]}(A)$ if a or b are in the spectrum of A, because in this case the integral over the contour $\mathscr{C}$ diverges in the vicinity of a and b. However, by choosing a contour symmetric with respect to the real axis and flattening it as shown in Fig. 4.3, we can show a formula valid for all $a, b \in \mathbb{R}$.

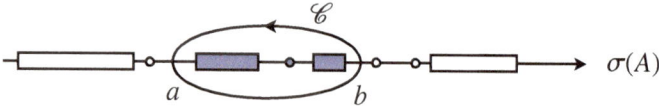

Fig. 4.2 Cauchy's formula (4.25) allows expressing the spectral projection $\mathbb{1}_{[a,b]}(A)$ with $a, b \notin \sigma(A)$ as the integral of the resolvent $(A - z)^{-1}$ over a contour $\mathscr{C}$ in the complex plane. © Mathieu Lewin 2021. All rights reserved

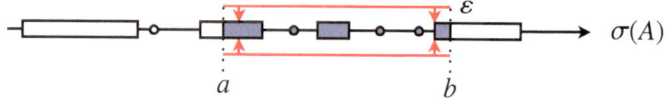

Fig. 4.3 Stone's formula (4.26) is a generalization to the case where a or b belongs to $\sigma(A)$. We take $\mathscr{C}$ equal to a rectangle that we flatten, and from which we remove the vertical sides. © Mathieu Lewin 2021. All rights reserved

Theorem 4.20 (Stone's Formula) *Let A be a self-adjoint operator and $a < b \in \mathbb{R}$. For all $v \in \mathfrak{H}$, we have*

$$\lim_{\varepsilon \to 0^+} \frac{1}{2i\pi} \int_a^b \left((A - s - i\varepsilon)^{-1} - (A - s + i\varepsilon)^{-1} \right) ds \, v$$

$$= \left(\mathbb{1}_{(a,b)}(A) + \frac{1}{2} \mathbb{1}_{\{a\}}(A) + \frac{1}{2} \mathbb{1}_{\{b\}}(A) \right) v. \qquad (4.26)$$

Proof Consider the function

$$f_\varepsilon(x) = \frac{1}{2i\pi} \int_a^b \left(\frac{1}{x - s - i\varepsilon} - \frac{1}{x - s + i\varepsilon} \right) ds$$

$$= \frac{1}{\pi} \arctan\left(\frac{b - x}{\varepsilon} \right) - \frac{1}{\pi} \arctan\left(\frac{a - x}{\varepsilon} \right)$$

which is uniformly bounded and converges to

$$\lim_{\varepsilon \to 0^+} f_\varepsilon(x) = \mathbb{1}_{(a,b)}(x) + \frac{1}{2} \mathbb{1}_{\{a\}}(x) + \frac{1}{2} \mathbb{1}_{\{b\}}(x)$$

for all $x \in \mathbb{R}$. The result follows from property (v) of Theorem 4.8 on functional calculus. ☐

Exercise 4.21 (Projectors $\mathbb{1}_{(-\infty,b]}(A)$) Show that we can take $a = -\infty$ in Formula (4.26), so that

$$\lim_{\varepsilon \to 0^+} \frac{1}{2i\pi} \int_{-\infty}^b \left((A - s - i\varepsilon)^{-1} - (A - s + i\varepsilon)^{-1} \right) ds \, v$$

$$= \left(\mathbb{1}_{(-\infty,b)}(A) + \frac{1}{2} \mathbb{1}_{\{b\}}(A) \right) v. \qquad (4.27)$$

By rotating the axes as in Fig. 4.4, also prove the formula

$$\lim_{\varepsilon \to 0^+} \frac{1}{2\pi} \int_{-\infty}^{-\varepsilon} \left((A - b - is)^{-1} - (A - b + is)^{-1} \right) ds \, v$$

$$= \left(\mathbb{1}_{(-\infty,b)}(A) + \frac{1}{2} \mathbb{1}_{\{b\}}(A) \right) v. \qquad (4.28)$$

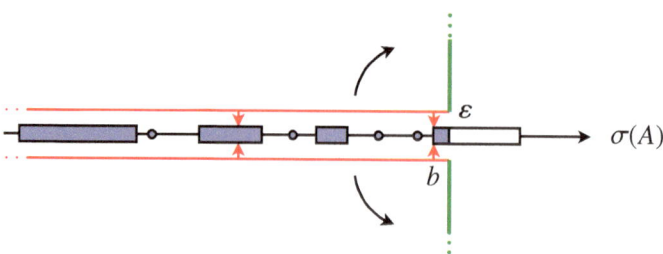

4.6   Powers

The following theorem is an immediate consequence of the spectral theorem.

Theorem 4.22 (Powers)   *Let $(A, D(A))$ be a self-adjoint operator and $k \geq 2$ an integer. The operator $(A^k, D(A^k))$ defined by (4.5) with $f(x) = x^k$ has the domain satisfying the recursion relation*

$$D(A^k) = \left\{ v \in D(A^{k-1}) \ : \ A^{k-1}v \in D(A) \right\} \tag{4.29}$$

with $A^k v = A(A^{k-1}v)$. The spectrum of the self-adjoint operator A^k is

$$\sigma(A^k) = \left\{ \lambda^k \ : \ \lambda \in \sigma(A) \right\}.$$

Proof Pass to the representation where A is a multiplication operator and notice that $\int_B (1 + |a|^{2k})|v|^2 \, d\mu$ converges if and only if $a^\ell v \in L^2(B, d\mu)$ for all $\ell = 0, \ldots, k$. $\qquad \square$

Example 4.23 (Square of Momenta on the Interval $I = (0, 1)$) Let $I = (0, 1)$ and consider the momentum operators

$$P_{\text{per},\theta} f = -if', \quad D(P_{\text{per},\theta}) = H^1_{\text{per},\theta}(I) = \left\{ f \in H^1(I) \ : \ f(1) = e^{i\theta} f(0) \right\},$$

which were studied in Sect. 2.8.1. Recall that these are the only possible self-adjoint realizations of the momentum on the interval $I = (0, 1)$. Unsurprisingly, we find that $P^2_{\text{per},\theta}$ is the Born-von Kármán Laplacian defined by $P^2_{\text{per}} f = -f''$ on the domain

$$D(P^2_{\text{per},\theta}) = \{ f \in H^1_{\text{per},\theta}(I) \ : \ f' \in H^1_{\text{per},\theta}(I) \}$$

$$= \{ f \in H^2(I) \ : \ f(1) = e^{i\theta} f(0), \ f'(1) = e^{i\theta} f'(0) \}.$$

More surprising is, perhaps, the fact that this proves that none of the other self-adjoint extensions of the Laplacian found in Sect. 2.8.3 (for example Dirichlet, Neumann and Robin) is the square of a momentum! As a side remark, recall that we saw in Exercise 3.27 that the Dirichlet and Neumann Laplacians can rather be written as $(P_0)^* P_0$ and $P_0 (P_0)^*$ where P_0 is however not self-adjoint.

Example 4.24 (Square of the Hydrogen Atom) Consider the operator describing the hydrogen atom

$$A = -\frac{\Delta}{2} - \frac{1}{|x|},$$

which is self-adjoint on $D(A) = H^2(\mathbb{R}^3)$ as we saw in Example 3.6. The singularity at 0 of the Coulomb potential is not sufficient to alter the self-adjoint domain of the Laplacian. One might think that the same is true for the domain of the powers A^k, but this is actually not the case. For example, the first eigenfunction $f(x) = \pi^{-1/2} e^{-|x|}$ found in Theorem 1.3 is not in $H^4(\mathbb{R}^3)$ while it is of course in the domain of A^k for all k. Thus, we have

$$D(A^2) \neq H^4(\mathbb{R}^3) = D(\Delta^2).$$

The reader can show as an exercise that

$$D(A^2) \cap H^4(\mathbb{R}^3) = \left\{ f \in H^4(\mathbb{R}^3) \ : \ f(0) = 0, \quad \nabla f(0) = 0 \right\}.$$

If the self-adjoint operator A has a non-negative spectrum, we can also define A^s for all real $s \geq 0$. The square root obtained for $s = 1/2$ is related to the quadratic form of A.

Theorem 4.25 (Square Root and Quadratic Form)   *Let $(A, D(A))$ be a non-negative self-adjoint operator, that is, with spectrum $\sigma(A) \subset \mathbb{R}_+$. Let $(\sqrt{A}, D(\sqrt{A}))$ be the operator defined by (4.5) with $f(x) = \sqrt{x}$. Then we have*

$$A = (\sqrt{A})^2, \qquad D(A) = \left\{ f \in D(\sqrt{A}) \ : \ \sqrt{A} f \in D(\sqrt{A}) \right\}.$$

The quadratic form q_A defined in Sect. 3.2 satisfies

$$\boxed{q_A(v) = \left\| \sqrt{A}\, v \right\|^2, \qquad Q(A) = D(\sqrt{A}).}$$

Proof The operator A is unitarily equivalent to a multiplication operator by a function a on a space $L^2(B, d\mu)$, with $a \geq 0$ μ-almost everywhere. In this representation, the quadratic form of A becomes

$$q_A(U^{-1} v) = \int_B a(x) |v(x)|^2 \, d\mu(x)$$

for all $v \in D(A)$. The space $Q(A)$ obtained by completion in Sect. 3.2 is therefore

$$Q(A) = U^{-1} \left\{ v \in L^2(B, d\mu) \ : \ \int_B a(x)|v(x)|^2 \, d\mu(x) < \infty \right\}$$

which is exactly $D(\sqrt{A})$ by definition. $\square$

Remark 4.26 (Interpolation) Consider two Hilbert spaces $\mathfrak{H}_1 \subset \mathfrak{H}_0$ with continuous embedding and $\mathfrak{H}_1$ dense in $\mathfrak{H}_0$. By the Riesz-Friedrichs Theorem 3.17, we can write $\mathfrak{H}_1 = Q(A)$ with a coercive self-adjoint operator $(A, D(A))$ on $\mathfrak{H}_0$. We then deduce the existence of a family of spaces

$$\mathfrak{H}_s := Q(A^s) = D(A^{\frac{s}{2}}), \qquad 0 \le s \le 1$$

which continuously interpolates between $\mathfrak{H}_0$ and $\mathfrak{H}_1$. They are non-increasing with s, that is to say $\mathfrak{H}_s \subset \mathfrak{H}_{s'}$ for $s \ge s'$ with continuous embedding. For example, for $A = -\Delta$ we have $\mathfrak{H}_1 = H^1(\mathbb{R}^d)$, $\mathfrak{H}_0 = L^2(\mathbb{R}^d)$ and $\mathfrak{H}_s = H^s(\mathbb{R}^d)$.

At this stage, it may be useful to revisit the domains $D(A)$ and $Q(A)$, and in particular the meaning that can be given to A on $Q(A)$.

When A is a self-adjoint operator, it is in particular closed and its domain $D(A)$ is a Hilbert space when it is equipped with the scalar product $\langle v, w \rangle_{D(A)} := \langle v, w \rangle + \langle Av, Aw \rangle$. As $D(A)$ is also included in $\mathfrak{H}$, with continuous embedding, it seems natural to look at its dual $D(A)'$ which is generally larger than $\mathfrak{H}$. By the spectral theorem, we can identify A with the multiplication operator by a locally bounded function a on a space $L^2(B, d\mu)$ and it follows that

$$D(A)' \simeq \left\{ f \in L^2_{\text{loc}}(B, d\mu) \ : \ \int_B \frac{|f(x)|^2}{1 + a(x)^2} \, d\mu(x) < \infty \right\}.$$

This is also a Hilbert space, associated with the bounded Borel measure $\mu/(1+a^2)$. Moreover, since a is locally bounded, we can without problem multiply any $f \in L^2(B, d\mu)$ by a. We obtain a function that is only in $L^2_{\text{loc}}(B, d\mu)$ and of course belongs to $L^2(B, d\mu/(1 + a^2))$. From this discussion we see that the operator A extends into a unique continuous operator on all $\mathfrak{H}$ with values in $D(A)'$. For any vector $v \in \mathfrak{H}$, $Av \in D(A)'$ is just the linear form $u \in D(A) \mapsto \langle v, Au \rangle$.

The argument is exactly similar for $Q(A)$ when $A \ge 0$. In this case A also extends into a unique bounded operator from $Q(A)$ into

$$Q(A)' \simeq \left\{ f \in L^2_{\text{loc}}(B, d\mu) \ : \ \int_B \frac{|f(x)|^2}{\sqrt{1 + a(x)^2}} \, d\mu(x) < \infty \right\}.$$

These observations are very useful when working with several equivalent quadratic forms but whose associated self-adjoint operators can be very different. For example, with the help of the KLMN Theorem 3.19 we know how to associate a unique self-adjoint operator C with a quadratic form $q_A + b$ when

$$|b(v)| \leq \eta q_A + \kappa \|v\|^2, \qquad \text{with } \eta < 1 \tag{4.30}$$

and A is coercive. But the operator C can have a very different domain from A and it may be impossible to define $B = C - A$ on a dense domain in $\mathfrak{H}$ (re-read on this Exercise 3.26). Thanks to the previous remark, we see that $B = C - A$ makes perfect sense as a bounded operator from $Q(A)$ into $Q(A)'$. Moreover,

$$\boxed{A^{-\frac{1}{2}} B A^{-\frac{1}{2}} \text{ is a bounded self-adjoint operator on } \mathfrak{H}.}$$

Indeed, $A^{-1/2}$ is bounded from $\mathfrak{H}$ into $Q(A)$, then B is bounded from $Q(A)$ into $Q(A)'$ and finally $A^{-1/2}$ is bounded from $Q(A)'$ into $\mathfrak{H}$ by the previous arguments. By replacing v with $A^{-1/2}v$ in (4.30), we even obtain the estimate

$$\left\| A^{-\frac{1}{2}} B A^{-\frac{1}{2}} \right\| \leq \eta + \kappa \left\| A^{-1} \right\| = \eta + \frac{\kappa}{\min \sigma(A)}.$$

This remark will be useful whenever we need to study perturbations of quadratic forms.

4.7 Schrödinger, Heat and Wave Equations

4.7.1 Schrödinger's Equation

Let A be a self-adjoint operator, with domain $D(A)$. In this section we show that the time-dependent Schrödinger equation

$$\begin{cases} i\partial_t v(t) = A\,v(t) \\ v(0) = v_0 \end{cases} \tag{4.31}$$

admits the unique solution $v(t) = e^{-itA}v_0$, where e^{-itA} is defined by functional calculus. A first difficulty consists in precisely defining the concept of solutions of (4.31). Here, we will restrict ourselves to solutions that are in $L^1\big((0, T), \mathfrak{H}\big)$ and satisfy (4.31) in the weak sense. The argument will be very similar for negative time solutions, on $(-T, 0)$.

There are subtleties concerning the definition of the space $L^1\big((0, T), \mathfrak{H}\big)$, particularly for the notion of measurability of functions with values in a Banach space.[1]

---

[1] The measurability of $t \mapsto v(t)$ can be defined in two ways, either by involving the Borel sets of $\mathfrak{H}$ (we then speak of strong measurability), or by rather requiring that the maps $t \mapsto \langle w, v(t) \rangle$

We do not wish to dwell on these difficulties here and just retain that the family $v(t)$ for $t \in (0, T)$ belongs to $L^1((0, T), \mathfrak{H})$ when

$$\int_0^T \|v(t)\| \, dt < \infty.$$

We saw in the previous section that it was possible to extend A to all of $\mathfrak{H}$ if we allow it to take its values in the dual $D(A)'$. This allows us to identify $A\mathfrak{H}$ with a subspace of $D(A)'$, with

$$\|Av\|_{D(A)'} = \sup_{u \in D(A)} \frac{|\langle v, Au \rangle|}{\|u\|_{D(A)}} \leq \|v\|.$$

Thus, we see that every $v(t) \in L^1((0, T), \mathfrak{H})$ is such that $Av(t) \in L^1((0, T), D(A)')$, which allows us to give a weak sense to the right-hand side of Eq. (4.31), as a linear form on $D(A)$. We also need to give a sense to the left term $\partial_t v(t)$, which we want to interpret as the derivative of $v(t)$ in the sense of distributions in $L^1((0, T), \mathfrak{H})$. Rather than talking about such derivatives, we will simply take the scalar product against a vector $w \in D(A)$ and talk about derivatives in the sense of distributions on $(0, T)$, using the relation

$$\frac{d}{dt} \langle w, v(t) \rangle = \langle w, \partial_t v(t) \rangle.$$

Definition 4.27 (Weak Solutions of Schrödinger's Equation) We say that $v \in L^1((0, T), \mathfrak{H})$ is a *weak solution* of Schrödinger's equation (4.31) if we have, in the sense of distributions on $(0, T)$,

$$\frac{d}{dt} \langle w, v(t) \rangle = -i \langle Aw, v(t) \rangle \tag{4.32}$$

with moreover

$$\lim_{t \to 0^+} \langle w, v(t) \rangle = \langle w, v_0 \rangle, \tag{4.33}$$

this for all $w \in D(A)$.

Our assumption that $v \in L^1((0, T), \mathfrak{H})$ implies that $f(t) = \langle w, v(t) \rangle \in L^1((0, T), \mathbb{C})$. Furthermore, Eq. (4.32) implies that $f' \in L^1((0, T), \mathbb{C})$ as well. By the properties of Sobolev spaces in one dimension recalled in Sect. A.2, this shows that $f \in C^0([0, T], \mathbb{C})$, that is, the left limit of (4.33) exists. For an interpretation

---

be measurable for all $w \in \mathfrak{H}$, which amounts to projecting $v(t)$ on a basis (we speak of weak measurability). We refer to [Eva10, App. E.5] for some details and references. In the case of a separable Hilbert space $\mathfrak{H}$ like us, the two notions turn out to be equivalent, see [RS72, Thm. IV.22].

of (4.32) using weak derivatives valued in the Banach space $D(A)$, see [Eva10, Sec. 5.9.2 & App. E.5] or [DL92, Chap. XVIII §1].

We will particularly discuss the *well-posedness* of the equation, which consists in asking, in addition to the existence and uniqueness of solutions, that the map $v_0 \mapsto v(t)$ be *continuous* in appropriate spaces. This means that the solution $v(t)$ will only change a little when the initial condition v_0 is slightly perturbed. The following result means that the Schrödinger equation is well-posed from $\mathfrak{H}$ to $C^0([0, T], \mathfrak{H})$.

Theorem 4.28 (Schrödinger Equation) *Let $(A, D(A))$ be a self-adjoint operator. Then for all $v_0 \in \mathfrak{H}$ and all $T > 0$, Schrödinger's Equation (4.31) admits a **unique weak solution** on $(0, T)$, given by*

$$v(t) = e^{-itA}v_0.$$

This solution is in fact defined on the whole of $\mathbb{R}$ and satisfies $v \in C^0(\mathbb{R}, \mathfrak{H})$, with the relation $\|v(t)\| = \|v_0\|$ for all $t \in \mathbb{R}$. If in addition $v_0 \in D(A)$, then $v \in C^1(\mathbb{R}, \mathfrak{H}) \cap C^0(\mathbb{R}, D(A))$ and solves (4.31) in the strong sense.

Proof By functional calculus, $v(t) = e^{-itA}v_0$ is indeed a strong solution of the equation if we add the additional assumption that $v_0 \in D(A)$. Let us now assume that $v_0 \in \mathfrak{H}$ without necessarily belonging to $D(A)$. We find

$$\frac{d}{dt}\left\langle w, e^{-itA}v_0 \right\rangle = \frac{d}{dt}\left\langle e^{itA}w, v_0 \right\rangle = -i\left\langle e^{itA}Aw, v_0 \right\rangle = -i\langle Aw, v(t) \rangle$$

for all $w \in D(A)$ and all $t \in \mathbb{R}$, which means that $v(t)$ is a weak solution on $\mathbb{R}$ if $v_0 \in \mathfrak{H}$. It remains to show uniqueness. We will use the fact that if $v \in L^1((0, T), \mathfrak{H})$ is a weak solution of the equation, then for all $w \in C^1([0, T], \mathfrak{H}) \cap C^0([0, T], D(A))$ we have

$$\frac{d}{dt}\langle w(t), v(t) \rangle = -i\langle Aw(t), v(t) \rangle + \langle \partial_t w(t), v(t) \rangle. \tag{4.34}$$

In other words, we can take w depending on time in (4.32) provided we remember to differentiate w. We leave the proof of (4.34) as an exercise. Let $v_0 \in \mathfrak{H}$ and $v(t)$ be any weak solution on $(0, T)$. Also let $w_0 \in D(A)$ and $w(t) = e^{-itA}w_0$, the latter being a strong solution of the equation, with $w(t) \in D(A)$ for all t. We thus obtain

$$\frac{d}{dt}\langle w(t), v(t) \rangle = -i\langle Aw(t), v(t) \rangle + i\langle Aw(t), v(t) \rangle = 0.$$

As $\varphi(t) := \langle w(t), v(t) \rangle$ is continuous on $[0, T]$ (it is integrable and has an integrable derivative), this shows that $\langle w(t), v(t) \rangle = \langle w_0, e^{itA}v(t) \rangle = \langle w_0, v_0 \rangle$ which is the limit at 0 of this same quantity. As we have this relation for all $t \in \mathbb{R}$ and all $w_0 \in D(A)$, we deduce by density of $D(A)$ that $e^{itA}v(t) = v_0$ for almost all $t \in \mathbb{R}$, that is $v(t) = e^{-itA}v_0$. $\square$

Example 4.29 (Laplacian) For $A = -\Delta$ on $D(A) = H^2(\mathbb{R}^d) \subset \mathfrak{H} = L^2(\mathbb{R}^d)$, the operator e^{-itA} is the multiplication by the function $e^{-it|k|^2}$ in Fourier and we can thus express the solution in direct space as

$$\left(e^{-itA}v_0\right)(x) = \frac{1}{(4\pi it)^{d/2}} \int_{\mathbb{R}^d} e^{i\frac{|x-y|^2}{4t}} v_0(y)\,dy.$$

This formula makes sense when $v_0 \in L^1(\mathbb{R}^d) \cap L^2(\mathbb{R}^d)$, for example.

Remark 4.30 (Beware of Series!) The formula $v(t) = e^{-itA}v_0$ for the unique solution of Schrödinger's equation shines with its simplicity, but it is important to keep in mind that the unitary e^{-itA} was defined by functional calculus (Theorem 4.8), which is not at all a trivial result. For example, one might think that one can write the solution in the form of a series

$$v(t) = e^{-itA}v_0 = \sum_{n\geq 0} \frac{(-it)^n}{n!} A^n v_0 \tag{4.35}$$

but this formula only makes sense for initial conditions v_0 belonging to $D(A^n)$ for all $n \geq 1$ and so that the series converges. Misuse of (4.35) can lead to false conclusions. Here is an example taken from [MTWB10, FLLØ16]. Consider the function $v_0 = e^{-|x|} \in \mathfrak{H} = L^2(\mathbb{R})$ and the operator $A = -d^2/dx^2$ which is self-adjoint on $D(A) = H^2(\mathbb{R})$. A calculation shows that

$$-v_0'' + v_0 = 2\delta_0 \tag{4.36}$$

in the sense of distributions on $\mathbb{R}$. Thus, we have $-v_0'' + v_0 = 0$ on $\mathbb{R} \setminus \{0\}$. The right-hand side of formula (4.35) therefore seems to provide

$$\sum_{n\geq 0} \frac{(it)^n}{n!} v_0^{(2n)} = v_0 \sum_{n\geq 0} \frac{(it)^n}{n!} = e^{it}v_0, \qquad \text{on } \mathbb{R} \setminus \{0\}.$$

This cannot be equal to $e^{-itA}v_0$ because otherwise v_0 would be an eigenvector of $A = -d^2/dx^2$, with eigenvalue -1, and we know that $\sigma(A) = [0, \infty)$ without any eigenvalue. In fact, $v_0(x) = e^{-|x|}$ does not belong to $H^2(\mathbb{R}) = D(A)$ since we have (4.36), which makes the use of the series (4.35) completely illicit, even outside the origin.

Exercise 4.31 Show that $v(t) = e^{-itA}v_0$ is k times differentiable at $t = 0$ in $\mathfrak{H}$ (thus for all $t \in \mathbb{R}$), if and only if $v_0 \in D(A^k)$, and then $v^{(k)}(0) = (-i)^k A^k v_0$. Give a condition on v_0 for the series (4.35) to converge and coincide with $v(t)$.

4.7.2 Heat Equation

The study carried out for Schrödinger's equation can be extended to other situations. For example, the heat equation takes the form

$$\begin{cases} \partial_t v(t) = -A\, v(t) \\ v(0) = v_0 \end{cases} \tag{4.37}$$

and can be studied with a similar method as before. For Schrödinger's equation (1.14) there was no big difference between positive and negative times; we say that the equation is reversible. This is not the case here. The formal solution is now given by $v(t) = e^{-tA} v_0$. This is a well-defined vector in $\mathfrak{H}$ only when $v_0 \in D(e^{-tA})$ for all $t \in (0, T)$. We have $D(e^{-tA}) = \mathfrak{H}$ for a $t > 0$ only when e^{-tA} is a bounded operator, that is, A is bounded-below. Thus, the problem is well posed in positive time only when A is bounded-below, and for all times $t \in \mathbb{R}$ only if A is a bounded operator. We state here a theorem that is shown in the same way as that for Schrödinger's equation, in the case of positive times only. Weak solutions are defined similarly to Definition 4.27 by of course requiring that

$$\frac{d}{dt} \langle w, v(t) \rangle = -\langle Aw, v(t) \rangle$$

for all $w \in D(A)$.

We leave the proof of the following theorem as an exercise.

Theorem 4.32 (Heat Equation) *Let* $(A, D(A))$ *be a bounded-below self-adjoint operator. Then for all* $v_0 \in \mathfrak{H}$ *and all* $T > 0$, *the heat Equation* (4.37) *admits a* **unique weak solution** $v \in L^1\big((0, T), \mathfrak{H}\big)$, *given by*

$$\boxed{v(t) = e^{-tA} v_0.}$$

It verifies $v \in C^0\big([0, +\infty), \mathfrak{H}\big)$, *with the estimate* $\|v(t)\| \le e^{-t \min \sigma(A)} \|v_0\|$ *for all* $t \ge 0$. *If in addition* $v_0 \in D(A)$, *then* $v \in C^1\big([0, +\infty), \mathfrak{H}\big) \cap C^0\big([0, +\infty), D(A)\big)$ *and solves* (4.37) *in the strong sense.*

Let f be any locally bounded Borel function. For Schrödinger's equation, $v_0 \in D(f(A))$ implies $v(t) \in D(f(A))$ for all $t \in \mathbb{R}$. In fact, $v(t) \in D(f(A))$ for a t if and only if $v_0 \in D(f(A))$ because we can apply the flow backwards. In other words, Schrödinger's equation propagates the "regularity with respect to A" of the initial condition without ever improving it. The situation is very different for the heat equation. For example, we have $v(t) \in D(e^{tA})$ for all $t > 0$, so in particular $v(t) \in D(A^k)$ for all $k \ge 1$, even if this property is false at $t = 0$.

Example 4.33 (Laplacian) For $A = -\Delta$ on $D(A) = H^2(\mathbb{R}^d) \subset \mathfrak{H} = L^2(\mathbb{R}^d)$, the operator e^{-tA} is the multiplication by the function $e^{-t|k|^2}$ in Fourier and we obtain

$$\left(e^{-tA}v_0\right)(x) = \frac{1}{(4\pi t)^{d/2}} \int_{\mathbb{R}^d} e^{-\frac{|x-y|^2}{4t}} v_0(y)\, dy.$$

This turns out to be C^∞ for all $t > 0$, for any (possibly quite rough function) v_0 in $L^2(\mathbb{R}^d)$.

4.7.3 Wave Equation

Among the many other interesting equations, we finally mention the wave equation

$$\begin{cases} \partial_{tt} v(t) = -A\, v(t) \\ v(0) = v_0 \\ \partial_t v(0) = v_1 \end{cases} \tag{4.38}$$

whose weak formulation is expressed in the form

$$\begin{cases} \dfrac{d^2}{dt^2} \langle w, v(t) \rangle = -\langle Aw, v(t) \rangle \\[2mm] \lim_{t \to 0^+} \langle w, v(t) \rangle = \langle w, v_0 \rangle \\[2mm] \lim_{t \to 0^+} \dfrac{d}{dt} \langle w, v(t) \rangle = \langle w, v_1 \rangle, \end{cases}$$

for all $w \in D(A)$. Now the function $f(t) = \langle w, v(t) \rangle$ has two integrable derivatives on $(0, T)$, so it is in $C^1([0, T], \mathbb{C})$. Like for Schrödinger's equation, the wave equation is reversible and does not make a big difference between positive and negative times.

The reader can show the following theorem as an exercise.

Theorem 4.34 (Wave Equation) *Let $(A, D(A))$ be a non-negative self-adjoint operator, that is, with spectrum $\sigma(A) \subset [0, +\infty)$. Then for all $v_0, v_1 \in \mathfrak{H}$ and $T > 0$, the wave Equation (4.38) admits a **unique weak solution** $v \in L^1((0, T), \mathfrak{H})$, given by*

$$v(t) = \cos(t\sqrt{A})v_0 + \frac{\sin(t\sqrt{A})}{\sqrt{A}} v_1$$

(with the convention $\sin(xt)/x = t$ *for* $x = 0$*). This solution is in fact defined on all* $\mathbb{R}$ *and verifies* $v \in C^0(\mathbb{R}, \mathfrak{H})$*, with the estimate*

$$\|v(t)\| \leq \|v_0\| + |t| \, \|v_1\|$$

for all $t \in \mathbb{R}$*. If in addition* $v_0 \in D(A)$ *and* $v_1 \in Q(A)$*, then* $v \in C^2(\mathbb{R}, \mathfrak{H}) \cap C^0(\mathbb{R}, D(A))$ *and solves (4.38) in the strong sense.*

4.8 Stone's Theorem and Symmetry Groups

4.8.1 Stone's Theorem

We saw in Sect. 4.7 that the unitary operator $U(t) = e^{-itA}$ is useful to solve the time-dependent Schrödinger equation. In this section we show that, conversely, any family of unitaries $(U(t))_{t \in \mathbb{R}}$ forming a strongly continuous one-parameter group, is necessarily in the form $U(t) = e^{-itA}$ with A a self-adjoint operator. This theorem, due to Stone [Sto30, Sto32b] and von Neumann [von32b], once again shows the importance of the notion of self-adjointness in quantum mechanics.

Theorem 4.35 (Stone) *Let* $(U(t))_{t \in \mathbb{R}}$ *be a family of bounded operators on a Hilbert space* $\mathfrak{H}$ *satisfying*

 (i) $U(t + s) = U(t)U(s) = U(s)U(t)$ *for all* $t, s \in \mathbb{R}$*,*
 (ii) $U(-t) = U(t)^*$ *for all* $t \in \mathbb{R}$*,*
 (iii) $U(0) = \mathbb{1}_{\mathfrak{H}}$*,*
 (iv) $t \mapsto U(t)$ *is strongly continuous at 0, that is, we have* $U(t)v \to v$ *in* $\mathfrak{H}$ *when* $t \to 0$*, for all fixed* $v \in \mathfrak{H}$*.*

Then there exists a unique self-adjoint operator $(A, D(A))$ *such that* $U(t) = e^{-itA}$ *for all* $t \in \mathbb{R}$*. The latter is called the* **infinitesimal generator** *of* $U(t)$*. Its domain is*

$$D(A) = \left\{ v \in \mathfrak{H} \text{ such that } t \mapsto U(t)v \text{ is differentiable at } t = 0 \right\} \qquad (4.39)$$

with

$$Av = i \frac{\mathrm{d}}{\mathrm{d}t} U(t)v \bigg|_{t=0}. \qquad (4.40)$$

Furthermore, if $\mathcal{D} \subset \mathfrak{H}$ *is a subspace such that*

- $\mathcal{D}$ *is dense in* $\mathfrak{H}$*,*
- $U(t)\mathcal{D} \subset \mathcal{D}$ *for all* $t \in \mathbb{R}$*,*
- $t \mapsto U(t)v$ *is differentiable at* $t = 0$ *for all* $v \in \mathcal{D}$ *(that is,* $\mathcal{D} \subset D(A)$*),*

then the restriction $A^{\min}$ *of* A *to the domain* $D(A^{\min}) = \mathcal{D}$ *is essentially self-adjoint, with* $\overline{A^{\min}} = A$.

The operators $U(t)$ in the statement are necessarily unitary, since $U(t)^*U(t) = U(-t)U(t) = U(t-t) = 1$. In particular, their operator norm is $\|U(t)\| = 1$ for all $t \in \mathbb{R}$.

Remark 4.36 The strong continuity at 0 in (iv) and the differentiability at 0 in (4.39) are equivalent to the same property at any $t_0 \in \mathbb{R}$, since $U(t + t_0) = U(t)U(t_0)$ by (i).

Proof We start by constructing a domain $\mathcal{D}$ on which $U(t)$ is differentiable, which will allow us to introduce $B := iU'(0)$ on $D(B) = \mathcal{D}$. Then we will show that B is symmetric and essentially self-adjoint, so the operator in the statement will be the closure $A = \overline{B}$. We introduce the space

$$\mathcal{D} := \left\{ \int_{\mathbb{R}} f(t)U(t)\, v\, dt, \qquad f \in C_c^\infty(\mathbb{R}), \ v \in \mathfrak{H} \right\}. \tag{4.41}$$

Here the integral is convergent in $\mathfrak{H}$ since $t \mapsto f(t)U(t)\,v$ is continuous according to (iv) and bounded by $|f(t)|\,\|v\|$ since $U(t)$ is unitary, which provides

$$\int_{\mathbb{R}} \|f(t)U(t)\,v\|\, dt = \|v\| \int_{\mathbb{R}} |f(t)|\, dt < \infty.$$

Let $\chi \in C_c^\infty(\mathbb{R})$ be supported in $[-1, 1]$ and such that $\int_{\mathbb{R}} \chi = 1$. As

$$\left\| n \int_{\mathbb{R}} \chi(nt)U(t)v\, dt - v \right\| \leq \int_{-1}^{1} \chi(t)\, \|U(t/n)v - v\|\, dt \xrightarrow[n \to \infty]{} 0$$

for all $v \in \mathfrak{H}$ by dominated convergence, we conclude that $\mathcal{D}$ is dense. Moreover,

$$U(s) \int_{\mathbb{R}} f(t)U(t)\, v\, dt = \int_{\mathbb{R}} f(t)U(t+s)\, v\, dt = \int_{\mathbb{R}} f(t-s)U(t)\, v\, dt \tag{4.42}$$

which demonstrates that $\mathcal{D}$ is invariant by the unitaries $U(s)$. Furthermore,

$$\frac{U(s) - 1}{s} \int_{\mathbb{R}} f(t)U(t)\, v\, dt = \int_{\mathbb{R}} \frac{f(t-s) - f(t)}{s} U(t)\, v\, dt \xrightarrow[s \to 0]{}$$
$$- \int_{\mathbb{R}} f'(t)U(t)\, v\, dt,$$

and thus $t \mapsto U(t)$ is differentiable on the dense domain $\mathcal{D}$ at $t = 0$.

In order to show at the same time the second part of the theorem, let us now consider a dense domain $\mathcal{D}$ (not necessarily the one introduced previously), invariant by all the $U(t)$ and on which U is differentiable at $t = 0$. Let us then show that $B = iU'(0)$ defined on $D(B) = \mathcal{D}$ is symmetric and essentially self-adjoint, with $U(t) = e^{-it\overline{B}}$. It is clear that B is a linear operator because the limit

$$\lim_{s \to 0} \frac{U(s) - 1}{s} v$$

depends linearly on v. Moreover, the differentiability at $t = 0$ implies that at all t, with

$$\frac{d}{dt} U(t)v = U(t)U'(0)v = -iU(t)Bv = -iBU(t)v \tag{4.43}$$

where we have used here the invariance of $\mathcal{D}$ by $U(t)$. Finally, we have

$$\left\langle w, \frac{U(s) - 1}{s} v \right\rangle = \left\langle \frac{U(s)^* - 1}{s} w, v \right\rangle = \left\langle \frac{U(-s) - 1}{s} w, v \right\rangle$$

which, by taking the limit $s \to 0$, shows that B is symmetric on $\mathcal{D}$. Let then B^* be the adjoint of B and suppose that $w \in \ker(B^* + i)$. According to (4.43), we have for all $t \in \mathbb{R}$

$$\frac{d}{dt} \langle w, U(t)v \rangle = \langle w, -iBU(t)v \rangle = \langle iB^*w, U(t)v \rangle = \langle w, U(t)v \rangle.$$

This is an ordinary differential equation whose solution is $\langle w, U(t)v \rangle = e^t \langle w, v \rangle$. But as $U(t)$ is unitary we have $|\langle w, U(t)v \rangle| \leq \|w\| \|v\|$ which therefore implies, by taking $t \to +\infty$, $\langle w, v \rangle = 0$. Thus, $w \in \mathcal{D}^\perp = \{0\}$ because $\mathcal{D}$ is supposed to be dense. The same argument for $-i$ and $t \to -\infty$ finally provides $\ker(B^* \pm i) = \operatorname{ran}(B \mp i)^\perp = \{0\}$. According to Exercise 2.29, this proves that B is essentially self-adjoint, with closure denoted $A = \overline{B}$.

It remains to show that $U(t) = e^{-itA}$. We have already seen that for all $v \in \mathcal{D}$, $v(t) := U(t)v$ is a solution of the equation

$$\begin{cases} i\partial_t v(t) = Av(t) \\ v(0) = v \end{cases}$$

(because A is an extension of B). We explained in Sect. 4.7.1 that this equation admits the unique solution $v(t) = e^{-itA}v$, so that $U(t)$ coincides with e^{-itA} on $\mathcal{D}$. By density they must coincide everywhere.

Finally, we discuss uniqueness. Suppose that $U(t) = e^{-itA'}$ for an operator A'. We have seen that $U(t)$ is differentiable at 0 on the domain $\mathcal{D}$ introduced in (4.41) at the beginning of the proof. However, by functional calculus we can see that $t \mapsto e^{-itA'}v$ is differentiable in $\mathfrak{H}$ if and only if $v \in D(A')$; in this case the derivative is $-iA'v$. This therefore shows that $\mathcal{D} \subset D(A')$, and that $A = A'$ on $\mathcal{D}$. Since A is essentially self-adjoint on $\mathcal{D}$ and A' is self-adjoint, we must have $A' = A$. □

In the following sections we will discuss the most classical examples of unitary groups, which all correspond to symmetry groups, and calculate their infinitesimal generator.

4.8.2   Group of Translations and Momentum

Let us work in $\mathfrak{H} = L^2(\mathbb{R})$ and consider the group of translations, which acts on functions by translating its variable as

$$(\mathcal{T}(t)v)(x) = v(x - t). \tag{4.44}$$

This group satisfies all the hypotheses (i)–(iv) of Theorem 4.35, so it can be written in the form of a Schrödinger propagator. Let us take $v \in C_c^\infty(\mathbb{R})$ and calculate the derivative explicitly

$$\frac{\mathrm{d}}{\mathrm{d}t}\mathcal{T}(t)v_{|t=0} = \frac{\mathrm{d}}{\mathrm{d}t}v(x - t)_{|t=0} = -v'(x).$$

Thus, on $\mathcal{D} = C_c^\infty(\mathbb{R})$ the generator coincides with the operator $P^{\min}$ defined on $D(P^{\min}) = C_c^\infty(\mathbb{R})$ by $P^{\min}v = -iv'$. We have already seen in Remark 2.14 and Theorem 2.36 that this operator is essentially self-adjoint and that the domain of its closure is $H^1(\mathbb{R})$.

Theorem 4.37 (Generator of Translations) *The translation operator (4.44) is*

$$\boxed{\mathcal{T}(t) = e^{-itP} \quad \text{where } P = -i\frac{\mathrm{d}}{\mathrm{d}x} \text{ on } D(P) = H^1(\mathbb{R}).}$$

Similarly, in dimension d we can translate a function by t in a direction $a \in \mathbb{R}^d$ by defining

$$(\mathcal{T}_a(t)v)(x) = v(x - ta) \tag{4.45}$$

and we get with the same reasoning that

$$\boxed{\mathcal{T}_a(t) = e^{-ita \cdot P}}$$

where the operator $a \cdot P = -i\sum_{j=1}^{d} a_j \partial_{x_j}$ is essentially self-adjoint on $C_c^\infty(\mathbb{R}^d)$. The domain of $a \cdot P$ is a Sobolev space in the direction a

$$D(a \cdot P) = \left\{ v \in L^2(\mathbb{R}^d) : a \cdot \nabla v \in L^2(\mathbb{R}^d) \right\}$$

which we have already encountered for P_j, obtained when $a = e_j$ (a vector of the canonical basis), in Theorem 2.17.

4.8.3 Rotation Group and Angular Momentum

Let us now work in dimension $d = 2$ in $L^2(\mathbb{R}^2)$ and consider the operation consisting of rotating the variable of functions around the origin. This amounts to the unitary transformation

$$(\mathcal{R}(\theta)v)(x_1, x_2) = v\big(\cos(\theta)x_1 + \sin(\theta)x_2, -\sin(\theta)x_1 + \cos(\theta)x_2\big) = v(R_{-\theta}x)$$
(4.46)

where

$$R_\theta = \begin{pmatrix} \cos\theta & -\sin\theta \\ \sin\theta & \cos\theta \end{pmatrix}$$

is the associated rotation matrix. Again, this family of operators satisfies the assumptions (i)–(iv) of Theorem 4.35 and it is possible to write $\mathcal{R}(\theta)$ as a Schrödinger propagator. To determine its generator, let us take $v \in C_c^\infty(\mathbb{R}^2)$ and calculate

$$\frac{d}{d\theta}\mathcal{R}(\theta)v_{|\theta=0} = (x_2\partial_{x_1} - x_1\partial_{x_2})v.$$

Theorem 4.38 (Generator of Rotations in 2D) *The symmetric operator*

$$L^{min} = -i(x_1\partial_{x_2} - x_2\partial_{x_1})$$

*is essentially self-adjoint on $D(L^{min}) = C_c^\infty(\mathbb{R}^2)$ and its closure $L = \overline{L^{min}}$, called the **angular momentum operator**, is the generator of rotations:*

$$\boxed{\mathcal{R}(\theta) = e^{-i\theta L}}$$

where $\mathcal{R}(\theta)$ was defined in (4.46). Furthermore, we have

$$D(L) = \left\{ v \in L^2(\mathbb{R}^2) \; : \; (x_1\partial_{x_2} - x_2\partial_{x_1})v \in L^2(\mathbb{R}^2) \right\}$$
(4.47)

where the term on the right is understood in the sense of distributions, and of course $L = -i(x_1\partial_{x_2} - x_2\partial_{x_1})$ on $D(L)$. In addition, $\sigma(L) = \mathbb{Z}$ where each eigenvalue is of infinite multiplicity.

Proof The operator defined on the domain $D(L)$ in (4.47) is closed and it is an extension of L^{min}. To see that it is the closure of L^{min}, we need to show that for every $v \in D(L)$ we can find a sequence $v_n \in C_c^\infty(\mathbb{R}^2)$ such that $v_n \to v$ and $L^{min} v_n = L v_n \to Lv$. This can be done by truncating and convolving with a regularizing kernel. More precisely, we can take

$$v_n(x) = n^2 \chi(x/n) \int_{\mathbb{R}^2} \chi(n(x-y)) v(y) \, dy = \chi(x/n)(\chi_n * v)(x) \tag{4.48}$$

where χ is a radial function of $C_c^\infty(\mathbb{R}^2)$ that equals 1 in a neighborhood of the origin and such that $\int_{\mathbb{R}^2} \chi = 1$. We have also introduced $\chi_n(x) = n^2 \chi(nx)$. We already know that $\chi_n * v$ is a C^∞ function that converges to v in $L^2(\mathbb{R}^2)$, and as $\chi(x/n) \to 1$ almost everywhere, we have $v_n \to v$ in $L^2(\mathbb{R}^2)$. Then, we notice that $L\chi = 0$ because χ is radial, which implies after a small calculation that

$$(L v_n)(x) = n^2 \chi(x/n) \int_{\mathbb{R}^2} \chi(n(x-y))(Lv)(y) \, dy$$

which converges to Lv for the same reason. Thus the closure of L^{min} is indeed the operator L introduced in the statement. By Stone's Theorem 4.35, we know that L is self-adjoint.

It remains to show that $\sigma(L) = \mathbb{Z}$. For this we note that $\mathcal{R}(2\pi) = \mathbb{1}_{\mathfrak{H}} = e^{-i2\pi L}$. By functional calculus this proves that the spectrum of L is contained in $\mathbb{Z}$ (after diagonalization, we have $e^{-2i\pi s} = 1$ for μ-almost every $s \in \sigma(L)$). In fact 0 is an eigenvalue whose eigenspace is of infinite dimension, because it contains all radial functions in $D(L)$, that is, invariant by rotations. We can also easily exhibit eigenfunctions for the other eigenvalues, for example

$$(x_1 + i x_2)^n \chi(x), \qquad n \in \mathbb{Z} \tag{4.49}$$

with $\chi \in C_c^\infty(\mathbb{R}^2 \setminus \{0\})$ radial. $\qquad\qquad\square$

It is common to write $L = x \wedge P$ where it is understood here that $X \wedge Y = \det(XY) = x_1 y_2 - x_2 y_1$. Another frequently encountered writing is $L = z\partial_z - \bar{z}\partial_{\bar{z}}$ where $z = x_1 + i x_2$, $\partial_z = \partial_{x_1} - i\partial_{x_2}$ and $\partial_{\bar{z}} = \partial_{x_1} + i\partial_{x_2}$.

Our study of the angular momentum operator can be generalized to dimension $d = 3$. By similar reasoning we find that the rotations $\mathcal{R}_\omega(\theta)$ of angle θ around an axis $\omega \in \mathbb{S}^2$ (a vector of the unit sphere) are given by

$$\mathcal{R}_\omega(\theta) = e^{-i\theta\omega \cdot L}, \qquad \omega \cdot L = \sum_{j=1}^{3} \omega_j L_j$$

where the three operators L_j correspond to rotations around the vectors of the canonical basis:

$$L = x \wedge (-i\nabla) = -i \begin{pmatrix} x_2 \partial_{x_3} - x_3 \partial_{x_2} \\ x_3 \partial_{x_1} - x_1 \partial_{x_3} \\ x_1 \partial_{x_2} - x_2 \partial_{x_1} \end{pmatrix} = \begin{pmatrix} L_1 \\ L_2 \\ L_3 \end{pmatrix}.$$

Each of the L_j is essentially self-adjoint on $C_c^\infty(\mathbb{R}^3)$ and self-adjoint on

$$D(L_j) = \left\{ v \in L^2(\mathbb{R}^3) \; : \; L_j v \in L^2(\mathbb{R}^3) \right\}.$$

As in dimension $d = 2$, we have $\sigma(L_j) = \mathbb{Z}$. An important operator is the square of the total angular momentum $|L|^2 := L_1^2 + L_2^2 + L_3^2$ which is also essentially self-adjoint on $C_c^\infty(\mathbb{R}^3)$. It is possible to show that

$$\sigma(|L|^2) = \left\{ \ell(\ell+1), \quad \ell \in \mathbb{N} \cup \{0\} \right\}.$$

This plays a particular role when studying Schrödinger operators with a radial potential. Indeed, by writing the Laplacian operator in spherical coordinates we find

$$-\Delta + V(|x|) = -\frac{1}{r^2}\frac{\partial}{\partial r} r^2 \frac{\partial}{\partial r} + \frac{|L|^2}{r^2} + V(r), \qquad r = |x|$$

which for example allows one to calculate the spectrum of the hydrogen atom explicitly. See [Tes09, Chap. 8 & 10] for more details on this subject.

4.8.4 Group of Dilations and Its Generator

Let us now consider another group acting on $L^2(\mathbb{R}^d)$ (in any dimension $d \geq 1$), consisting of dilating a function. As the dilation factor must be a positive number, it is common to write it in the form e^t, which reveals the additive group structure. Let us therefore define the family of unitaries

$$\left(\mathcal{D}(t)v\right)(x) = e^{\frac{-dt}{2}} v(e^{-t}x) \tag{4.50}$$

which satisfies properties (i)–(iv) of Theorem 4.35. In order to determine the associated generator, we take $v \in C_c^\infty(\mathbb{R}^d)$ and calculate

$$\frac{\mathrm{d}}{\mathrm{d}t}\mathcal{D}(t)v_{|t=0} = -\frac{d}{2}v(x) - x \cdot \nabla v(x) = -\frac{1}{2}(\nabla \cdot x + x \cdot \nabla)v(x).$$

Theorem 4.39 (Generator of Dilations in $\mathbb{R}^d$) *The symmetric operator*

$$A^{\min} = -\frac{i}{2}(\nabla \cdot x + x \cdot \nabla)$$

is essentially self-adjoint on $D(A^{\min}) = C_c^\infty(\mathbb{R}^d)$ and its closure $A = \overline{A^{\min}}$ is the generator of dilations:

$$\boxed{\mathcal{D}(t) = e^{-itA}}$$

where $\mathcal{D}(t)$ was defined in (4.50). Furthermore, we have

$$D(A) = \left\{ v \in L^2(\mathbb{R}^d) \; : \; \nabla \cdot x \, v \in L^2(\mathbb{R}^d) \right\}$$

$$= \left\{ v \in L^2(\mathbb{R}^d) \; : \; \nabla \cdot x \, v \text{ and } x \cdot \nabla v \in L^2(\mathbb{R}^d) \right\} \tag{4.51}$$

and of course $A = -i(\nabla \cdot x + x \cdot \nabla)/2$ on $D(A)$. In addition, $\sigma(A) = \mathbb{R}$ and contains no eigenvalues.

Proof The proof is essentially the same as that of Theorem 4.38. For the closure, we take the same function v_n as in (4.48) (with the factor n^d instead of n^2), but there is less cancellation because we do not have $A\chi = 0$ (0 is not an eigenvalue of A). A calculation shows that

$$(x \cdot \nabla)v_n = \xi\left(\frac{\cdot}{n}\right)(\chi_n * v) + \chi\left(\frac{\cdot}{n}\right)\eta_n * v + \chi\left(\frac{\cdot}{n}\right)\chi_n * (x \cdot \nabla v) \tag{4.52}$$

where $\xi = x \cdot \nabla\chi$ and $\eta_n = n^d(\nabla \cdot x\chi)(n\cdot)$. It is useful to choose a function χ that is constant equal to 1 in a small neighborhood of the origin. In this case the function $\xi = x \cdot \nabla\chi$ vanishes in a neighborhood of 0 and thus $\xi(x/n) \to 0$ everywhere. Furthermore, $\chi_n * v \to v$ strongly in $L^2(\mathbb{R}^d)$ and we deduce by dominated convergence that the first term of (4.52) tends to 0. The second term also tends to 0, this time because

$$\eta_n \rightharpoonup \left(\int_{\mathbb{R}^d} \nabla \cdot (x\chi)\right) \delta_0 = 0$$

in the sense of measures. Finally, the last term of (4.52) converges as desired to $x \cdot \nabla v$ in $L^2(\mathbb{R}^d)$.

To determine the spectrum we notice that the homogeneous functions $|x|^s$ are formal eigenfunctions of A, but none is in $L^2(\mathbb{R}^d)$. More precisely, we have

$$-\frac{i}{2}(\nabla \cdot x + x \cdot \nabla)|x|^s = \left(-i\frac{d}{2} - is\right)|x|^s,$$

in the sense of distributions on $\mathbb{R}^d \setminus \{0\}$. This suggests choosing s in the form $s = -d/2 + i\lambda$. We can regularize the corresponding function at 0 and at infinity by setting, for example,

$$u_\varepsilon = |x|^{-\frac{d}{2}+i\lambda} \frac{|x|^\varepsilon}{1+|x|^{2\varepsilon}}$$

with $\varepsilon > 0$, so that $u_\varepsilon \in D(A)$. We then find

$$i(A-\lambda)u_\varepsilon(x) = |x|^{-\frac{d}{2}+i\lambda} \, x \cdot \nabla \frac{|x|^\varepsilon}{1+|x|^{2\varepsilon}} = |x|^{-\frac{d}{2}+i\lambda}\varepsilon \frac{|x|^\varepsilon}{1+|x|^{2\varepsilon}} \left(1 - 2\frac{|x|^{2\varepsilon}}{1+|x|^{2\varepsilon}}\right)$$

and so $\left|(A-\lambda)u_\varepsilon(x)\right| \le \varepsilon |u_\varepsilon(x)|$. We obtain $\|(A-\lambda)u_\varepsilon\| \le \varepsilon \|u_\varepsilon\|$, which shows that $u_\varepsilon/\|u_\varepsilon\|$ is a Weyl sequence, and finally that $\lambda \in \sigma(A)$ for all $\lambda \in \mathbb{R}$, by Theorem 2.30.

Finally, suppose that $u \in \ker(A - \lambda)$ for a $\lambda \in \mathbb{R}$. Then we have $\mathcal{D}(t)u = e^{-itA}u = e^{-it\lambda}u$, which means that $u(x) = e^{-t(d/2-i\lambda)}u(e^{-t}x)$ for all t, almost everywhere in x. This is impossible for a non-zero function of $L^2(\mathbb{R}^d)$. For example, for $t = -\log(2)$ we get $|u(x)|^2 = 2^d|u(2x)|^2$, which implies after change of variable that the integrals of $|u|^2$ over the annuli $\{2^k \le |x| < 2^{k+1}\}$ are all equal for $k \in \mathbb{Z}$. As the sum must be equal to $\int_{\mathbb{R}^d} |u|^2$ which is finite, these integrals are all zero and therefore $u = 0$ almost everywhere. Thus, A has no eigenvalues. □

The operator A appears in the proof of the Virial identity, already encountered in (1.33) for the hydrogen atom.

Theorem 4.40 (Virial/Pohožaev Identity) *Let $V \in L^p(\mathbb{R}^d) + L^\infty(\mathbb{R}^d)$ be a real-valued function, with p satisfying*

$$\begin{cases} p = 2 & \text{if } d \in \{1, 2, 3\}, \\ p > 2 & \text{if } d = 4, \\ p = \frac{d}{2} & \text{if } d \ge 5. \end{cases} \tag{4.53}$$

Let u be a function of $H^2(\mathbb{R}^d)$ such that

$$-\Delta u(x) + V(x)u(x) = \lambda u(x). \tag{4.54}$$

If in addition $x \cdot \nabla V(x) \in L^p(\mathbb{R}^d) + L^\infty(\mathbb{R}^d)$ then we have

$$\int_{\mathbb{R}^d} |\nabla u(x)|^2 \, dx = \frac{1}{2} \int_{\mathbb{R}^d} |u(x)|^2 x \cdot \nabla V(x) \, dx \tag{4.55}$$

which provides in particular the relation

$$\lambda = \int_{\mathbb{R}^d} |u(x)|^2 \left(V(x) + \frac{1}{2} x \cdot \nabla V(x) \right) dx.$$ (4.56)

If $V(x) + \frac{1}{2} x \cdot \nabla V(x) < 0$ (for example for $V(x) = -|x|^{-s}$ with $0 < s < 2$) we see that $-\Delta + V$ cannot have a non-negative eigenvalue. There are many improvements of this result, see [RS78, Sec. XIII.13].

Proof The idea is to multiply the equation by $\overline{Au}$ where A is the generator of dilations, then to integrate by parts. Unfortunately, we do not know that u is in the domain of A. Taking advantage of the fact that $u \in H^2(\mathbb{R}^d)$, we will rather multiply by the regularization of $\overline{Au}$ in the form

$$i \frac{d}{2} \overline{u(x)} + i \frac{x}{1 + \varepsilon |x|^2} \cdot \overline{\nabla u(x)} \in H^1(\mathbb{R}^d),$$

which gives

$$\frac{d}{2} \left(\int_{\mathbb{R}^d} |\nabla u(x)|^2 dx + \int_{\mathbb{R}^d} (V(x) - \lambda)|u(x)|^2 dx \right)$$

$$= \Re \int_{\mathbb{R}^d} \Delta u(x) \, \overline{\nabla u(x)} \cdot \frac{x}{1 + \varepsilon |x|^2} dx - \Re \int_{\mathbb{R}^d} \frac{(V(x) - \lambda) x}{1 + \varepsilon |x|^2} \cdot \overline{\nabla u(x)} u(x) dx.$$ (4.57)

Noting that $\Re(u \nabla \overline{u}) = \nabla |u|^2 / 2$ and using the assumption on $x \cdot \nabla V$, we can integrate by parts the second integral, which gives

$$\Re \int_{\mathbb{R}^d} \frac{(V(x) - \lambda) x}{1 + \varepsilon |x|^2} \cdot \overline{\nabla u(x)} u(x) dx$$

$$= -\frac{1}{2} \int_{\mathbb{R}^d} |u(x)|^2 \mathrm{div} \frac{(V(x) - \lambda) x}{1 + \varepsilon |x|^2} dx$$

$$= -\frac{d}{2} \int_{\mathbb{R}^d} |u(x)|^2 \frac{V(x) - \lambda}{1 + \varepsilon |x|^2} dx - \frac{1}{2} \int_{\mathbb{R}^d} |u(x)|^2 x \cdot \nabla \frac{V(x) - \lambda}{1 + \varepsilon |x|^2} dx$$

$$= -\frac{d}{2} \int_{\mathbb{R}^d} (V(x) - \lambda)|u(x)|^2 dx - \frac{1}{2} \int_{\mathbb{R}^d} |u(x)|^2 x \cdot \nabla V(x) dx + o(1)_{\varepsilon \to 0}.$$

Similarly, we have for the first term

$$\Re \int_{\mathbb{R}^d} \Delta u(x) \, \overline{\nabla u(x)} \cdot \frac{x}{1 + \varepsilon |x|^2} dx$$

$$= \sum_{i,j=1}^{d} \Re \int_{\mathbb{R}^d} \partial_{jj} u(x) \frac{x_i}{1 + \varepsilon |x|^2} \overline{\partial_i u(x)} dx$$

$$= - \sum_{i,j=1}^{d} \Re \int_{\mathbb{R}^d} \left(\frac{\delta_{ij} - 2\varepsilon x_i x_j (1 + \varepsilon |x|^2)^{-1}}{1 + \varepsilon |x|^2} \right) \overline{\partial_i u(x)} \partial_j u(x) \, dx$$

$$+ \frac{1}{2} \int_{\mathbb{R}^d} |\nabla u(x)|^2 \text{div} \frac{x}{1 + \varepsilon |x|^2} \, dx$$

$$= \left(-1 + \frac{d}{2} \right) \int_{\mathbb{R}^d} |\nabla u(x)|^2 \, dx + o(1)_{\varepsilon \to 0}.$$

In the second equality, we first integrated by parts one of the derivatives ∂_j, then we integrated by parts ∂_i for the term $\Re \left(\partial_j u \, \overline{\partial_{ij} u} \right) = \partial_i |\partial_j u|^2 / 2$, which provides the divergence of $x(1 + \varepsilon |x|^2)^{-1}$. By inserting into (4.57) and taking the limit $\varepsilon \to 0$ we arrive at (4.55). □

4.9 Commutators and Conserved Quantities

If A and B are two bounded self-adjoint operators, we can define their **commutator** by $[A, B] = AB - BA$. If this commutator vanishes, we say that A and B commute and we find that the observable B is constant along the trajectories of the Schrödinger flow generated by A, that is to say:

$$\frac{d}{dt} \left\langle e^{-itA} v, B e^{-itA} v \right\rangle = \left\langle -i A e^{-itA} v, B e^{-itA} v \right\rangle + \left\langle e^{-itA} v, -i B A e^{-itA} v \right\rangle$$

$$= i \left\langle e^{-itA} v, [A, B] e^{-itA} v \right\rangle = 0$$

for all $v \in \mathfrak{H}$. In other words, we have $e^{itA} B e^{-itA} = B$, or equivalently $[e^{itA}, B] = 0$. This being valid for all t, we easily conclude that $[f(A), B] = 0$ for any bounded function f. Reasoning in the same way for B, we arrive at $[f(A), g(B)] = 0$.

The generalization of this argument to unbounded operators is not obvious due to the difficulty in defining the commutator $[A, B]$ on a good domain. One might think to ask for example that $[A, B]$ makes sense and vanishes on a dense subspace $\mathcal{D}$ of $\mathfrak{H}$, but this notion does not have the expected properties [RS72, Sec. VIII.5]. Here is a result that clarifies how to approach this question differently.

Theorem 4.41 (Commutator) *Let A and B be two self-adjoint operators on their respective domain $D(A)$, $D(B) \subset \mathfrak{H}$. The following assertions are equivalent:*

(i) $[f(A), g(B)] = 0$ for all Borel functions f, g bounded on a neighborhood of $\sigma(A) \cup \sigma(B)$;

(ii) $[e^{itA}, e^{isB}] = 0$ for all $t, s \in \mathbb{R}$;

(iii) $[(A - z)^{-1}, (B - z')^{-1}] = 0$ for all $z, z' \in \mathbb{C} \setminus \mathbb{R}$;

(iv) $[\mathbb{1}_I(A), \mathbb{1}_{I'}(B)] = 0$ for all intervals $I, I' \subset \mathbb{R}$;

(v) $D(A)$ is stable by e^{itB} for all $t \in \mathbb{R}$ and we have $e^{itB}A = Ae^{itB}$ on $D(A)$;

(vi) $D(B)$ is stable by e^{itA} for all $t \in \mathbb{R}$ and we have $e^{itA}B = Be^{itA}$ on $D(B)$.

If these equivalent conditions are met, we say that A and B commute.

In particular, if A and B commute and $v_0 \in D(B)$, we conclude that $v(t) = e^{-itA}v_0$ belongs to $D(B)$ for all t and that the average value of the observable B is constant over time: $\langle v(t), Bv(t) \rangle = \langle v_0, e^{itA}Be^{-itA}v_0 \rangle = \langle v_0, Bv_0 \rangle$.

Proof We leave as an exercise the equivalence between (i)–(iv) which is shown as for the functional calculus in Sect. 4.4. We will only prove that, for example, (vi) is equivalent to (i). If (i) is true, then e^{itA} commutes with $B(1 + \varepsilon B^2)^{-1} = f(B)$ where $f(x) = x/(1 + \varepsilon x^2)$ is bounded, so

$$\left\| B(1 + \varepsilon B^2)^{-1}e^{itA}v \right\| = \left\| e^{itA}B(1 + \varepsilon B^2)^{-1}v \right\| = \left\| B(1 + \varepsilon B^2)^{-1}v \right\|$$

for all $t \in \mathbb{R}$. As we have by the spectral theorem and functional calculus

$$w \in D(B) \iff \limsup_{\varepsilon \to 0^+} \left\| B(1 + \varepsilon B^2)^{-1}w \right\| < \infty$$

with in this case $\lim_{\varepsilon \to 0} B(1 + \varepsilon B^2)^{-1}v = Bv$, we conclude that $e^{itA}v \in D(B)$ for all $v \in D(B)$, and that $Be^{itA}v = e^{itA}Bv$.

Conversely, if $Be^{itA} = e^{itA}B$ on $D(B)$, we deduce that $e^{-itA}Be^{itA} = B$ and thus $e^{-itA}(B - z)e^{itA} = B - z$ on $D(B)$ for all $t \in \mathbb{R}$ and all $z \in \mathbb{C} \setminus \mathbb{R}$. But

$$\underbrace{e^{-itA}(B - z)e^{itA}}_{=B-z}\, e^{-itA}(B - z)^{-1}e^{itA} = e^{-itA}(B - z)(B - z)^{-1}e^{itA} = 1\!\!1_{\mathfrak{H}}$$

which shows that $e^{-itA}(B - z)^{-1}e^{itA} = (B - z)^{-1}$ or, written differently,

$$\left[e^{-itA}, (B - z)^{-1} \right] = 0, \qquad \forall t \in \mathbb{R},\ \forall z \in \mathbb{C} \setminus \mathbb{R}.$$

By the same argument as for (i)–(iv), we conclude that $[f(A), g(B)] = 0$ for all f, g Borel bounded.   □

Example 4.42 Consider a Schrödinger operator in the form $H = -\Delta + V(x)$ where $V \in L^2(\mathbb{R}^3) + L^\infty(\mathbb{R}^3)$ is a real-valued function. Then H is self-adjoint on $H^2(\mathbb{R}^3)$ by Theorem 3.4. Furthermore, we have $e^{-ia\cdot L}H^2(\mathbb{R}^3) \subset H^2(\mathbb{R}^3)$ where L is the angular momentum operator, for all $a \in \mathbb{R}^3$. This is because $e^{-ia\cdot L}$ just consists of applying a rotation of axis $a/|a|$ and angle $|a|$ on the functions, as we saw in Sect. 4.8.3. If V is a radial function, which means that $V(Rx) = V(x)$ for any rotation $R \in SO(3)$, we then have

$$e^{ia\cdot L}He^{-ia\cdot L} = H$$

for all $a \in \mathbb{R}^3$, on $D(H) = H^2(\mathbb{R}^3)$. We conclude that H commutes with $a \cdot L$ for all $a \in \mathbb{R}^3$. In particular, we have by Theorem 4.41

$$e^{itH} L_j e^{-itH} = L_j, \qquad \forall t \in \mathbb{R}$$

and the angular momentum is therefore preserved along the trajectories.

When two self-adjoint operators commute in the sense of Theorem 4.41, it is possible to "diagonalize them simultaneously". The precise statement goes as follows.

Theorem 4.43 (Simultaneous Diagonalization of Two Commuting Operators)
Let A and B be two self-adjoint operators on their respective domain $D(A), D(B) \subset \mathfrak{H}$. We assume that they commute in the sense of Theorem 4.41. Then there exists $d \geq 1$, a Borel set $\Omega \subset \mathbb{R}^d$, a locally finite Borel measure μ on Ω, two real-valued locally bounded functions $a, b \in L^\infty_{\text{loc}}(\Omega, \mathrm{d}\mu)$, and an isomorphism $U : \mathfrak{H} \to L^2(\Omega, \mathrm{d}\mu)$ such that

$$\boxed{UAU^{-1} = M_a, \qquad UBU^{-1} = M_b.}$$

More precisely, it is possible to take $d = 3$, $\Omega = \sigma(A) \times \sigma(B) \times \mathbb{N} \subset \mathbb{R}^3$, μ a finite measure on Ω and

$$a(s_1, s_2, n) = s_1, \qquad b(s_1, s_2, n) = s_2.$$

The support of the cylindrical projection $\tilde{\mu} = \sum_n \mu(\cdot, \cdot, n)$ on $\mathbb{R}^2$ is called the **joint spectrum** of the pair (A, B). It is a closed subset of $\sigma(A) \times \sigma(B) \subset \mathbb{R}^2$.

The result can obviously be generalized to any number of commuting operators.

Proof We only sketch the proof which follows a similar reasoning as the one we used for the spectral theorem of one operator in Sect. 4.4. More details can for instance be read in [Sim15, Sec. 5.6]. The first step is to define a functional calculus for the pair (A, B), in the sense that we associate a bounded operator $F(A, B)$ to any bounded function $F : \mathbb{R}^2 \to \mathbb{C}$ (continuous will be enough, similarly as in Theorem 4.11). In the case of a finite sum of tensor products, $F(x, y) = \sum_{i=1}^N f_i(x)g_i(y)$ with f_i, g_i bounded functions, we simply let $F(A, B) = \sum_{i=1}^N f_i(A)g_i(B)$. It is important that the operators commute, so that we do not have to worry about the order in which we place the operators $f_i(A)$ and $g_i(B)$, by Theorem 4.41. The goal is to extend the definition to any function F. We give ourselves an $F \in C_0^0(\mathbb{R}^2, \mathbb{C})$ and approximate it by step functions in the form

$$F_n(x, y) = \sum_{m, \ell \in \mathbb{Z}} F(\tau_m, \tau_\ell) \, \mathbb{1}_{I_m}(x) \, \mathbb{1}_{I_\ell}(y)$$

where $I_m = [\tau_m, \tau_{m+1})$ and $\tau_m = 2^{-n}m$. This is not a continuous function, but it nevertheless converges to F uniformly. Then we define as before

$$F_n(A, B) = \sum_{m,\ell \in \mathbb{Z}} F(\tau_m, \tau_\ell)\, \mathbb{1}_{I_m}(A)\, \mathbb{1}_{I_\ell}(B).$$

Using that $\mathbb{1}_{I_m}(A)\mathbb{1}_{I_\ell}(A) = \mathbb{1}_{I_m \cap I_\ell}(A) = \delta_{m,\ell}\mathbb{1}_{I_m}(A)$ by (iv) in Theorem 4.16 and that the spectral projections of A and B commute by Theorem 4.41, we find

$$F_n(A, B)^* F_n(A, B) = \sum_{m,\ell \in \mathbb{Z}} |F(\tau_m, \tau_\ell)|^2 \mathbb{1}_{I_m}(A)\, \mathbb{1}_{I_\ell}(B) = |F_n|^2(A, B).$$

This implies

$$\|F_n(A, B)v\|^2 = \langle v, |F_n|^2(A, B)v\rangle \le \max_{m,\ell} |F(\tau_m, \tau_\ell)|^2\, \|v\|^2 \le \max_{\mathbb{R}^2} |F|^2\, \|v\|^2.$$

In the first inequality we used that all the terms are non-negative since

$$\langle v, \mathbb{1}_{I_m}(A)\, \mathbb{1}_{I_\ell}(B)v\rangle = \langle \mathbb{1}_{I_\ell}(B)v, \mathbb{1}_{I_m}(A)\, \mathbb{1}_{I_\ell}(B)v\rangle \ge 0,$$

and that $\sum_m \mathbb{1}_{I_m}(A) = \sum_\ell \mathbb{1}_{I_\ell}(B) = 1$. Thus we have $\|F_n(A, B)\| \le \|F\|_\infty$. A similar estimate proves that $F_n(A, B)$ is a Cauchy sequence, hence converges in norm to a limit, which is $F(A, B)$ by definition. The so-defined functional calculus satisfies similar properties as in Theorem 4.11. It is a continuous morphism of C^*-algebras with $F(A, B) = f(A)g(B)$ whenever $F(x, y) = f(x)g(y)$, for any two continuous functions $f, g \in C_0^0(\mathbb{R}, \mathbb{C})$. The rest of the argument is then exactly like in Sect. 4.4 and we leave the details as an exercise. Namely, for any fixed non-zero $v \in \mathfrak{H}$, we obtain the existence of a unique finite Borel measure μ_v on $\mathbb{R}^2$ such that $\langle v, F(A, B)v\rangle = \int_{\mathbb{R}^2} F(x, y)\, d\mu_v(x, y)$. Next we define the cyclic subspace X_v similarly as in (4.14) with $F(A, B)$ instead of $f(A)$ and go on like in Sect. 4.4. $\square$

Complementary Exercises

Exercise 4.44 (Helffer-Sjöstrand Formula and Density of $\mathcal{A}$) Let $f \in C_c^\infty(\mathbb{R})$ be a C^∞ function with compact support. We introduce a "quasi-analytic" extension of f on $\mathbb{C}$ by

$$\tilde{f}(z) = \big(f(x) + iyf'(x)\big)\chi\left(\frac{y}{\sqrt{1+x^2}}\right)$$

where $z = x + iy$ and χ is a C^∞ function with compact support in $[-2, 2]$ that is constant equal to 1 on $[-1, 1]$. We note that $\tilde{f}_{|\mathbb{R}} = f$. Show the formula

$$f(x) = -\frac{1}{\pi} \int_\mathbb{C} \frac{\partial \tilde{f}(z)}{\partial \bar{z}} \frac{dz}{x - z}, \qquad \text{where} \qquad \frac{\partial \tilde{f}(z)}{\partial \bar{z}} := \frac{\partial \tilde{f}(z)}{\partial x} + i \frac{\partial \tilde{f}(z)}{\partial y}.$$

(4.58)

Deduce that $\mathrm{span}\{(x - z)^{-1}, \ z \in \mathbb{C} \setminus \mathbb{R}\}$ is a dense subspace in $C_0^0(\mathbb{R})$. For a proof of Theorem 4.11 based on Formula (4.58), see [Dav95].

Exercise 4.45 (Invariant Spaces) Let A be a self-adjoint operator on its domain $D(A) \subset \mathfrak{H}$. We say that a closed subspace $\mathcal{V} \subset \mathfrak{H}$ is A-**invariant** if $(A - z)^{-1}\mathcal{V} \subset \mathcal{V}$ for all $z \in \mathbb{C} \setminus \mathbb{R}$.

1. Show then that $f(A)\mathcal{V} \subset \mathcal{V}$ for all $f \in C_0^0(\mathbb{R}, \mathbb{C})$.
2. Show that $\mathcal{V}^\perp$ is also A-invariant.
3. Show that $\mathcal{V}$ is A-invariant if and only if A commutes with the orthogonal projector $\Pi_\mathcal{V}$ on $\mathcal{V}$, in the sense of Sect. 4.9.

See Problem B.2 for more properties of invariant spaces.

Exercise 4.46 (Laplacian and Cyclic Vectors) Consider the Laplace operator $A = -d^2/dx^2$ which is self-adjoint on $D(A) = H^2(\mathbb{R}) \subset \mathfrak{H} = L^2(\mathbb{R})$. Using the fact that $L^2(\mathbb{R}) = L_{\text{even}}^2(\mathbb{R}) \oplus L_{\text{odd}}^2(\mathbb{R})$ (orthogonal sum of the space of even and odd functions), show that A admits no cyclic vector. However, show that there exist two vectors v_1 and $v_2 \in \mathfrak{H}$ such that $L^2(\mathbb{R}) = \mathcal{X}_{v_1} \oplus \mathcal{X}_{v_2}$, where $\mathcal{X}_v$ is the cyclic subspace defined in (4.14) in the proof of the spectral theorem. This shows that, in the representation of the Laplacian as a multiplication operator, we can use only two copies of $\sigma(A) = \mathbb{R}_+$ in $\mathbb{R}^2$. What about the Laplacian in higher dimensions?

Exercise 4.47 (Convergence(s) of Operators) We recall that a sequence of bounded operators (B_n) on $\mathfrak{H}$ converges to a bounded operator B *in norm* when $\|B_n - B\| \to 0$, *strongly* when $\|B_n v - Bv\| \to 0$ for all $v \in \mathfrak{H}$, and *weakly* when $B_n v \rightharpoonup Bv$ weakly in $\mathfrak{H}$ for all $v \in \mathfrak{H}$, that is $\langle w, B_n v \rangle \to \langle w, Bv \rangle$ for all $v, w \in \mathfrak{H}$. The Banach-Steinhaus theorem implies that if $B_n \to B$ strongly, then $\|B_n\|$ is bounded.

1. Let B_n be the operator of multiplication by the function $b_n(x) = 1/(1 + x^2/n)$ on the space $\mathfrak{H} = L^2(\mathbb{R})$. Study the convergence of B_n according to the three previous notions. Determine the spectrum of B_n and that of its limit B. What can we conclude?
2. Let $w \in \mathfrak{H}$ and $v_n \rightharpoonup 0$ be a sequence that converges weakly to 0 in $\mathfrak{H}$, with $\|v_n\| = 1$ for all n. We set $B_n h := \langle w, h \rangle v_n$. Show that B_n is bounded and study its convergence according to the three previous notions.

In the following, we consider a sequence $(A_n, D(A_n))$ of *self-adjoint* operators on $\mathfrak{H}$, as well as a self-adjoint operator $(A, D(A))$. We study the convergence of $f(A_n)$ to $f(A)$ for various bounded functions.

3. Show that for all $z, z' \in \mathbb{C} \setminus \mathbb{R}$, we have the relation

$$(A_n - z)^{-1} - (A - z)^{-1} = \frac{A_n - z'}{A_n - z} \left((A_n - z')^{-1} - (A - z')^{-1} \right) \frac{A - z'}{A - z}.$$
$$(4.59)$$

4. Show the equivalence of the three propositions:

   - $(A_n + i)^{-1}$ converges to $(A + i)^{-1}$ in norm;
   - $(A_n - z)^{-1}$ converges to $(A - z)^{-1}$ in norm for all $z \in \mathbb{C} \setminus \mathbb{R}$;
   - $f(A_n)$ converges to $f(A)$ in norm for all functions $f \in C_0^0(\mathbb{R})$ (continuous on $\mathbb{R}$ and tending to 0 at infinity).

   Then show the same equivalence if we replace "in norm" with "strongly".
5. Suppose that $(A_n + i)^{-1} \to (A + i)^{-1}$ strongly. Show that if $\lambda \in \sigma(A)$, then there exists a sequence $\lambda_n \in \sigma(A_n)$ such that $\lambda_n \to \lambda$. Can we hope for more?
6. We assume that $(A_n + i)^{-1} \to (A + i)^{-1}$ in norm. Show that if $[a, b] \subset \rho(A)$, then $[a, b] \subset \rho(A_n)$ for n large enough. What can we conclude from this?

Exercise 4.48 (RAGE Theorem) Let μ be a finite Borel measure on $\mathbb{R}$ and $A_\mu = \{x \in \mathbb{R} : \mu(\{x\}) > 0\}$ be the set of its atoms. We introduce its Fourier transform

$$\widehat{\mu}(t) := \frac{1}{\sqrt{2\pi}} \int_{\mathbb{R}} e^{-ixt} \, d\mu(x)$$

which we recall is a bounded continuous function on $\mathbb{R}$.

1. Show that

$$\frac{1}{T} \int_0^T |\widehat{\mu}(t)|^2 \, dt = \frac{1}{2T} \int_{-T}^T |\widehat{\mu}(t)|^2 \, dt = \iint_{\mathbb{R}^2} K_T(x - y) \, d\mu(x) \, d\mu(y)$$

   for a function K_T that is even, to be determined.
2. Show that for all $x \in \mathbb{R}$,

$$\lim_{T \to \infty} \int_{\mathbb{R}} K_T(x - y) \, d\mu(y) = \frac{\mu(\{x\})}{2\pi}.$$

3. Deduce a theorem of Wiener:

$$\lim_{T \to \infty} \frac{1}{T} \int_0^T |\widehat{\mu}(t)|^2 \, dt = \frac{1}{2\pi} \sum_{x \in A_\mu} \mu(\{x\})^2.$$

Now let A be a self-adjoint operator on its domain $D(A) \subset \mathfrak{H}$.

4. By applying the spectral theorem, show that the set of its eigenvalues is countable.

5. Show that there exists an orthonormal system $\{u_j\}_{j\geq 1}$ of eigenvectors of A, with $u_j \in D(A)$ and $Au_j = \lambda_j u_j$ for all j, such that $A_{|\text{span}(u_j, \, j\geq 1)^{\perp}}$ has no eigenvalues.
6. Let v be any vector in $\mathfrak{H}$. Show that for any operator R of finite rank

$$\lim_{T\to\infty} \frac{1}{T} \int_0^T \left\langle e^{-itA}v, \, Re^{-itA}v \right\rangle dt = \sum_{j\geq 1} |\langle v, u_j\rangle|^2 \langle u_j, Ru_j\rangle.$$

This result, due to Ruelle, Amrein, Georgescu and Enss [Rue69, AG74, Ens78], means that only the projection of a quantum state onto the point spectrum survives in the long time limit, on average. See also [AW15, Sec. 2.4].

Chapter 5
Spectrum of Self-adjoint Operators

In this chapter, we study more precisely the spectrum of self-adjoint operators.

5.1 Perturbation Theory

In this first section, we study how the spectrum of a self-adjoint operator A is modified when we add a small operator B in an appropriate sense. To simplify, we start with the case where the perturbation B is just bounded, before examining that of relatively bounded perturbations. These estimates on the spectrum will then allow us to determine the regularity of spectral projectors and eigenvalues with respect to the perturbation.

5.1.1 *Bounded Perturbations*

The following result indicates that the spectrum of $A + B$ is close to that of A when $\|B\|$ is small.

Lemma 5.1 (Bounded Perturbations) *Let A be a self-adjoint operator on its domain $D(A) \subset \mathfrak{H}$ and B a bounded self-adjoint operator. Then the spectrum of $C := A + B$ is close to that of A in the sense of Hausdorff distance:*

$$\sup_{\lambda \in \sigma(A)} d(\lambda, \sigma(C)) \leq \|B\|, \qquad \sup_{\lambda' \in \sigma(C)} d(\lambda', \sigma(A)) \leq \|B\|. \qquad (5.1)$$

Here $d(x, \sigma(A)) = \min_{y \in \sigma(A)} |x - y|$ is the distance from x to the spectrum of A. We recall that the operator $C = A + B$ is self-adjoint on $D(A)$, by the Rellich-Kato

M. Lewin, *Spectral Theory and Quantum Mechanics*, Universitext, https://doi.org/10.1007/978-3-031-66878-4_5

Theorem 3.1, since B is bounded. The estimate (5.1) means that for every $\lambda \in \sigma(A)$ there exists a $\lambda' \in \sigma(C)$ at a distance at most $\|B\|$ and vice versa. Thus, we have $\sigma(A + B) \subset \sigma(A) + [-\|B\|, \|B\|]$; each element of the spectrum is moved by at most $\|B\|$ under the action of the perturbation.

Proof Let $\lambda' \in \sigma(C)$ and suppose by contradiction that $[\lambda' - \|B\|, \lambda' + \|B\|]$ does not intersect $\sigma(A)$, that is, it is included in $\rho(A)$. Then there exists $d > \|B\|$ such that $[\lambda' - d, \lambda' + d] \subset \rho(A)$ because the resolvent set is open. We have

$$\left\| B(A - \lambda')^{-1} \right\| \leq \|B\| \left\| (A - \lambda')^{-1} \right\| = \frac{\|B\|}{d(\lambda', \sigma(A))} \leq \frac{\|B\|}{d} < 1$$

where we used Corollary 4.6 to calculate $\|(A - \lambda')^{-1}\|$. By writing

$$C - \lambda' = A - \lambda' + B = \left(1 + B(A - \lambda')^{-1}\right)(A - \lambda')$$

we see that $\lambda' \in \rho(C)$, a contradiction. Thus we have shown that $d(\lambda', \sigma(A)) \leq \|B\|$. The other inequality is obtained by exchanging the roles of A and C. □

5.1.2 Relatively Bounded Perturbations

The assumption that B is bounded does not cover all practical cases because we often encounter perturbations that are *relatively bounded* with respect to A, that is, such that $B(A + i)^{-1}$ is bounded. The following theorem contains an estimate of the same type as that of Lemma 5.1, but which is more complicated because of the dependence on the point a considered in the spectrum.

Theorem 5.2 (Relatively Bounded Perturbations) *Let A be a self-adjoint operator on its domain $D(A) \subset \mathfrak{H}$ and $a \in \rho(A) \cap \mathbb{R}$. Let B be a symmetric operator on $D(A)$ such that $B(A - a)^{-1}$ is a bounded operator satisfying*

$$\left\| B(A - a)^{-1} \right\| < 1. \tag{5.2}$$

Then the operator $C := A + B$ is self-adjoint on $D(A)$ and $a \notin \sigma(C)$. More precisely,

$$(a - \eta, a + \eta) \subset \rho(C) \quad \text{for} \quad \eta := d(a, \sigma(A))\left(1 - \left\| B(A - a)^{-1} \right\|\right). \tag{5.3}$$

Finally, the two spectra are locally *close in the sense of Hausdorff distance: for all*

$$d(a, \sigma(A)) \leq R < \frac{\eta}{\left\| B(A - a)^{-1} \right\|} \tag{5.4}$$

we have

$$\sup_{\lambda \in \sigma(A) \cap [a-R,a+R]} d\big(\lambda, \sigma(C)\big) \leq \frac{R^2 \, \big\| B(A-a)^{-1} \big\|}{\eta - R \, \big\| B(A-a)^{-1} \big\|},$$

$$\sup_{\lambda' \in \sigma(C) \cap [a-R,a+R]} d\big(\lambda', \sigma(A)\big) \leq \frac{R^2 \, \big\| B(A-a)^{-1} \big\|}{\eta - R \, \big\| B(A-a)^{-1} \big\|}. \tag{5.5}$$

The inequality (5.5) provides a *local* estimate of the distance between the spectra of A and $A + B$, on an interval $[a - R, a + R]$ of length $2R$ around a. The error on the right depends on the considered R and deteriorates as R increases. Because of the condition (5.4), no information is provided beyond $a \pm \eta / \| B(A-a)^{-1} \|$. Note that the second part of the theorem requires $\big\| B(A-a)^{-1} \big\| < 1/2$.

Proof The hypothesis (5.2) implies

$$\| Bv \| = \big\| B(A-a)^{-1}(A-a)v \big\| \leq \big\| B(A-a)^{-1} \big\| \Big(\| Av \| + |a| \, \| v \| \Big),$$

which shows that $A + B$ is self-adjoint on $D(A)$, by the Rellich-Kato Theorem 3.1. Moreover,

$$(A + B - a) = \Big(1 + B(A-a)^{-1} \Big)(A-a) \tag{5.6}$$

is invertible with a bounded inverse according to (5.2), so $a \in \rho(A + B)$. Then we notice that $(a - \eta, a + \eta) \subset \rho(A)$ because $\eta < d(a, \sigma(A))$ by definition. This allows us to write as before

$$(A + B - z) = \Big(1 + B(A-z)^{-1} \Big)(A-z) \tag{5.7}$$

for all $z \in (a - \eta, a + \eta)$. To show that $z \in \rho(A + B)$ it is therefore sufficient to prove that $\| B(A-z)^{-1} \| < 1$. As

$$\big\| B(A-z)^{-1} \big\| \leq \big\| B(A-a)^{-1} \big\| \left\| \frac{A-a}{A-z} \right\|$$

with, by Corollary 4.6,

$$\left\| \frac{A-a}{A-z} \right\| = \left\| 1 + \frac{z-a}{A-z} \right\| \leq 1 + \frac{|z-a|}{d(z, \sigma(A))},$$

we see that it is sufficient to show that

$$\big\| B(A-a)^{-1} \big\| \left(1 + \frac{|z-a|}{d(z, \sigma(A))} \right) < 1, \qquad \forall z \in (a - \eta, a + \eta). \tag{5.8}$$

For $b \in \sigma(A)$ we have

$$|b - z| \geq |b - a| - |a - z| \geq d(a, \sigma(A)) - \eta \geq d(a, \sigma(A)) \left\| B(A - a)^{-1} \right\|$$

which implies $d(z, \sigma(A)) \geq d(a, \sigma(A)) \left\| B(A - a)^{-1} \right\|$ and thus we obtain

$$\left\| B(A - a)^{-1} \right\| \left(1 + \frac{|z - a|}{d(z, \sigma(A))}\right)$$

$$< \left\| B(A - a)^{-1} \right\| \left(1 + \frac{\eta}{d(a, \sigma(A)) \left\| B(A - a)^{-1} \right\|}\right) = 1,$$

as desired. It remains to estimate the Hausdorff distance between the two spectra. The relation (5.6) means that

$$(A + B - a)^{-1} - (A - a)^{-1} = (A - a)^{-1} \left(\left(1 + B(A - a)^{-1}\right)^{-1} - 1\right)$$

so that

$$\left\| (A + B - a)^{-1} - (A - a)^{-1} \right\| \leq \frac{\left\| B(A - a)^{-1} \right\|}{d(a, \sigma(A)) \left(1 - \left\| B(A - a)^{-1} \right\|\right)} =: \delta.$$

$$(5.9)$$

The two resolvents are bounded self-adjoint operators, whose spectrum is

$$\sigma\left((A + B - a)^{-1}\right) = \overline{\left\{\frac{1}{\lambda - a}, \ \lambda \in \sigma(A + B)\right\}},$$

$$\sigma\left((A - a)^{-1}\right) = \overline{\left\{\frac{1}{\lambda - a}, \ \lambda \in \sigma(A)\right\}},$$

by Lemma 2.11 or by functional calculus. The closure only adds the point 0 if they are not bounded. Lemma 5.1 provides that the two spectra are at a Hausdorff distance smaller than δ. In particular, for every $\lambda \in \sigma(A)$ we have

$$\inf_{\lambda' \in \sigma(A + B)} \left| \frac{1}{\lambda - a} - \frac{1}{\lambda' - a} \right| \leq \delta.$$

If we add the assumption that $|\lambda - a| \leq R$ (such an R is necessarily at least equal to $d(a, \sigma(A))$), then an optimum λ' must satisfy

$$\frac{1}{|\lambda' - a|} \geq \frac{1}{|\lambda - a|} - \left| \frac{1}{\lambda - a} - \frac{1}{\lambda' - a} \right| \geq \frac{1 - R\delta}{R},$$

that is $|\lambda' - a| \le R/(1 - R\delta)$. As $|\lambda - \lambda'| \le |\lambda - a| |\lambda' - a| \delta$ we finally obtain

$$|\lambda - \lambda'| \le \frac{R^2 \delta}{1 - R\delta},$$

which concludes the proof. $\qquad\square$

5.1.3 Bounded Perturbations in the Sense of Quadratic Forms

We finally consider the case where A is a bounded-below operator and where B is a small perturbation in the sense of quadratic forms, but not in the sense of operators. We will see that the previous results remain all true, with similar estimates involving $\||A - a|^{-\frac{1}{2}} B |A - a|^{-\frac{1}{2}}\|$ everywhere instead of $\|B(A - a)^{-1}\|$. To simplify the argument, we will limit ourselves to the case where a is located below the spectrum of A. The main tool is then the comparison of operators using their quadratic form.

Definition 5.3 (Comparing Operators in Quadratic Form Sense) Let $(A, D(A))$ and $(C, D(C))$ be two bounded-below self-adjoint operators. We say that $A \le C$ when $Q(C) \subset Q(A)$ and $q_A(v) \le q_C(v)$ for all $v \in Q(C)$.

It is often useful in practice that only the domains of quadratic forms come into play, since we recall that $D(A)$ and $D(C)$ can be very different. The following result will be useful for deducing a comparison between resolvents from an estimate between quadratic forms.

Theorem 5.4 (Inverse and Inequalities Between Operators) *Let $(A, D(A))$ and $(C, D(C))$ be two coercive self-adjoint operators. Then we have $A \le C$ if and only if $C^{-1} \le A^{-1}$.*

We say that the inverse function $x \mapsto x^{-1}$ is **operator-monotone**. It is generally not true that every monotone function is operator-monotone [Bha97]. For example, $0 \le A \le C$ does indeed imply $\sqrt{A} \le \sqrt{C}$ (the function $x \mapsto \sqrt{x}$ is operator-monotone) whereas in general $A^2 \nleq C^2$ (the function $x \mapsto x^2$ is not operator-monotone), see the counter-example in Exercise 5.5 below.

Proof The inequality $q_A \le q_C$ implies that for all $z \in \mathfrak{H}$

$$-\langle z, A^{-1} z \rangle = \inf_{v \in Q(A)} \{q_A(v) - 2\Re\langle v, z \rangle\}$$

$$\le \inf_{v \in Q(C)} \{q_A(v) - 2\Re\langle v, z \rangle\}$$

$$\le \inf_{v \in Q(C)} \{q_C(v) - 2\Re\langle v, z \rangle\} = -\langle z, C^{-1} z \rangle$$

and therefore $q_{C^{-1}} \le q_{A^{-1}}$ on all $\mathfrak{H}$. We have here used Theorem 3.14 which specifies that the infimum is reached for $v = A^{-1}z$ (resp. $v = C^{-1}z$). We have also used the fact that $Q(C) \subset Q(A)$ so the infimum on $Q(A)$ is lower than that on $Q(C)$.

For the converse part of the statement, we note that, in a similar way,

$$\inf_{v \in \mathfrak{H}} \left\{ \left\langle v, A^{-1}v \right\rangle - 2\Re \langle v, z \rangle \right\} = \begin{cases} -\left\| A^{\frac{1}{2}}z \right\|^2 = -q_A(z) & \text{if } z \in Q(A), \\ -\infty & \text{otherwise.} \end{cases}$$

The first case is shown by completing the square as in the proof of Theorem 3.14 and the second with the help of the sequence $v_n = A\mathbb{1}(|A| \leq n)z$. The inequality $C^{-1} \leq A^{-1}$ on $\mathfrak{H}$ therefore immediately implies that $Q(C) \subset Q(A)$ and $q_A \leq q_C$.

$\square$

Exercise 5.5 (The Square Function Is Not Operator-Monotone) We work in $\mathfrak{H} = \mathbb{C}^2$ and consider the two matrices

$$A = \begin{pmatrix} 1 & 1 \\ 1 & 1 \end{pmatrix}, \qquad B = \begin{pmatrix} 2 & 1 \\ 1 & 1 \end{pmatrix}.$$

Show that $A \leq B$ but that $A^2 \nleq B^2$.

By comparing the resolvents we can obtain information about the entire spectrum, when two quadratic forms are close. An example of such a result is provided in the following statement.

Theorem 5.6 (Spectra for q_A Close to q_C) *Let $(A, D(A))$ and $(C, D(C))$ be two coercive self-adjoint operators such that*

$$(1 - \varepsilon)A \leq C \leq (1 + \varepsilon)A \tag{5.10}$$

for some $0 < \varepsilon < 1$. Then we have

$$\left\| A^{-1} - C^{-1} \right\| \leq \varepsilon \left\| C^{-1} \right\| \leq \frac{\varepsilon \left\| A^{-1} \right\|}{1 - \varepsilon}. \tag{5.11}$$

The spectra of A and C are locally close in the sense of the Hausdorff distance: for all $\|A^{-1}\|^{-1} \leq R < \eta/\varepsilon$ with $\eta := \frac{1-\varepsilon}{\|A^{-1}\|}$, we have

$$\sup_{\lambda \in \sigma(A) \cap [0, R]} d(\lambda, \sigma(C)) \leq \frac{\varepsilon R^2}{\eta - R\varepsilon}, \qquad \sup_{\lambda' \in \sigma(C) \cap [0, R]} d(\lambda', \sigma(A)) \leq \frac{\varepsilon R^2}{\eta - R\varepsilon}. \tag{5.12}$$

Proof Theorem 5.4 provides $(1 - \varepsilon)C^{-1} \leq A^{-1} \leq (1 + \varepsilon)C^{-1}$ so that

$$-\varepsilon \left\| C^{-1} \right\| \leq A^{-1} - C^{-1} \leq \varepsilon \left\| C^{-1} \right\|$$

and

$$(1 - \varepsilon) \left\| C^{-1} \right\| \leq \left\| A^{-1} \right\|.$$

This gives (5.6). We then apply Lemma 5.1 and argue as in the proof of Theorem 5.2.

$\square$

The estimate (5.12) is exactly the same as (5.5) in Theorem 5.2 with $a = 0$, but $\|BA^{-1}\|$ has been replaced by $\|A^{-1/2}BA^{-1/2}\|$. Recall that $B = C - A$ makes sense as a bounded operator from $Q(A)$ to $Q(A)'$ as discussed at the end of Sect. 4.6. The hypothesis (5.10) can then be rewritten as

$$\|A^{-\frac{1}{2}}BA^{-\frac{1}{2}}\| \leq \varepsilon.$$

The approach presented in this section relies on the fact that $a = 0$ is below the spectrum of A. This is the situation we will most often encounter for Schrödinger operators $-\Delta + V$. If we want to deal with the case of an $a \in \rho(A)$ in the middle of the spectrum (for example when V is periodic as in Chap. 7), we must argue differently. One solution is to write

$$A + B - a = \frac{A - a}{|A - a|} |A - a|^{\frac{1}{2}} \left(1 + \frac{|A - a|}{A - a} |A - a|^{-\frac{1}{2}} B |A - a|^{-\frac{1}{2}} \right) |A - a|^{\frac{1}{2}}.$$

$$(5.13)$$

By functional calculus, $\frac{|A-a|}{A-a}$ is a unitary operator, therefore of norm 1 and it can be seen that $a \in \rho(A + B)$ when

$$\left\| |A - a|^{-\frac{1}{2}} B |A - a|^{-\frac{1}{2}} \right\| < 1.$$

We can then argue as in the proof of Theorem 5.2, but we leave the details as an exercise.

5.1.4 Analyticity of Spectral Projections and Eigenvalues

The estimates on the spectrum will now allow us to study the regularity of the spectral projections $\mathbb{1}_{[a,b]}(A + B)$ with respect to the perturbation B, as long as a and b remain outside the spectrum. To simplify the statement we will rather fix a relatively bounded operator B, consider a perturbation in the form εB with ε small enough and then discuss the regularity with respect to the real parameter ε.

Recall that if B is a symmetric operator such that $B(A + i)^{-1}$ is bounded, then $B(A - a)^{-1}$ is bounded for all $a \in \rho(A) \cap \mathbb{R}$ with the estimate

$$\left\| B(A - a)^{-1} \right\| \leq \left\| B(A + i)^{-1} \right\| \left\| \frac{A + i}{A - a} \right\| \leq \left\| B(A + i)^{-1} \right\| \left(1 + \frac{|a| + 1}{\mathrm{d}(a, \sigma(A))} \right).$$

Theorem 5.7 (Regularity of Spectral Projectors) *Let A be a self-adjoint operator on its domain $D(A) \subset \mathfrak{H}$ and $a, b \in \rho(A) \cap \mathbb{R}$ with $a < b$. Let $B \neq 0$ be a symmetric operator on $D(A)$ such that $B(A + i)^{-1}$ is a bounded operator. We set*

$$\varepsilon_0 = \min \left(\frac{1}{\|B(A - a)^{-1}\|}, \frac{1}{\|B(A - b)^{-1}\|} \right). \tag{5.14}$$

Then:

(i) *a and b are in $\rho(A + \varepsilon B)$ for all $\varepsilon \in (-\varepsilon_0, \varepsilon_0)$.*

(ii) *The spectral projector*

$$P(\varepsilon) := \mathbb{1}_{[a,b]}(A + \varepsilon B)$$

is real analytic on $(-\varepsilon_0, \varepsilon_0)$, that is, given by a normally convergent power series $P(\varepsilon) = P(0) + \sum_{n \geq 1} \varepsilon^n b_n$ where the b_n's are bounded operators such that $\sum_{n \geq 1} |\varepsilon|^n \|b_n\| < \infty$ for all $\varepsilon \in (-\varepsilon_0, \varepsilon_0)$. In particular, $\varepsilon \mapsto P(\varepsilon)$ is C^∞ and of constant rank (finite or infinite) on $(-\varepsilon_0, \varepsilon_0)$.

(iii) *If A has a finite number of eigenvalues in the interval $[a, b]$ whose sum of multiplicities is $k < \infty$, then the same is true for $A + \varepsilon B$ for all $\varepsilon \in (-\varepsilon_0, \varepsilon_0)$.*

(iv) *If A has a unique simple eigenvalue in $[a, b]$ then $A + \varepsilon B$ also has a unique simple eigenvalue $\lambda(\varepsilon)$ which is a real analytic function on $(-\varepsilon_0, \varepsilon_0)$.*

If A is bounded-below and B is A-bounded in the sense of quadratic forms, we have the same result for the Friedrichs realization of $A + \varepsilon B$ by replacing ε_0 with

$$\varepsilon_0 = \min \left(\frac{1}{\left\| |A - a|^{-\frac{1}{2}} B |A - a|^{-\frac{1}{2}} \right\|}, \frac{1}{\left\| |A - b|^{-\frac{1}{2}} B |A - b|^{-\frac{1}{2}} \right\|} \right).$$

The theorem specifies that the spectrum located between a and b cannot change in nature when we insert the perturbation εB and ensure that nothing can cross a and b. For example, if $[a, b]$ contains only a finite number of eigenvalues of finite multiplicities, no eigenvalue can appear or disappear when ε varies. Moreover, if there is a unique eigenvalue, it is real analytic with respect to ε.

Proof Property (i) that the spectrum of $A + \varepsilon B$ does not contain a and b for all $\varepsilon \in (-\varepsilon_0, \varepsilon_0)$ follows immediately from Theorem 5.2. We can then express the projector $P(\varepsilon)$ using Cauchy's formula

$$P(\varepsilon) = -\frac{1}{2i\pi} \oint_{\mathscr{C}} (A + \varepsilon B - z)^{-1} \, dz,$$

by Theorem 4.19 from the previous chapter, with a closed loop $\mathscr{C}$ in the complex plane that crosses the real axis only at a and b and surrounds the interval $[a, b]$. We

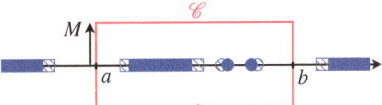

Fig. 5.1 Rectangle $\mathscr{C}$ used in the proof of Theorem 5.7 to study the spectral projector $P(\varepsilon) = \mathbb{1}_{[a,b]}(A + \varepsilon B)$ using Cauchy's formula. We take $|\varepsilon| < \varepsilon_0$ to ensure that the spectrum of $A + \varepsilon B$ (hatched in the figure) does not approach a and b. © Mathieu Lewin 2021. All rights reserved

take, for example, a rectangle whose sides are parallel to the real and imaginary axis and of height $2M$ as displayed in Fig. 5.1.

Next, we write

$$A + \varepsilon B - z = \left(1 + \varepsilon B(A - z)^{-1}\right)(A - z),$$

and seek to control the norm of the operator $B(A - z)^{-1}$ for $z \in \mathscr{C}$. When z belongs to the left vertical side of the rectangle $\mathscr{C}$, we write

$$\left\| B(A - z)^{-1} \right\| \leq \left\| B(A - a)^{-1} \right\| \left\| \frac{A - a}{A - z} \right\| \leq \left\| B(A - a)^{-1} \right\|.$$

Indeed, by functional calculus we have

$$\left\| \frac{A - a}{A - z} \right\| \leq \sup_{x \in \mathbb{R} \setminus \{0\}} \left| \frac{x}{x - i\tau} \right| = 1$$

for $z = a + i\tau$ with $\tau \in [-M, M]$. The same estimate is valid on the right vertical side of the rectangle, with b instead of a. When z belongs to a horizontal side and has a real part less than $(a + b)/2$, we rather use

$$\frac{A - a}{A - z} = \frac{A - \mathfrak{R}(z)}{A - z} + \frac{\mathfrak{R}(z) - a}{A - z}$$

which this time provides

$$\left\| \frac{A - a}{A - z} \right\| \leq 1 + \frac{b - a}{2M}.$$

We argue similarly with b instead of a, when $\mathfrak{R}(z) \geq (a + b)/2$. According to the definition (5.14) of ε_0, we thus obtain

$$\left\| \varepsilon B(A - z)^{-1} \right\| \leq \begin{cases} \frac{|\varepsilon|}{\varepsilon_0} & \text{for } z \text{ on a vertical side of } \mathscr{C}, \\ \left(1 + \frac{b-a}{2M}\right) \frac{|\varepsilon|}{\varepsilon_0} & \text{for } z \text{ on a horizontal side of } \mathscr{C}. \end{cases} \tag{5.15}$$

Let $|\varepsilon| < \varepsilon_0$ and M be large enough so that

$$\left(1 + \frac{b-a}{2M}\right)\frac{|\varepsilon|}{\varepsilon_0} < 1.$$

In this case the resolvent is given by the convergent series

$$(A + \varepsilon B - z)^{-1} = (A - z)^{-1}\left(1 + \varepsilon B(A - z)^{-1}\right)^{-1}$$
$$= (A - z)^{-1}\sum_{n \geq 0}(-1)^n \varepsilon^n \left(B(A - z)^{-1}\right)^n,$$

for all $z \in \mathscr{C}$. After exchanging the integral and the sum, we obtain as desired

$$P(\varepsilon) = P(0) + \sum_{n \geq 1}\varepsilon^n b_n$$

where the operator

$$b_n := \frac{(-1)^{n+1}}{2i\pi}\oint_{\mathscr{C}}(A - z)^{-1}\left(B(A - z)^{-1}\right)^n dz \qquad (5.16)$$

satisfies the estimate

$$\|b_n\| \leq \frac{1}{2\pi}\left(\frac{2M}{d(a, \sigma(A))} + \frac{2M}{d(b, \sigma(A))} + \frac{2(b-a)}{M}\left(1 + \frac{b-a}{2M}\right)^n\right)\varepsilon_0^{-n}. \qquad (5.17)$$

Here we have used that

$$\|(A - z)^{-1}\| = \frac{1}{d(z, \sigma(A))} \leq \begin{cases} \frac{1}{d(a,\sigma(A))} & \text{for } z \text{ on the left side,} \\ \frac{1}{d(b,\sigma(A))} & \text{for } z \text{ on the right side,} \\ \frac{1}{M} & \text{for } z \text{ on a horizontal side,} \end{cases}$$

according to Corollary 4.6. This clearly shows that $P(\varepsilon)$ is real analytic, hence C^∞, on the interval $(-\varepsilon_0, \varepsilon_0)$.

In reality, the operators b_n from (5.16) are independent of the height M of the rectangle. Indeed, they are the terms in the series expansion of $P(\varepsilon)$, which is independent of M. This also follows from Cauchy's formula, since A has no spectrum outside the real axis so the integral over a curve outside the axis is zero. We can therefore choose a different M for each b_n in the estimate (5.17) of its norm.

By taking for example $M = n/2$ we obtain the explicit estimate

$$\|b_n\| \leq \frac{1}{2\pi} \left(\frac{n}{d(a, \sigma(A))} + \frac{n}{d(b, \sigma(A))} + \frac{4(b-a)e^{b-a}}{n} \right) \varepsilon_0^{-n} \qquad (5.18)$$

for $n \geq 1$.

As the rank of an orthogonal projector is always an integer when it is finite, it is quite intuitive that it cannot change along a continuous curve of such projectors. This is confirmed by the following lemma.

Lemma 5.8 (Pairs of Projectors) *Let P and P' be two orthogonal projectors on a Hilbert space $\mathfrak{H}$, such that $\|P - P'\| < 1$. Then $\mathrm{rank}(P) = \mathrm{rank}(P')$.*

Proof (of the Lemma) If both projectors are of infinite rank, there is nothing to prove. Suppose for example that P is of rank $k < \infty$. If the image of P' is of dimension $\geq k + 1$, then it must intersect the orthogonal of the image of P, that is $\ker P$. But for $v \in \mathrm{Im}(P') \cap \ker(P)$ we have $(P - P')v = -v$ therefore $\|(P - P')v\| = \|v\|$ which contradicts the assumption that $\|P - P'\| < 1$. Thus $\mathrm{rank}(P') \leq k = \mathrm{rank}(P)$. In particular, P' is of finite rank and the other inequality is found by swapping P and P'. $\qquad\square$

The lemma allows us to deduce that $\mathrm{rank}\,P(\varepsilon) = \mathrm{rank}\,P(0)$ for all $\varepsilon \in (-\varepsilon_0, \varepsilon_0)$ by a continuation argument, since the map $\varepsilon \mapsto P(\varepsilon)$ is continuous on this interval. This concludes the proof of (ii).

Next, we use the following auxiliary result, that is a simple consequence of the spectral theorem and immediately implies (iii).

Lemma 5.9 *Let A be a self-adjoint operator on its domain $D(A) \subset \mathfrak{H}$ and two real numbers $a < b$. Then $\mathbb{1}_{(a,b)}(A)$ has finite rank $k \in \mathbb{N}$ if and only if $\sigma(A) \cap (a, b)$ consists of finitely many eigenvalues whose sum of multiplicities is equal to k.*

Proof (of the Lemma) If $\sigma(A) \cap (a, b)$ consists of finitely many eigenvalues of finite multiplicity, then $\mathbb{1}_{(a,b)}(A)$ is the orthogonal projection onto the direct sum of the corresponding eigenspaces, as we have seen in Sect. 4.5. Hence its rank equals the sum of the multiplicities, as claimed. Let us now assume that $\mathbb{1}_{(a,b)}(A)$ has finite rank $k \in \mathbb{N}$. By the spectral Theorem 4.4 we have $A = U^{-1}M_f U$ where $f(s, n) = s$ on $L^2(\sigma(A) \times \mathbb{N}, d\mu)$ so that

$$\mathrm{rank}\left(\mathbb{1}_{(a,b)}(A)\right) = \dim L^2((a, b) \times \mathbb{N}, d\mu)$$

as we have already seen in (4.17). This space being of finite dimension k by assumption, we can already deduce that the measure μ can charge at most a finite number of points in $(a, b) \times \mathbb{N}$. This is because the dimension of the space of functions supported on these points is equal to the number of points:

$$\dim L^2\left((a, b) \times \mathbb{N}, \sum_{j=1}^{d} v_j \delta_{x_j}\right) = d, \qquad \text{for } v_j > 0 \text{ and } x_j \neq x_k \text{ for } j \neq k.$$

$$(5.19)$$

Let us now show that μ cannot contain anything else than these Dirac deltas. We pick a $\lambda \in (a, b) \cap \sigma(A)$ such that $\mu(\{\lambda\} \times \mathbb{N}) = 0$ and recall that $\mu((\lambda - \varepsilon, \lambda + \varepsilon) \times \mathbb{N}) > 0$ for all $\varepsilon > 0$ by definition of the spectrum. Hence there must exist a decreasing sequence $\varepsilon_n \searrow 0$ such that $\mu((\lambda - \varepsilon_n, \lambda + \varepsilon_n) \times \mathbb{N})$ is strictly decreasing and converges to $\mu(\{\lambda\} \times \mathbb{N}) = 0$. We can also assume that $\varepsilon_1 < \min(|\lambda - a|, |\lambda - b|)$ so that $(\lambda - \varepsilon_n, \lambda + \varepsilon_n) \subset (a, b)$. Then, we have $\mu((\lambda - \varepsilon_{n-1}, \lambda - \varepsilon_n] \cup [\lambda + \varepsilon_n, \lambda + \varepsilon_{n-1}) \times \mathbb{N}) > 0$ for all n. Picking any function $f_n \neq 0$ supported in this set, we obtain an infinite sequence of orthogonal functions in $L^2((a, b) \times \mathbb{N}, d\mu)$. This cannot exist in a space of finite dimension and we thus conclude that μ must be a finite combination Dirac delta's. Their number has to be equal to k by (5.19). □

We finally turn to (iv). When $k = 1$, (iii) implies that $A + \varepsilon B$ has exactly one simple eigenvalue for all $\varepsilon \in (-\varepsilon_0, \varepsilon_0)$ when this is the case for A. Let ε_1 be any real number in $(-\varepsilon_0, \varepsilon_0)$ and v_1 a normalized eigenvector for the operator $A + \varepsilon_1 B$. Then $\varepsilon \mapsto \langle v_1, P(\varepsilon) v_1 \rangle$ is a real analytic function that equals 1 at $\varepsilon = \varepsilon_1$ by continuity of P. An adaptation of the proof of (ii) provides that the operator $\varepsilon \mapsto (A + \varepsilon B) P(\varepsilon) = \lambda(\varepsilon) P(\varepsilon)$ is also real analytic on $(-\varepsilon_0, \varepsilon_0)$. Thus,

$$\lambda(\varepsilon) = \frac{\langle v_1, (A + \varepsilon B) P(\varepsilon) v_1 \rangle}{\langle v_1, P(\varepsilon) v_1 \rangle}$$

is real analytic in the neighborhood of ε_1 defined by the condition that $\langle v_1, P(\varepsilon) v_1 \rangle \neq 0$. This shows that $\lambda(\varepsilon)$ is real analytic over the entire interval $(-\varepsilon_0, \varepsilon_0)$, which concludes the proof of Theorem 5.7 in the case where B is A-bounded.

When A is bounded-below and B is only A-bounded in the sense of quadratic forms, we must use the same decomposition as in (5.13) for $A + \varepsilon B - z$ and therefore estimate the norm of the operator b_n from (5.16) by

$$\|b_n\| \leq \frac{1}{2\pi} \oint_{\mathscr{C}} \left\||A - z|^{-\frac{1}{2}}\right\|^2 \left\||A - z|^{-\frac{1}{2}} B |A - z|^{-\frac{1}{2}}\right\|^n |dz|.$$

The reader can then verify that we can use the same inequalities as before, taking roots everywhere. □

5.2 Point Spectrum, Continuous Spectrum, Essential Spectrum, Discrete Spectrum

5.2.1 Definitions

We have already introduced the concept of **eigenvalue** which corresponds to the elements λ of the spectrum $\sigma(A)$ such that $\ker(A - \lambda) \neq \{0\}$.

Definition 5.10 (Point Spectrum, Continuous Spectrum)   Let $(A, D(A))$ be a self-adjoint operator. We call **point spectrum** and we note

$$\sigma_{\mathrm{pt}}(A) := \left\{ \lambda \in \sigma(A) \ : \ \ker(A - \lambda) \neq \{0\} \right\}$$

the set of all the **eigenvalues** of A. We call **continuous spectrum** and we note $\sigma_{\mathrm{cont}}(A)$ its complement in $\sigma(A)$.

By the spectral theorem, a real number $\lambda \in \sigma(A)$ belongs to the point spectrum when

$$\mu\big(\{\lambda\} \times \mathbb{N}\big) = \sum_{n \geq 0} \mu(\{(\lambda, n)\}) > 0,$$

in the representation where A becomes the multiplication operator by the function $a(s, n) = s$ on $L^2(\sigma(A) \times \mathbb{N}, d\mu)$. This corresponds to asking that μ charges some of the points (λ, n), which justifies the name "point spectrum".

The separation of the spectrum between point and continuous is not a very good notion because it is **unstable under small perturbations**, as illustrated in the following example.

Example 5.11   Let $I = (0, 1)$. Consider the operator $A = 0$ on $\mathfrak{H} = L^2(I)$ and the multiplication operator $B = M_f$ by the function $f(x) = x$. Then $\sigma(A) = \{0\}$ with $\ker(A) = \mathfrak{H}$, that is, there is only point spectrum. On the other hand, $\sigma(A + \varepsilon B) = \sigma(\varepsilon B) = [0, \varepsilon]$ without any eigenvalue, according to Theorem 4.3.

In Theorem 5.7 (iii) we saw that, on the contrary, the set of isolated eigenvalues of finite multiplicity was **stable under small perturbations**. This leads us to consider a better classification of the spectrum.

Definition 5.12 (Discrete Spectrum, Essential Spectrum)   Let $(A, D(A))$ be a self-adjoint operator. We call **discrete spectrum** and denote $\sigma_{\mathrm{disc}}(A)$ the set of **isolated eigenvalues of finite multiplicity**. We call **essential spectrum** and denote $\sigma_{\mathrm{ess}}(A)$ its complement in $\sigma(A)$.

Recall that a point $\lambda \in \sigma(A)$ of the spectrum is called isolated when there exists $\varepsilon > 0$ such that $(\lambda - \varepsilon, \lambda + \varepsilon) \cap \sigma(A) = \{\lambda\}$. The following is a consequence of Lemma 5.9.

Lemma 5.13 (Essential Spectrum and Spectral Projections) *Let A be a self-adjoint operator on its domain $D(A) \subset \mathfrak{H}$. Then we have*

(i) *$\lambda \in \sigma_{\mathrm{ess}}(A)$ if and only if $\mathbb{1}_{(\lambda-\varepsilon,\lambda+\varepsilon)}(A)$ has infinite rank for all $\varepsilon > 0$;*

(ii) *$\lambda \in \sigma_{\mathrm{disc}}(A)$ if and only if $\lambda \in \sigma(A)$ and there exists $\varepsilon > 0$ such that $\mathbb{1}_{(\lambda-\varepsilon,\lambda+\varepsilon)}(A)$ has finite rank.*

Proof The two statements are equivalent, hence we only prove the second. If $\lambda \in \sigma_{\mathrm{disc}}(A)$ then for $\varepsilon > 0$ small enough such that $(\lambda - \varepsilon, \lambda + \varepsilon) \cap \sigma(A) = \{\lambda\}$, we have that $\mathbb{1}_{(\lambda-\varepsilon,\lambda+\varepsilon)}(A) = \mathbb{1}_{\{\lambda\}}(A)$. As we have seen in Sect. 4.5, the latter is the projection onto $\ker(A - \lambda)$, which is finite-dimensional by definition. Conversely, if $\lambda \in \sigma(A)$ and $\mathbb{1}_{(\lambda-\varepsilon,\lambda+\varepsilon)}(A)$ has finite rank $k \in \mathbb{N}$ for some $\varepsilon > 0$, then by Lemma 5.9 the spectrum in $(\lambda - \varepsilon, \lambda + \varepsilon)$ is composed of finitely many eigenvalues, whose sum of multiplicities is k. Any eigenvalue in this open interval is then isolated from the rest of the spectrum. Since λ is assumed to be in the spectrum, the result follows.   □

The characterization in terms of spectral projections implies the following.

Lemma 5.14 ($\sigma_{\mathrm{ess}}(A)$ Is Closed) *The essential spectrum of a self-adjoint operator is closed.*

Proof Take a sequence $\lambda_n \in \sigma_{\mathrm{ess}}(A)$ converging to some λ. We already know that $\lambda \in \sigma(A)$ because the spectrum is closed. Let $\varepsilon > 0$. For n large enough we have $\lambda_n \in (\lambda - \varepsilon/2, \lambda + \varepsilon/2)$, so that $(\lambda_n - \varepsilon/2, \lambda_n + \varepsilon/2) \subset (\lambda - \varepsilon, \lambda + \varepsilon)$. Now, $\mathbb{1}_{(\lambda_n-\varepsilon/2,\lambda_n+\varepsilon/2)}(A)$ has infinite rank by Lemma 5.13, hence so does the larger projection $\mathbb{1}_{(\lambda-\varepsilon,\lambda+\varepsilon)}(A)$. Since this is valid for all $\varepsilon > 0$, this proves that $\lambda \in \sigma_{\mathrm{ess}}(A)$.   □

5.2.2 Weyl's Characterization of the Essential Spectrum

We provide here a characterization of the essential spectrum $\sigma_{\mathrm{ess}}(A)$ using Weyl sequences. Recall that $\lambda \in \sigma(A)$ if and only if there exists a sequence $(v_n) \in D(A)^{\mathbb{N}}$ such that $\|v_n\| = 1$ and $(A - \lambda)v_n \to 0$ (Theorem 2.30).

Theorem 5.15 (Weyl's Characterization of the Essential Spectrum) *Let A be a self-adjoint operator on its domain $D(A) \subset \mathfrak{H}$. Then we have $\lambda \in \sigma_{\mathrm{ess}}(A)$ if and only if there exists a sequence $(v_n) \in D(A)^{\mathbb{N}}$ such that*

- $\|v_n\| = 1$;
- $(A - \lambda)v_n \to 0$;
- $v_n \rightharpoonup 0$ *weakly in $\mathfrak{H}$.*

Such a sequence is sometimes called a **singular Weyl sequence**. It follows from the proof below that we can even assume that (v_n) is an orthonormal system.

Proof We argue like in Remark 4.18. If $\lambda \in \sigma_{\mathrm{ess}}(A)$, we construct the sequence v_n by induction, taking v_n in the range of the spectral projection $\mathbb{1}_{(\lambda-1/n,\lambda+1/n)}(A)$ with the additional assumption that $v_n \in \mathrm{vect}(v_1, \ldots, v_{n-1})^{\perp}$. As $\mathbb{1}_{(\lambda-1/n,\lambda+1/n)}(A)$ is infinite rank by Lemma 5.13, we can always find such a vector. Then $v_n \rightharpoonup 0$ by construction and moreover $\|(A - \lambda)v_n\|^2 \leq \|v_n\|^2/n^2 \to 0$ by (4.18) in Remark 4.18, as we wanted. Conversely, let (v_n) be a sequence as in the statement. From (4.20), we know that

$$v_n - \mathbb{1}_{(\lambda-\varepsilon,\lambda+\varepsilon)}(A)v_n = \mathbb{1}_{\mathbb{R}\setminus(\lambda-\varepsilon,\lambda+\varepsilon)}(A)v_n \xrightarrow[n\to\infty]{} 0.$$

Table 5.1 Characterization of the elements of the spectrum of a self-adjoint operator using Weyl sequences

$\lambda \in \sigma(A)$	There exists a sequence $(v_n) \in D(A)^{\mathbb{N}}$ such that $\|v_n\| = 1$ and $(A - \lambda)v_n \to 0$ strongly (Weyl sequence)
$\lambda \in \sigma_{\mathrm{ess}}(A)$	There exists a Weyl sequence such that $v_n \rightharpoonup 0$ weakly
$\lambda \in \sigma_{\mathrm{disc}}(A)$	All Weyl sequences have at least one strongly convergent sub-sequence in $\mathfrak{H}$
$\lambda \in \sigma_{\mathrm{cont}}(A)$	All Weyl sequences (v_n) satisfy $v_n \rightharpoonup 0$ weakly
$\lambda \in \sigma_{\mathrm{pt}}(A)$	There exists a Weyl sequence that admits a weak limit different from 0. Equivalently, $\ker(A - \lambda) \neq \{0\}$

Letting $w_n := \mathbb{1}_{(\lambda-\varepsilon,\lambda+\varepsilon)}(A)v_n$, we see that $w_n \rightharpoonup 0$ weakly and $\|w_n\| \to 1$. This is only possible if the range of $\mathbb{1}_{(\lambda-\varepsilon,\lambda+\varepsilon)}(A)$ is infinite-dimensional. This concludes the proof by Lemma 5.13. $\qquad\square$

Table 5.1 provides a summary of the characterization of the elements of the spectrum using Weyl sequences.

Remark 5.16 (Weyl's Criterion for Non-self-adjoint Operators) There is no unambiguous definition of the essential and discrete spectra for a non-self-adjoint operator [RS78, Sec. XIII.4]. For example, in the non-self-adjoint case, an isolated point of the spectrum is not always an eigenvalue. However, the theory works very similarly if we add the assumption that the operator is diagonalizable, that is, unitarily equivalent to a multiplication operator by a locally bounded function.

If A is a self-adjoint operator, consider for example the resolvent $(A - z)^{-1}$ with $z \in \rho(A)$. It is a bounded operator but is not self-adjoint when $z \in \mathbb{C} \setminus \mathbb{R}$. However, it is unitarily equivalent to a multiplication operator, by the spectral theorem applied to A. Then its isolated eigenvalues (in $\mathbb{C}$) of finite multiplicity are exactly given by

$$\sigma_{\mathrm{disc}}\big((A - z)^{-1}\big) = \Big\{(\lambda - z)^{-1}, \ \lambda \in \sigma_{\mathrm{disc}}(A)\Big\},$$

because the function $x \mapsto (x - z)^{-1}$ is bijective and continuous on a neighborhood of $\sigma(A)$. Read again Lemma 2.11 on this. The essential spectrum is then equal to

$$\sigma_{\mathrm{ess}}\big((A - z)^{-1}\big) = \Big\{(\lambda - z)^{-1}, \ \lambda \in \sigma_{\mathrm{ess}}(A)\Big\} \cup \begin{cases} \emptyset & \text{if } A \text{ is bounded,} \\ \{0\} & \text{if } A \text{ is unbounded.} \end{cases}$$

The essential spectrum $\sigma_{\mathrm{ess}}\big((A - z)^{-1}\big)$ is characterized by Weyl sequences exactly as in Theorem 5.15, with a similar proof.

There exists a formula based on Weyl sequences, providing the bottom of the essential spectrum under the additional assumption that the operator is bounded from below.

Theorem 5.17 (Formula for $\min \sigma_{\text{ess}}(A)$**)** *Let A be a bounded-below self-adjoint operator on its domain $D(A) \subset \mathfrak{H}$. Then we have*

$$\min \sigma_{\text{ess}}(A) = \min_{\substack{(v_n)\in D(A)^{\mathbb{N}} \\ \|v_n\|=1 \\ v_n \rightharpoonup 0}} \liminf_{n\to\infty} \langle v_n, A v_n \rangle = \min_{\substack{(v_n)\in Q(A)^{\mathbb{N}} \\ \|v_n\|=1 \\ v_n \rightharpoonup 0}} \liminf_{n\to\infty} q_A(v_n), \qquad (5.20)$$

with the convention that the three terms are $+\infty$ *when* $\sigma_{\text{ess}}(A) = \emptyset$.

The minimum in (5.20) is taken over all sequences (v_n) in the indicated space. The weak convergence $v_n \rightharpoonup 0$ is assumed to hold in $\mathfrak{H}$. In the following, we will use the notation

$$\Sigma(A) := \begin{cases} \min \sigma_{\text{ess}}(A) & \text{if } \sigma_{\text{ess}}(A) \neq \emptyset, \\ +\infty & \text{if } \sigma_{\text{ess}}(A) = \emptyset \end{cases} \qquad (5.21)$$

for the bottom of the essential spectrum. Let us recall that $\Sigma(A)$ belongs to $\sigma_{\text{ess}}(A)$ if this set is non-empty, according to Lemma 5.14. When $\Sigma(A) = +\infty$, Formula (5.20) simply means that $q_A(v_n) \to +\infty$ for any sequence $(v_n) \in Q(A)^{\mathbb{N}}$ such that $\|v_n\| = 1$ and $v_n \rightharpoonup 0$.

Proof The equality of the two terms on the right of (5.20) follows from the density of $D(A)$ in $Q(A)$ for the associated norm. Suppose first $\sigma_{\text{ess}}(A) \neq \emptyset$. Since $\Sigma(A)$ belongs to the essential spectrum, there exists a Weyl sequence satisfying $\|v_n\| = 1$, $(A - \Sigma(A))v_n \to 0$ and $v_n \rightharpoonup 0$. Taking the scalar product against v_n we find that $\langle v_n, (A - \Sigma(A))v_n \rangle \to 0$, that is $\langle v_n, A v_n \rangle \to \Sigma(A)$. Conversely, if $v_n \in D(A)$ is such that $\liminf_{n\to\infty} q_A(v_n) < \infty$, then we can write

$$v_n = \mathbb{1}_{(-\infty, \Sigma(A)-\varepsilon]}(A)v_n + \mathbb{1}_{(\Sigma(A)-\varepsilon, +\infty)}(A)v_n := v_n^1 + v_n^2.$$

From the spectral theorem we know that A is bounded in the range of $\mathbb{1}_{(-\infty, \Sigma(A)-\varepsilon]}(A)$ and that $v_n^2 \in D(A)$. By definition of the spectral projections we have $\|v_n^1\|^2 + \|v_n^2\|^2 = \|v_n\|^2 = 1$ as well as

$$q_A(v_n) = \langle v_n^1, A v_n^1 \rangle + \langle v_n^2, A v_n^2 \rangle \geq \langle v_n^1, A v_n^1 \rangle + (\Sigma(A) - \varepsilon)\|v_n^2\|^2. \qquad (5.22)$$

The weak convergence $v_n \rightharpoonup 0$ implies $v_n^1 \rightharpoonup 0$ because the spectral projections are bounded. In fact, the range of $\mathbb{1}_{(-\infty, \Sigma(A)-\varepsilon]}(A)$ is finite-dimensional since $(-\infty, \Sigma(A) - \varepsilon]$ contains a finite number of eigenvalues of finite multiplicity, by definition of $\Sigma(A)$. In this space weak convergence must thus be strong and we conclude that $v_n^1 \to 0$. This implies $\|v_n^2\| \to 1$. We conclude that $\langle v_n^1, A v_n^1 \rangle \to 0$ since A is bounded in the corresponding spectral subspace. This proves that

$$\liminf_{n\to\infty} q_A(v_n) \geq \Sigma(A) - \varepsilon$$

and the result follows by taking $\varepsilon \to 0$. The proof is similar if $\sigma_{\mathrm{ess}}(A) = \emptyset$, using the spectral projection of the interval $(-\infty, M]$ and taking $M \to \infty$ at the end. $\square$

5.3   Compact Operators

Let us recall that an operator K is called compact when the image of the unit ball is compact or, in other words, when $K v_n \to 0$ strongly for any sequence $v_n \rightharpoonup 0$ weakly. Compact operators form a closed ideal $\mathcal{K}(\mathfrak{H})$ of the algebra $\mathcal{B}(\mathfrak{H})$ of bounded operators, in which finite rank operators are dense. The closure of $\mathcal{K}(\mathfrak{H})$ means that if $\| K_n - B \| \to 0$ where all the K_n are compact, then B is automatically compact. Saying that it is an ideal means that BK and KB belong to $\mathcal{K}(\mathfrak{H})$ for all $K \in \mathcal{K}(\mathfrak{H})$ and all $B \in \mathcal{B}(\mathfrak{H})$. A finite rank operator is in the form

$$R = \sum_{j=1}^{J} |w_j\rangle\langle v_j|,$$

a notation with ket's and bra's which means $Rv = \sum_{j=1}^{J} \langle v_j, v\rangle w_j$ where the v_j, w_j are vectors of $\mathfrak{H}$. We also sometimes write $R = \sum_{j=1}^{J} w_j(v_j)^*$. By diagonalizing the self-adjoint operator $R^* R$ we can also reduce to the case where the v_j form an orthonormal system, in which case we obtain $w_j = Rv_j$, which are pairwise orthogonal because $\langle Rv_j, Rv_k\rangle = \langle v_j, R^* Rv_k\rangle$.

5.3.1   Diagonalization

Here is a consequence of Theorem 5.15.

Corollary 5.18 (Compact Operators)   *Suppose* $\dim(\mathfrak{H}) = +\infty$. *A bounded self-adjoint operator A is compact if and only if $\sigma_{\mathrm{ess}}(A) = \{0\}$. In particular, A is then diagonalizable in an orthonormal basis, with eigenvalues λ_j tending to 0. The non-zero eigenvalues are all of finite multiplicity.*

We obtain that every compact self-adjoint operator decomposes in the form $A = \sum_j \lambda_j |e_j\rangle\langle e_j|$ where the e_j are the eigenvectors associated with the λ_j's. If A is not self-adjoint, we can apply the preceding result to $A^* A$, whose eigenvalues are denoted μ_j. By introducing $\lambda_j := \sqrt{\mu_j}$ which are called the *singular values of* A, we find that $A = \sum_j \lambda_j |f_j\rangle\langle e_j|$ where the e_j are the eigenvectors of $A^* A$ and $f_j := Ae_j / \| Ae_j \| = Ae_j / \lambda_j$ also form an orthonormal system. This is the best we can do for a non-self-adjoint compact operator.

Proof   If A is compact and $\lambda \in \sigma_{\mathrm{ess}}(A)$, then there exists a Weyl sequence (v_n) normalized such that $(A - \lambda)v_n \to 0$ and $v_n \rightharpoonup 0$. Then $Av_n \to 0$, which implies

$\lambda v_n \to 0$ and is only possible if $\lambda = 0$, because $\|v_n\| = 1$ by hypothesis. Therefore $\sigma_{\text{ess}}(A) \subset \{0\}$. But for a bounded operator we have $\sigma(A) \subset [-\|A\|, \|A\|]$. Any accumulation point of eigenvalues belongs to $\sigma_{\text{ess}}(A)$ so the hypothesis $\sigma_{\text{ess}}(A) = \emptyset$ implies that the spectrum of A is composed only of a finite number of eigenvalues of finite multiplicity. But then $\dim(\mathfrak{H}) = \operatorname{ran}(\mathbb{1}_{\mathbb{R}}(A)) < \infty$ by Lemma 5.9. Therefore $\sigma_{\text{ess}}(A) \neq \emptyset$ and finally $\sigma_{\text{ess}}(A) = \{0\}$.

Conversely, if $\sigma_{\text{ess}}(A) = \{0\}$, we can write $A = A\mathbb{1}_{[-\varepsilon,\varepsilon]}(A) + A\mathbb{1}_{\mathbb{R}\setminus[-\varepsilon,\varepsilon]}(A)$. As A is bounded, its spectrum is also bounded and the assumption $\sigma_{\text{ess}}(A) = \{0\}$ then implies that $\sigma(A) \cap (\mathbb{R}\setminus[-\varepsilon,\varepsilon])$ is composed of a finite number of eigenvalues of finite multiplicities. In particular, the operator $A\mathbb{1}_{\mathbb{R}\setminus[-\varepsilon,\varepsilon]}(A)$ has finite rank by Lemma 5.9. As $\|A\mathbb{1}_{[-\varepsilon,\varepsilon]}(A)\| \leq \varepsilon$ we see that A is a norm limit of a sequence of finite rank operators, and is therefore compact.   □

In the following sections we give several examples of compact operators.

5.3.2 Hilbert-Schmidt Operators

We consider $\mathfrak{H} = L^2(B, d\mu)$ for any measurable set $B \subset \mathbb{R}^d$ and μ a locally finite Borel measure (for example $B = \mathbb{R}^d$ and μ the Lebesgue measure).

Definition 5.19 (Hilbert-Schmidt)   An operator A on $L^2(B, d\mu)$ is called **Hilbert-Schmidt** when it is given by an integral kernel $a \in L^2(B \times B, d\mu \otimes \mu)$, that is to say by the formula

$$(Au)(x) := \int_B a(x, y)\, u(y)\, d\mu(y). \qquad (5.23)$$

We have not specified the domain of definition of A because a Hilbert-Schmidt operator is always bounded and therefore defined on all $\mathfrak{H}$. Indeed, by the Cauchy-Schwarz inequality, we have

$$|(Au)(x)|^2 \leq \|u\|^2_{L^2(B,d\mu)} \int_B |a(x, y)|^2\, d\mu(y) \qquad (5.24)$$

which implies, by integrating with respect to x, that A is bounded with

$$\|A\| \leq \|a\|_{L^2(B \times B, d\mu \otimes \mu)}.$$

It is easy to verify that the adjoint of A is also Hilbert-Schmidt, with the integral kernel $\overline{a(y, x)}$. Thus, A is self-adjoint if and only if the function a satisfies $\overline{a(y, x)} = a(x, y)$ almost everywhere. We should think of the function $a(x, y)$ as the continuous equivalent of the coefficients M_{ij} of a matrix in finite dimension. The formula (5.23) is then the continuous equivalent of the formula $(Mv)_i = \sum_j M_{ij} v_j$ where the sum has been replaced by an integral.

Proposition 5.20 (Compactness) *A Hilbert-Schmidt operator is compact.*

Proof By Fubini's theorem, the function $y \mapsto a(x, y)$ belongs to $L^2(B, d\mu)$ for almost all x. Thus, if $u_n \rightharpoonup 0$ weakly in $L^2(B, d\mu)$, we have

$$(Au_n)(x) = \int_B a(x, y)\, u_n(y)\, d\mu(y) \to 0$$

for almost all x, by definition of the weak convergence of u_n in the variable y. Therefore $Au_n \to 0$ almost everywhere. Moreover, the estimate (5.24) provides a domination independent of n because (u_n) is bounded in $L^2(B, d\mu)$. By dominated convergence we then have $\|Au_n\|_{L^2(B, d\mu)} \to 0$.   □

A finite-rank operator is always Hilbert-Schmidt. Indeed, using the previous notations, we find that the operator $A = |v\rangle\langle w|$ has the kernel $a(x, y) = v(x)\overline{w(y)}$. It is possible to characterize the Hilbert-Schmidt operators among the compact operators on $\mathfrak{H} = L^2(B, d\mu)$, from their singular values, that is, the eigenvalues of A^*A.

Theorem 5.21 (Characterization of Hilbert-Schmidt Operators) *Let A be a compact operator on $\mathfrak{H} = L^2(B, d\mu)$, with B and μ as before. Then A is Hilbert-Schmidt if and only if its singular values $\mu_j(A) := \sqrt{\lambda_j(A^*A)}$ are square-summable, and in this case we have*

$$a(x, y) = \sum_j (Au_j)(x)\overline{u_j(y)} \tag{5.25}$$

where the sum is convergent in $L^2(B \times B, d\mu \otimes \mu)$ and where (u_j) is an orthonormal basis of eigenvectors for the compact self-adjoint operator A^*A. Moreover,

$$\sum_j \mu_j(A)^2 = \int_B \int_B |a(x, y)|^2\, d\mu(x)\, d\mu(y).$$

Proof Let A be a compact operator. After diagonalizing A^*A, we can write $A = \sum_j |Au_j\rangle\langle u_j|$ where the sum $\sum_j |Au_j\rangle\langle u_j|$ is operator norm convergent. As we mentioned, each of the operators $|Au_j\rangle\langle u_j|$ is Hilbert-Schmidt, with the integral kernel $F_j(x, y) := (Au_j)(x)\overline{u_j(y)}$. We note that the F_j are pairwise orthogonal in $L^2(B \times B, d\mu \otimes \mu)$ because the u_j are, and that $\|F_j\|^2_{L^2(B \times B)} = \|Au_j\|^2_{L^2(B)} = \mu_j(A)^2$.

If $\sum_j \mu_j(A)^2 = \sum_j \|F_j\|^2 < \infty$, the function $a(x, y) := \sum_j F_j(x, y)$ is well defined in $L^2(B \times B, d\mu \otimes \mu)$. This defines a Hilbert-Schmidt operator $\tilde{A}$ with kernel a. A calculation shows that $\tilde{A}$ coincides with A on the basis of the u_j, and they must therefore be equal everywhere. In particular, A is Hilbert-Schmidt.

Conversely, if A is Hilbert-Schmidt we can write by the Cauchy-Schwarz inequality

$$\left\| \sum_{j=1}^{J} F_j \right\|_{L^2(B\times B)}^2 = \sum_{j=1}^{J} \mu_j(A)^2 = \sum_{j=1}^{J} \langle Au_j, Au_j \rangle$$

$$= \sum_{j=1}^{J} \int_B \int_B \overline{(Au_j)(x)} a(x,y) u_j(y)\, d\mu(x)\, d\mu(y)$$

$$= \int_{B\times B} \overline{\sum_{j=1}^{J} F_j(x,y)} a(x,y)\, d\mu \otimes \mu(x,y)$$

$$\le \|a\|_{L^2(B\times B)} \left\| \sum_{j=1}^{J} F_j \right\|_{L^2(B\times B)}.$$

So

$$\sum_{j=1}^{J} \mu_j(A)^2 = \left\| \sum_{j=1}^{J} F_j \right\|_{L^2(B\times B)}^2 \le \|a\|_{L^2(B\times B)}^2$$

and the sum converges.   □

The set of Hilbert-Schmidt operators on the space $L^2(B, d\mu)$ is denoted $\mathfrak{S}^2(B, d\mu)$. It is a Hilbert space that is isometric to $L^2(B \times B, d\mu \otimes \mu)$ when it is equipped with the norm $\|A\|_{\mathfrak{S}^2(B,d\mu)} := \|a\|_{L^2(B\times B, d\mu\otimes\mu)}$ where a is the integral kernel of A. We should be careful that $\mathfrak{S}^2(B, d\mu)$ is not closed for the operator norm. In fact, $\mathfrak{S}^2(B, d\mu)$ contains all finite rank operators, so its closure is $\mathcal{K}(\mathfrak{H})$. It turns out that $\mathfrak{S}^2(B, d\mu)$ is also an ideal of $\mathcal{B}(\mathfrak{H})$, that is, $AM \in \mathfrak{S}^2(B, d\mu)$ for any bounded operator M, when $A \in \mathfrak{S}^2(B, d\mu)$.

Exercise 5.22 ($\mathfrak{S}^2(B, d\mu)$ Is an Ideal of the Algebra $\mathcal{B}(\mathfrak{H})$) Let A be a bounded operator on $\mathfrak{H} = L^2(B, d\mu)$. Let (e_n) and (f_n) be any two orthonormal bases of $\mathfrak{H}$. Show the equality

$$\sum_{n\ge 1} \|Ae_n\|^2 = \sum_{n,m\ge 1} |\langle f_m, Ae_n \rangle|^2 = \sum_{m\ge 1} \|A^* f_m\|^2, \qquad (5.26)$$

where the terms can be finite or infinite. Deduce that A is a Hilbert-Schmidt operator if and only if these series are all convergent, with

$$\|A\|_{\mathfrak{S}^2(B,d\mu)}^2 := \sum_{n\ge 1} \|Ae_n\|^2$$

for any orthonormal basis. Show in this way that AM and MA are Hilbert-Schmidt when A is and M is any bounded operator.

Here is a particularly important example of a Hilbert-Schmidt operator.

Proposition 5.23 *When $f, g \in L^2(\mathbb{R}^d)$, the operators*

$$A_1 = f(x)g(-i\nabla), \qquad A_2 = g(-i\nabla)f(x) \tag{5.27}$$

defined first on $C_c^\infty(\mathbb{R}^d)$ are closable and their closures are Hilbert-Schmidt on all $\mathfrak{H} = L^2(\mathbb{R}^d)$, with

$$\left\|\overline{A_1}\right\|_{\mathfrak{S}^2} = \left\|\overline{A_2}\right\|_{\mathfrak{S}^2} = (2\pi)^{-\frac{d}{2}} \|f\|_{L^2(\mathbb{R}^d)} \|g\|_{L^2(\mathbb{R}^d)} \tag{5.28}$$

and the corresponding integral kernels

$$a_1(x, y) = (2\pi)^{-\frac{d}{2}} f(x)\check{g}(x - y), \qquad a_2(x, y) = (2\pi)^{-\frac{d}{2}} \check{g}(x - y)f(y).$$

We recall that $g(-i\nabla)$ is the operator of multiplication by the function $g(k)$ in Fourier, which equals the convolution by the function $(2\pi)^{-d/2}\check{g}$ in direct space. In (5.27) we interpret $f(x)$ as the operator M_f of multiplication by the function f. The operators A_1 and A_2 have already been considered in Sect. 1.5.5.

As the operators $f(x)$ and $g(-i\nabla)$ are not generally bounded, therefore not defined on all $L^2(\mathbb{R}^d)$, we must as in the statement start by defining A_1 and A_2 on an appropriate subspace and then show that they admit a unique bounded extension to all $L^2(\mathbb{R}^d)$. We therefore see that it can happen that a product of unbounded operators is finally bounded!

In reality, the operator of multiplication by $f(x)$ is well defined on all $L^2(\mathbb{R}^d)$ but it naturally takes its values in $L^1(\mathbb{R}^d)$, by the Cauchy-Schwarz inequality. The obtained function then has a bounded Fourier transform, which allows us to multiply by $g(k)$ and thus fall back into $L^2(\mathbb{R}^d)$. The product $g(-i\nabla)f(x)$ is therefore well defined on all $L^2(\mathbb{R}^d)$ with this interpretation. The situation is similar for the product in the other direction. To simplify the notations, we will therefore call in the following $f(x)g(-i\nabla)$ and $g(-i\nabla)f(x)$ the Hilbert-Schmidt operators defined on the whole space $L^2(\mathbb{R}^d)$.

Proof For any function $u \in C_c^\infty(\mathbb{R}^d)$, we have $fu \in L^1(\mathbb{R}^d) \cap L^2(\mathbb{R}^d)$ because $u \in L^2(\mathbb{R}^d) \cap L^\infty(\mathbb{R}^d)$. Its Fourier transform verifies $\widehat{fu} \in L^2(\mathbb{R}^d) \cap L^\infty(\mathbb{R}^d)$. This guarantees that $fu \in D(g(-i\nabla))$, that is $g\,\widehat{fu} \in L^2(\mathbb{R}^d)$, since $\widehat{fu} \in L^\infty(\mathbb{R}^d)$. Thus A_2 is well defined on $C_c^\infty(\mathbb{R}^d)$. An explicit calculation then shows that

$$(A_2 u)(x) = (2\pi)^{-\frac{d}{2}} \int_{\mathbb{R}^d} \check{g}(x - y)f(y)u(y)\,dy$$

for all $u \in C_c^\infty(\mathbb{R}^d)$. However, the integral kernel $a_2(x, y) = (2\pi)^{d/2}\check{g}(x - y)f(y)$ belongs to $L^2(\mathbb{R}^d \times \mathbb{R}^d)$ and defines a Hilbert-Schmidt operator. As this operator is bounded, by Theorem 5.21, it is the closure of A_2, previously defined on $C_c^\infty(\mathbb{R}^d)$. The argument is the same for A_1. $\qquad\square$

5.3.3 Operators $f(x)g(-i\nabla)$ for $f, g \in L^p(\mathbb{R}^d)$

We have shown that the operators $f(x)g(-i\nabla)$ and $g(-i\nabla)f(x)$ were compact (in fact Hilbert-Schmidt) on $L^2(\mathbb{R}^d)$ as soon as $f, g \in L^2(\mathbb{R}^d)$. The exponent 2 for f and g is not related to the fact that we are working in the Hilbert space $L^2(\mathbb{R}^d)$. It turns out that these operators are well defined for all $f, g \in L^p(\mathbb{R}^d)$ with $p \in [2, +\infty]$ and they are compact for $p < +\infty$.

Theorem 5.24 (Operators $f(x)g(-i\nabla)$ and $g(-i\nabla)f(x)$) *When*

$$f, g \in L^p(\mathbb{R}^d) \quad with \quad 2 \le p \le \infty,$$

the operators

$$A_1 = f(x)g(-i\nabla) \quad and \quad A_2 = g(-i\nabla)f(x)$$

first defined on $C_c^\infty(\mathbb{R}^d)$ are closable and their closures are bounded on the Hilbert space $L^2(\mathbb{R}^d)$ with

$$\|\overline{A_1}\| = \|\overline{A_2}\| \le (2\pi)^{-\frac{d}{p}} \|f\|_{L^p(\mathbb{R}^d)} \|g\|_{L^p(\mathbb{R}^d)}. \tag{5.29}$$

Moreover, these operators are compact when $p \in [2, +\infty)$ or when $p = +\infty$ and both f and g tend to 0 at infinity.

The operators A_1 and A_2 have the respective formal integral kernels

$$a_1(x, y) = (2\pi)^{-\frac{d}{2}} f(x)\check{g}(x - y), \qquad a_2(x, y) = (2\pi)^{-\frac{d}{2}} \check{g}(x - y)f(y)$$

but one must be careful that $\check{g}$ is in principle a tempered distribution. For instance, for $g \equiv 1$ we obtain $\check{g} = (2\pi)^{d/2}\delta_0$.

Proof Let $u \in C_c^\infty(\mathbb{R}^d)$ and $p \in [2, +\infty]$. Then, by Hölder's inequality, we have $fu \in L^q(\mathbb{R}^d)$ with $\|fu\|_{L^q(\mathbb{R}^d)} \le \|f\|_{L^p(\mathbb{R}^d)} \|u\|_{L^2(\mathbb{R}^d)}$ and $1/q = 1/2 + 1/p$. As $q = 2p/(p + 2) \in [1, 2)$, this implies that $\widehat{fu} \in L^{q'}(\mathbb{R}^d)$ where $1/q' + 1/q = 1$. We recall indeed that if $\varphi \in L^r(\mathbb{R}^d)$ with $1 \le r \le 2$, then $\widehat{\varphi} \in L^{r'}(\mathbb{R}^d)$ where $1/r + 1/r' = 1$ with

$$\|\widehat{\varphi}\|_{L^{r'}(\mathbb{R}^d)} \le (2\pi)^{-\frac{d(2-r)}{2r}} \|\varphi\|_{L^r(\mathbb{R}^d)} \tag{5.30}$$

(Hausdorff-Young inequality). We therefore have

$$\|\widehat{fu}\|_{L^{q'}(\mathbb{R}^d)} \le (2\pi)^{-\frac{d}{p}} \|f\|_{L^p(\mathbb{R}^d)} \|u\|_{L^2(\mathbb{R}^d)}$$

because $(2 - q)/2q = 1/p$. Furthermore, we also have $\widehat{fu} \in L^2(\mathbb{R}^d)$ because $fu \in L^p(\mathbb{R}^d) \cap L^q(\mathbb{R}^d) \subset L^2(\mathbb{R}^d)$. Thus $fu \in D\big(g(-i\nabla)\big)$ because, again by Hölder's inequality,

$$\|g\,\widehat{fu}\|_{L^2(\mathbb{R}^d)} \leq \|g\|_{L^p(\mathbb{R}^d)}\|\widehat{fu}\|_{L^{q'}(\mathbb{R}^d)} \leq (2\pi)^{-\frac{d}{p}} \|f\|_{L^p(\mathbb{R}^d)}\|g\|_{L^p(\mathbb{R}^d)}\|u\|_{L^2(\mathbb{R}^d)},$$

noting that $1/2 = 1/p + 1/q'$. In conclusion, we have shown that A_2 is well defined on $C_c^\infty(\mathbb{R}^d)$ and satisfies the estimate

$$\|A_2 u\|_{L^2(\mathbb{R}^d)} \leq (2\pi)^{-\frac{d}{p}} \|f\|_{L^p(\mathbb{R}^d)} \|g\|_{L^p(\mathbb{R}^d)} \|u\|_{L^2(\mathbb{R}^d)}$$

on this space. This shows that A_2 is closable and that its closure is a bounded operator on all $L^2(\mathbb{R}^d)$, which verifies as we wanted

$$\|\overline{A_2}\| \leq (2\pi)^{-\frac{d}{p}} \|f\|_{L^p(\mathbb{R}^d)} \|g\|_{L^p(\mathbb{R}^d)}.$$

The argument is valid for all $2 \leq p \leq \infty$ and it is similar for A_1.

Let us now show that $\overline{A_2}$ is compact when $2 \leq p < \infty$ or if $p = +\infty$ but we add the additional assumption that $f, g \to 0$ at infinity. For this, it is enough to notice that in all these cases we can approach f and g by sequences (f_k) and (g_k) of functions in $C_c^\infty(\mathbb{R}^d)$ (resp. $L_c^\infty(\mathbb{R}^d)$ if $p = \infty$) that converge to f and g in $L^p(\mathbb{R}^d)$. We then have, by the previous argument,

$$\|g(-i\nabla)f(x) - g_k(-i\nabla)f_k(x)\|$$

$$\leq \|(g - g_k)(-i\nabla)f(x)\| + \|g_k(-i\nabla)(f - f_k)(x)\|$$

$$\leq (2\pi)^{-\frac{d}{p}} \big(\|f\|_{L^p(\mathbb{R}^d)} \|g - g_k\|_{L^p(\mathbb{R}^d)} + \|f - f_k\|_{L^p(\mathbb{R}^d)} \|g_k\|_{L^p(\mathbb{R}^d)}\big) \to 0.$$

However, each of the operators $g_k(-i\nabla)f_k(x)$ is Hilbert-Schmidt therefore compact by Proposition 5.23. Hence, the operator $g(-i\nabla)f(x)$ is also compact, as the limit in norm of a sequence of compact operators. For A_1 we can use for example that $A_1 = \mathcal{F}^{-1} f(-i\nabla)g(x)\mathcal{F}$ which is compact because the Fourier transform is unitary. $\square$

Remark 5.25 Even if the function $x \mapsto |x|^{-d/p}$ is (just) not in $L^p(\mathbb{R}^d)$, it turns out that the operator $f(x)g(-i\nabla)$ remains compact if we replace f or g by $|x|^{-d/p}$. This follows from the Hardy-Littlewood-Sobolev inequality [LL01] which stipulates that

$$\left\| f * |x|^{-s} \right\|_{L^p(\mathbb{R}^d)} \leq C \|f\|_{L^q(\mathbb{R}^d)}, \qquad \text{for} \quad 1 < p, q < \infty, \quad 1 + \frac{1}{p} = \frac{1}{q} + \frac{s}{d}, \tag{5.31}$$

and which allows us to replace the Hausdorff-Young inequality used in the proof.

5.4   Operators with Compact Resolvent

Definition 5.26 (Operators with Compact Resolvent) We say that a self-adjoint operator A has **compact resolvent** if $(A + i)^{-1}$ is compact.

For all $z \in \rho(A)$, we have

$$(A - z)^{-1} = (A + i)^{-1} \frac{A + i}{A - z} = (A + i)^{-1} \underbrace{\left(1 + (z + i)(A - z)^{-1} \right)}_{\in \mathcal{B}(\mathfrak{H})} \tag{5.32}$$

which is therefore compact when $(A + i)^{-1}$ is. This argument shows that it is equivalent to ask that $(A - z)^{-1}$ is compact for all or for one $z \in \rho(A)$. Here is then a corollary of the same type as for compact operators.

Corollary 5.27 (Operators with Compact Resolvent) *Let A be a self-adjoint operator on its domain $D(A) \subset \mathfrak{H}$, with $\dim(\mathfrak{H}) = +\infty$. Then A has compact resolvent if and only if $\sigma_{\text{ess}}(A) = \emptyset$. In this case, A is unbounded and its spectrum is composed of a sequence (λ_n) of eigenvalues of finite multiplicity which verify*

$$\lim_{n \to \infty} |\lambda_n| = +\infty.$$

In particular, A is diagonalizable in an orthonormal basis.

Proof Suppose that $(A + i)^{-1}$ is compact and consider $\lambda \in \sigma_{\text{ess}}(A)$, with an associated Weyl sequence $v_n \rightharpoonup 0$. Then

$$0 \leftarrow (A - \lambda)v_n = (A + i)v_n - (\lambda + i)v_n$$

so that, by continuity of $(A+i)^{-1}$, we find that $v_n - (\lambda+i)(A+i)^{-1}v_n \to 0$ strongly. But $v_n \rightharpoonup 0$ and the compactness of $(A + i)^{-1}$ then implies that $(A + i)^{-1}v_n \to 0$ strongly, which implies $v_n \to 0$ and contradicts $\|v_n\| = 1$. Thus, $\sigma_{\text{ess}}(A) = \emptyset$.

Conversely, if $\sigma_{\text{ess}}(A) = \emptyset$ the spectrum of A is composed only of isolated eigenvalues of finite multiplicity, which must necessarily accumulate at $\pm\infty$ otherwise $\mathfrak{H}$ would be of finite dimension by the spectral theorem. Let in this case $a \in \mathbb{R} \setminus \sigma(A)$. By functional calculus and Remark 5.16, the spectrum of the self-adjoint operator $(A - a)^{-1}$ is composed of a sequence of eigenvalues of finite multiplicity, which tend to 0. This means that $\sigma_{\text{ess}}((A - a)^{-1}) = \{0\}$ and therefore, according to Corollary 5.18, that $(A - a)^{-1}$ is compact. $\square$

5.4.1   Application: Laplacian(s) on the Interval $I = (0, 1)$

Recall that the Laplacian A_V on $I = (0, 1)$ was constructed in Sect. 2.8.3, for all $V \subset \mathbb{C}^4$ an isotropic subspace of the matrix (2.27) of dimension 2. It is defined by $A_V v = -v''$ on the domain

$$D(A_V) = \left\{ v \in H^2(I) \; : \; (v(0), v'(0), v(1), v'(1)) \in V \right\}.$$

More precisely, we have seen in Theorem 2.44 that

$$\boxed{D(A_V) = H_0^2(I) + \text{vect}(v_1, v_2)} \tag{5.33}$$

where $v_1, v_2 \in H^2(I)$ are such that $(v_i(0), v_i'(0), v_i(1), v_i'(1))$ form a basis of V.

Theorem 5.28 (Spectrum of the Laplacian on $I = (0, 1)$) *Let $V \subset \mathbb{C}^4$ be an isotropic subspace of the matrix (2.27) with $\dim(V) = 2$ and A_V the corresponding self-adjoint Laplacian. Then, A_V has a compact resolvent and its spectrum is composed of a sequence of eigenvalues of finite multiplicity tending to $+\infty$. Moreover, A_V has at most two strictly negative eigenvalues (counted with multiplicity), that is, $\mathbb{1}_{(-\infty,0)}(A_V)$ is of rank less than or equal to 2.*

Proof Consider a sequence $v_n \rightharpoonup 0$ weakly in $L^2(I)$ and let $w_n := (A_V + i)^{-1} v_n$ which we want to show converges strongly to 0. As $(A_V + i)^{-1}$ is a bounded operator, w_n is bounded and we even have $w_n \rightharpoonup 0$ weakly. But $-w_n'' + i w_n = v_n$, which shows that w_n'' is bounded in $L^2(I)$. By Lemma A.4 (elliptic regularity on an interval), we deduce that w_n is bounded in $H^2(I)$ and, finally, that $w_n \to 0$ strongly in $\mathfrak{H} = L^2(I)$, according to the compactness of the embedding $H^2(I) \hookrightarrow L^2(I)$ (Rellich-Kondrachov Theorem A.18). This proves the compactness of $(A_V + i)^{-1}$.

Finally, let us show that $\mathbb{1}_{(-\infty,0)}(A_V)$ has rank at most equal to 2, which will imply *a fortiori* that A_V is bounded from below, and therefore that the eigenvalues must accumulate at $+\infty$. By contradiction, if $\mathbb{1}_{(-\infty,0)}(A_V)$ is of rank three or more, then the same is true for $\mathbb{1}_{(-M,0)}(A_V)$ when M is large enough. But in this case the image of $\mathbb{1}_{(-M,0)}(A_V)$ (which is included in $D(A_V)$ because M is finite) must intersect $H_0^2(I)$ which is of co-dimension 2 in $D(A_V)$, according to (5.33). Let $v \neq 0$ be in this intersection. As $v \in H_0^2(I)$ we have seen that the boundary terms vanish in the integration by parts, so that

$$\langle v, A_V v \rangle = \int_0^1 |v'(t)|^2 \, dt \geq 0.$$

This contradicts the fact that $v \in \mathbb{1}_{(-M,0)}(A_V)\mathfrak{H}$ because on this space, we have $\langle v, A_V v \rangle < 0$ for all $v \neq 0$.   □

The behavior of the eigenvalues of the Laplacian with Robin boundary condition (depending on the associated parameter θ) is studied below in Exercise 5.56.

5.4.2 Application: Confining Potential

We have a similar result in the case of a potential that tends to infinity at infinity.

Theorem 5.29 (Spectrum of Schrödinger Operators with Confining Potential)
*Let V be a real-valued measurable function with $V_- \in L^p(\mathbb{R}^d, \mathbb{R}) + L^\infty(\mathbb{R}^d, \mathbb{R})$
where p satisfies*

$$
\begin{cases}
p = 1 & \text{if } d = 1, \\
p > 1 & \text{if } d = 2, \\
p = \frac{d}{2} & \text{if } d \geq 3
\end{cases}
\tag{5.34}
$$

and $V_+ \in L^1_{\text{loc}}(\mathbb{R}^d)$ is such that

$$
\lim_{|x| \to \infty} V_+(x) = +\infty.
$$

Let $H = -\Delta + V$ be the Friedrichs self-adjoint realization constructed in Theorem 3.22. Then H has a compact resolvent. Its spectrum is composed of eigenvalues of finite multiplicity, which tend to $+\infty$. In particular, H is diagonalizable in an orthonormal basis.

Proof As we have little information about the domain of $-\Delta + V$, we rather work with the quadratic form. According to Lemma 1.10, the operator $-\Delta/2 - V_-$ is bounded-below. So we can choose a constant C such that $-\Delta/2 - V_- \geq -C + 1$ and obtain

$$
H + C \geq -\frac{\Delta}{2} + V_+ + 1 \geq 1.
\tag{5.35}
$$

In particular, $-C \in \rho(H)$. Let v_n be a sequence in $L^2(\mathbb{R}^d)$ that converges weakly to 0 and introduce $w_n = (H + C)^{-1} v_n$ which we want to show converges strongly to 0. As $(H + C)^{-1}$ is a bounded operator, we already know that $w_n \rightharpoonup 0$ in $L^2(\mathbb{R}^d)$. Moreover, $w_n \in D(H) \subset H^1(\mathbb{R}^d)$ and $(H + C)w_n = v_n$. By taking the scalar product with w_n and using (5.35), we obtain

$$
\frac{1}{2} \int_{\mathbb{R}^d} |\nabla w_n|^2 + \int_{\mathbb{R}^d} V_+ |w_n|^2 + \int_{\mathbb{R}^d} |w_n|^2 \leq \langle w_n, (H + C)w_n \rangle
$$

$$
= \int_{\mathbb{R}^d} \overline{w_n} v_n \leq \|w_n\| \, \|v_n\| = O(1).
$$

The sequence (w_n) is therefore bounded in the energy space $\mathcal{V}$ introduced in (3.26). In particular (w_n) is bounded in $H^1(\mathbb{R}^d)$ hence converges strongly locally to 0 by the Rellich-Kondrachov Theorem A.18. Furthermore, $\sqrt{V_+}w_n$ is bounded in $L^2(\mathbb{R}^d)$. By writing

$$\int_{\mathbb{R}^d} |w_n|^2 = \int_{B_R} |w_n|^2 + \int_{(B_R)^c} |w_n|^2 \le \int_{B_R} |w_n|^2 + \frac{1}{\inf_{(B_R)^c} V_+} \int_{(B_R)^c} V_+ |w_n|^2,$$

we deduce that $w_n \to 0$ strongly. Indeed, the first term tends to 0 for any fixed R and the second is small when R is large enough, since $V_+ \to \infty$ by hypothesis. $\quad\square$

Example 5.30 (Diagonalization of the Harmonic Oscillator)   Here we provide some hints for calculating the spectrum of the harmonic oscillator $H = -d^2/dx^2 + \omega^2 x^2$ where $\omega > 0$, but leave the details as an exercise. The domain of H was determined earlier in (3.25) and the ground state energy in Exercise 1.28. The idea is to notice that (at least formally [RS75, Thm. X.25]),

$$H = -\frac{d^2}{dx^2} + \omega^2 x^2 = \left(\frac{d}{dx} + \omega x\right)\left(-\frac{d}{dx} + \omega x\right) - \omega$$

$$= \left(-\frac{d}{dx} + \omega x\right)\left(\frac{d}{dx} + \omega x\right) + \omega. \tag{5.36}$$

From an energy perspective, this is written as

$$q_H(u) = \int_{\mathbb{R}} |u'(x) + \omega x u(x)|^2 \, dx + \omega \int_{\mathbb{R}} |u(x)|^2 \, dx \tag{5.37}$$

where each term makes sense in $Q(H)$, as we have already seen in Exercise 1.28. The first term of (5.37) vanishes exactly for the solutions of the equation $u'(x) + \omega x u(x) = 0$, that is, the multiples of

$$f_0(x) = \left(\frac{\omega}{\pi}\right)^{\frac{1}{4}} e^{-\frac{\omega x^2}{2}}.$$

The multiplicative factor was, as usual, chosen so that f_0 is normalized in $L^2(\mathbb{R})$. Thus the first eigenvalue is $\min \sigma(H) = \omega$ and it is non-degenerate. Moreover, by induction we can show using the relation (5.36) that the function

$$f_n(x) = \frac{1}{\sqrt{n!(2\omega)^{\frac{n}{2}}}} \left(-\frac{d}{dx} + \omega x\right)^n f_0(x) = \frac{\omega^{\frac{1}{4}}}{\pi^{\frac{1}{4}}\sqrt{n!(2\omega)^{\frac{n}{2}}}} P_n(\sqrt{\omega}x) e^{-\frac{\omega x^2}{2}}$$

is a normalized eigenvector of H, with eigenvalue $(2n + 1)\omega$. The functions $P_n(x) = e^{x^2}\frac{d^n}{dx^n}e^{x^2}$ are the Hermite polynomials, which form an orthogonal basis of $L^2(\mathbb{R}, e^{-x^2}dx)$. The (f_n) therefore form an orthonormal basis of $L^2(\mathbb{R})$ and we have diagonalized H with

$$\boxed{\sigma(H) = (2\mathbb{N} + 1)\omega.}$$

In dimension $d \geq 2$, we obtain a basis of $L^2(\mathbb{R}^d)$ by forming the tensor products of the f_n. More precisely, $f_{n_1}(x_1) \cdots f_{n_d}(x_d)$ is an eigenvector associated with the eigenvalue $2(n_1 + \cdots + n_d)\omega + d\omega$. The spectrum is therefore equal to

$$\boxed{\sigma(H) = (2\mathbb{N} + d)\omega.}$$

The multiplicity of the eigenvalue $(2n + 1)\omega$ is the number of ways that n can be written as a sum of integers $n = n_1 + \cdots + n_d$, which is $\frac{(n+d-1)!}{n!(d-1)!}$.

5.4.3 Application: Laplacian on a Bounded Open Set*

The argument is similar for the Robin Laplacian on any bounded open set $\Omega \subset \mathbb{R}^d$, introduced earlier in Sect. 3.3.4.

Theorem 5.31 (Spectrum of the Laplacian on $\Omega \subset \mathbb{R}^d$) *Let Ω be a bounded open set of $\mathbb{R}^d$ whose boundary is piecewise Lipschitz. The Robin Laplacian $(-\Delta)_{\mathrm{Rob},\theta}$ defined in Sect. 3.3.4 has a compact resolvent for all $\theta \in [0, 1)$ and its spectrum is composed of a sequence of eigenvalues of finite multiplicity, which diverge to $+\infty$.*

Proof Recall that the quadratic form of the Robin Laplacian is given by (3.37). As before, let $w_n := ((-\Delta)_{\mathrm{Rob},\theta} + C)^{-1}v_n$ where $v_n \rightharpoonup 0$ in $L^2(\Omega)$ and with C large enough. Then $-\Delta w_n + C w_n = v_n$ which, when taking the scalar product against $w_n \in H^1(\Omega)$, implies

$$\int_\Omega |\nabla w_n|^2 + \frac{1}{\tan(\pi\theta)} \int_{\partial\Omega} |w_n|^2 + C \int_\Omega |w_n|^2 \leq \|v_n\| \|w_n\| = O(1).$$

Thanks to inequality (A.15), this proves that w_n is bounded in $H^1(\Omega)$. This provides the expected compactness according to the Rellich-Kondrachov Theorem A.18. $\square$

5.5 Weyl's Theory on the Invariance of the Essential Spectrum

In this section we study under what condition on B we have $\sigma_{\mathrm{ess}}(A + B) = \sigma_{\mathrm{ess}}(A)$.

5.5.1   Perturbations Leaving the Essential Spectrum Invariant

We have seen in Chap. 3 the notion of (infinitesimally) relatively bounded perturbations. We introduce here a stronger notion, under which the essential spectrum will be unchanged.

Definition 5.32 (Relatively Compact Perturbations)   Let A be a self-adjoint operator on $D(A) \subset \mathfrak{H}$ and B a symmetric operator on $D(A)$. We say that B is A-**compact** when the operator $B(A + i)^{-1}$ is compact (thus bounded) in $\mathfrak{H}$.

For all $z \in \rho(A)$ we can write according to (5.32)

$$B(A - z)^{-1} = B(A + i)^{-1} \left(1 + (z + i)(A - \lambda)^{-1} \right)$$

so that B is A-compact if $B(A - z)^{-1}$ is compact for one or for all $z \in \rho(A)$.

Lemma 5.33 (A-compact $\Longrightarrow$ Infinitesimally A-bounded)   *If B is a symmetric A-compact operator, then B is infinitesimally A-bounded, which means that for all ε, there exists C_ε such that*

$$\|Bv\| \leq \varepsilon\|Av\| + C_\varepsilon\|v\|, \qquad \forall v \in D(A).$$

Proof   We write

$$B(A + i\mu)^{-1} = B(A + i)^{-1} \frac{A + i}{A + i\mu}.$$

For $\mu \geq 1$, we introduce $f_\mu(x) = (x + i)/(x + i\mu)$, and note that

$$\forall x \in \mathbb{R}, \qquad |f_\mu(x)| = \left| \frac{x + i}{x + i\mu} \right| = \left(\frac{x^2 + 1}{x^2 + \mu^2} \right)^{1/2} \leq 1$$

so that, by functional calculus (assertion (ii) of Theorem 4.8),

$$\|f_\mu(A)\| = \left\| \frac{A + i}{A + i\mu} \right\| \leq 1.$$

Moreover, the function $f_\mu(x)$ converges to 0 for all fixed x, when $\mu \to \infty$. This shows, again by functional calculus (assertion (v) of Theorem 4.8), that

$$\lim_{\mu \to \infty} f_\mu(A)\, v = \lim_{\mu \to \infty} \overline{f_\mu}(A)\, v = 0$$

for all fixed $v \in \mathfrak{H}$. As $B(A + i)^{-1}$ is compact, we can write $B(A + i)^{-1} = C + R$ where $\|C\| \leq \varepsilon$ and $R = \sum_{j=1}^{J} |w_j\rangle\langle v_j|$ has finite rank. Then

$$Rf_\mu(A) = \sum_{j=1}^{J} |w_j\rangle\langle \overline{f_\mu}(A)v_j|$$

so that

$$\left\| R\frac{A+i}{A+i\mu} \right\| \le \sum_{j=1}^{J} \|w_j\| \, \|\overline{f_\mu}(A)v_j\|$$

where $\|\overline{f_\mu}(A)v_j\| \to 0$ for all $j = 1, \dots, J$, as explained previously. Moreover

$$\left\| C\frac{A+i}{A+i\mu} \right\| \le \|C\| \le \varepsilon.$$

By passing to the limit we obtain

$$\limsup_{\mu\to\infty} \left\| B(A+i\mu)^{-1} \right\| = \limsup_{\mu\to\infty} \left\| B(A+i)^{-1}\frac{A+i}{A+i\mu} \right\| \le \|C\| \le \varepsilon.$$

Taking finally $\varepsilon \to 0$ provides

$$\lim_{\mu\to\infty} \left\| B(A+i\mu)^{-1} \right\| = 0. \tag{5.38}$$

As

$$\|Bv\| = \left\| B(A+i\mu)^{-1}(A+i\mu)v \right\| \le \left\| B(A+i\mu)^{-1} \right\| (\|Av\| + \mu \|v\|),$$

where the coefficient of $\|Av\|$ tends to 0 when $\mu \to \infty$, this implies the result. $\square$

Here is now a result that provides the invariance of the essential spectrum when B is A-compact.

Theorem 5.34 (Weyl I) *Let A be a self-adjoint operator on its domain $D(A) \subset \mathfrak{H}$ and B a symmetric operator that is A-compact. Then $A + B$ is self-adjoint on $D(A)$ and $\sigma_{\mathrm{ess}}(A + B) = \sigma_{\mathrm{ess}}(A)$.*

Proof The self-adjointness of $A + B$ on $D(A)$ follows from the Rellich-Kato Theorem 3.1 and Lemma 5.33. If $\lambda \in \sigma_{\mathrm{ess}}(A)$, there exists a sequence $v_n \in D(A)$ such that $(A - \lambda)v_n \to 0$ and $v_n \rightharpoonup 0$. Then $(A + B - \lambda)v_n = (A - \lambda)v_n - B(A + i)^{-1}(A+i)v_n$ converges to 0 strongly because $(A+i)v_n = (A - \lambda)v_n + (\lambda + i)v_n$ tends to 0 weakly and $B(A + i)^{-1}$ is compact. Thus $\lambda \in \sigma_{\mathrm{ess}}(A + B)$ and we have shown that $\sigma_{\mathrm{ess}}(A) \subset \sigma_{\mathrm{ess}}(A + B)$.

The reverse inclusion follows by swapping the roles of A and $A + B$ but we must first show that $-B$ is $A + B$ compact. This follows from the relation $A + B + i\mu = \left(1 + B(A + i\mu)^{-1}\right)(A + i\mu)$ which implies that

$$B(A + B + i\mu)^{-1} = B(A + i\mu)^{-1}\left(1 + B(A + i\mu)^{-1}\right)^{-1}$$

is indeed compact. The operator on the right is invertible as soon as $\|B(A + i\mu)^{-1}\| < 1$, which is the case for μ large enough according to (5.38). $\square$

Remark 5.35 Self-adjointness is important for the stability of the essential spectrum. In Exercise 5.53 we give the example of a non-self-adjoint bounded operator, to which we add a finite rank operator (thus compact), but which greatly modifies the essential spectrum.

There are several other versions of the same type as Theorem 5.34. For example, instead of the compactness of $B(A + i)^{-1}$, we can ask for that of $(A + B + i)^{-1} - (A + i)^{-1}$. As

$$(A + B + i)^{-1} - (A + i)^{-1} = -(A + B + i)^{-1}B(A + i)^{-1}$$

where $(A + B + i)^{-1}$ is bounded, it is in general a weaker assumption.

Theorem 5.36 (Weyl II) *Let A and A' be two arbitrary self-adjoint operators on their respective domains $D(A), D(A') \subset \mathfrak{H}$. If there exists $z \in \rho(A) \cap \rho(A')$ such that $(A - z)^{-1} - (A' - z)^{-1}$ is compact, then $\sigma_{\mathrm{ess}}(A) = \sigma_{\mathrm{ess}}(A')$.*

Proof The essential spectra of $(A - z)^{-1}$ and of $(A' - z)^{-1}$ were calculated in terms of those of A and A' in Remark 5.16 and it is characterized with Weyl sequences as in Theorem 5.15. The proof is then the same as in Theorem 5.34. If $\lambda \in \sigma_{\mathrm{ess}}(A)$, there exists a sequence (v_n) such that $\|v_n\| = 1$ and $((A - z)^{-1} - (\lambda - z)^{-1})v_n \to 0$. The compactness of $(A - z)^{-1} - (A' - z)^{-1}$ implies

$$((A' - z)^{-1} - (\lambda - z)^{-1})v_n \to 0,$$

which shows that $(\lambda - z)^{-1} \in \sigma_{\mathrm{ess}}((A' - z)^{-1})$, and therefore that $\lambda \in \sigma_{\mathrm{ess}}(A')$. We obtain the other inclusion by swapping the roles of A and A'. $\square$

There is also a slightly more complicated version involving only quadratic forms.

Theorem 5.37 (Weyl III) *Let A be a coercive self-adjoint operator on its domain $D(A) \subset \mathfrak{H}$. Let b be a quadratic form on $Q(A)$, such that*

$$|b(v)| \leq \eta\, q_A(v) + \kappa\, \|v\|^2, \qquad \forall v \in Q(A), \tag{5.39}$$

for a real $0 \leq \eta < 1$. Let C be the unique self-adjoint operator associated with the closed quadratic form $v \mapsto q_A(v) + b(v)$ on $Q(A)$ (KLMN Theorem 3.19). If b is continuous for the weak topology of $Q(A)$, that is

$$\lim_{n \to \infty} b(v_n) = b(v) \tag{5.40}$$

for any sequence $v_n \rightharpoonup v$ weakly in the Hilbert space $(Q(A), \varphi_A)$, then $\sigma_{\mathrm{ess}}(C) = \sigma_{\mathrm{ess}}(A)$.

Proof Let us set $B = C - A$, an operator that is bounded from $Q(A)$ to $Q(A)'$, as discussed at the end of Sect. 4.6. The operator $K = A^{-1/2}BA^{-1/2}$ is well defined and bounded on $\mathfrak{H}$. The operator $A^{-1/2}$ is bounded from $Q(A)'$ to $\mathfrak{H}$, since $Q(A) = D(\sqrt{A})$. After polarization, the assumption (5.40) can be rewritten as $\lim_{n\to\infty} b(v'_n, v_n) = 0$ for all sequences $v_n, v'_n \rightharpoonup 0$ in $Q(A)$, where $b(\cdot, \cdot)$ denotes by abuse of notation the sesquilinear form associated with b. These sequences can precisely be written $v_n = A^{-1/2}w_n$ and $v'_n = A^{-1/2}w'_n$ where $w_n, w'_n \rightharpoonup 0$ in $\mathfrak{H}$. Thus, the hypothesis (5.40) means that

$$\lim_{n\to\infty} \langle w'_n, Kw_n \rangle = \lim_{n\to\infty} \langle w'_n, A^{-\frac{1}{2}}BA^{-\frac{1}{2}}w_n \rangle = 0$$

for all sequences $w_n, w'_n \rightharpoonup 0$ in $\mathfrak{H}$. As K is bounded we can take $w'_n = Kw_n \rightharpoonup 0$, which implies that $\|Kw_n\| \to 0$ for any sequence $w_n \rightharpoonup 0$, therefore K is compact. The hypothesis (5.40) is thus a reformulation of the fact that $K = A^{-1/2}BA^{-1/2}$ is compact. Moreover, we have for α, β large enough

$$\frac{1}{\beta}\left\|(A+\alpha)^{\frac{1}{2}}v\right\|^2 \le q_A(v) + b(v) + \alpha\|v\|^2 = \left\|(C+\alpha)^{\frac{1}{2}}v\right\|^2 \le \beta\left\|(A+\kappa)^{\frac{1}{2}}v\right\|^2,$$

for all $v \in Q(A) = Q(A')$, according to the hypothesis (5.39). Written differently, this means that $(A+\alpha)^{1/2}(C+\alpha)^{-1/2}$ and $(C+\alpha)^{1/2}(A+\alpha)^{-1/2}$ are bounded operators. In conclusion, the operator

$$(C+\alpha)^{-1} - (A+\alpha)^{-1}$$

$$= (A+\alpha)^{-1}B(C+\alpha)^{-1}$$

$$= \underbrace{\frac{\sqrt{A}}{A+\alpha}}_{\in\mathcal{B}(\mathfrak{H})} \underbrace{A^{-\frac{1}{2}}BA^{-\frac{1}{2}}}_{=K\,\in\mathcal{K}(\mathfrak{H})} \underbrace{\frac{\sqrt{A}}{\sqrt{A+\kappa}}}_{\in\mathcal{B}(\mathfrak{H})} \underbrace{(A+\alpha)^{\frac{1}{2}}(C+\alpha)^{-\frac{1}{2}}}_{\in\mathcal{B}(\mathfrak{H})} \underbrace{(C+\alpha)^{-\frac{1}{2}}}_{\in\mathcal{B}(\mathfrak{H})}$$

is compact, which concludes the proof, by Theorem 5.36. $\square$

5.5.2 Essential Spectrum of Schrödinger Operators

We have already proven in Lemma 1.10 that any potential $V \in L^p(\mathbb{R}^d, \mathbb{R}) + L^\infty_\varepsilon(\mathbb{R}^d, \mathbb{R})$ negligible at infinity and satisfying the hypothesis

$$\begin{cases} p = 1 & \text{if } d = 1, \\ p > 1 & \text{if } d = 2, \\ p = \frac{d}{2} & \text{if } d \ge 3. \end{cases} \tag{5.41}$$

provided an energy

$$u \in H^1(\mathbb{R}^d) \mapsto \int_{\mathbb{R}^d} V(x)|u(x)|^2 \, dx$$

that is weakly continuous for the norm $H^1(\mathbb{R}^d)$. By Theorem 5.37, this immediately implies the following result.

Corollary 5.38 (Essential Spectrum of Schrödinger Operators) *Let* $V \in L^p(\mathbb{R}^d, \mathbb{R}) + L_\varepsilon^\infty(\mathbb{R}^d, \mathbb{R})$ *be negligible at infinity with* p *satisfying Assumption* (5.41). *Let* $-\Delta + V$ *be the Friedrichs realization obtained in Corollary 3.20. Then*

$$\boxed{\sigma_{\mathrm{ess}}(-\Delta + V) = \sigma_{\mathrm{ess}}(-\Delta) = \mathbb{R}_+.}$$

Example 5.39 For the hydrogen atom, we obtain $\sigma_{\mathrm{ess}}(-\Delta/2 - 1/|x|) = \mathbb{R}_+$ in $L^2(\mathbb{R}^3)$, as announced in Chap. 1, Fig. 1.5.

5.6   Discrete Spectrum and Courant-Fischer Formula

After studying the essential spectrum we now discuss the presence or absence of a discrete spectrum.

5.6.1   Courant-Fischer Formula

We start with the Courant-Fischer formula [Cou20, Fis05], which generalizes a similar formula for Hermitian matrices.

Theorem 5.40 (Courant-Fischer) *Let* A *be a self-adjoint operator bounded from below on its domain* $D(A) \subset \mathfrak{H}$ *and* $\Sigma(A) := \min \sigma_{\mathrm{ess}}(A) \in \mathbb{R} \cup \{+\infty\}$ *the bottom of its essential spectrum. Then*

$$\boxed{\mu_k(A) := \inf_{\substack{W \subset D(A) \\ \dim(W) = k}} \max_{\substack{v \in W \\ \|v\|_{\mathfrak{H}} = 1}} \langle v, Av \rangle = \inf_{\substack{W \subset Q(A) \\ \dim(W) = k}} \max_{\substack{v \in W \\ \|v\|_{\mathfrak{H}} = 1}} q_A(v)} \tag{5.42}$$

is equal to

- *the* k-*th eigenvalue of* A *counted with multiplicity if* $\mathbb{1}_{(-\infty, \Sigma(A))}(A)$ *has rank equal to* k *or more;*
- $\Sigma(A)$ *otherwise.*

If $\mu_k(A) < \Sigma(A)$, *the infimum of* (5.42) *is exactly reached for the spaces W that are generated by k eigenvectors of A, whose associated eigenvalues are all less than or equal to* $\mu_k(A)$. *Another formula for* $\mu_k(A)$ *is given by*

$$\mu_k(A) = \sup_{\substack{W \subset D(A) \\ \dim(W^\perp) = k-1}} \inf_{\substack{v \in W \\ \|v\|_{\mathfrak{H}} = 1}} \langle v, Av \rangle = \sup_{\substack{W \subset Q(A) \\ \dim(W^\perp) = k-1}} \inf_{\substack{v \in W \\ \|v\|_{\mathfrak{H}} = 1}} q_A(v). \qquad (5.43)$$

The Courant-Fischer formula (5.42) is often called *Rayleigh-Ritz* [Str71, Rit09] in physics and *Hylleraas-Undheim-McDonald (HUM)* [HU30, Mac33] in quantum chemistry, named after those who first used it to calculate an approximation of the eigenvalues of a self-adjoint operator. Various other authors have in fact used similar formulas before, such as Weber [Web69] and Poincaré [Poi90] in the nineteenth century. The formula implies that

$$\mu_k(A) \leq \inf_{\substack{W \subset D \\ \dim(W) = k}} \max_{\substack{v \in W \\ \|v\|_{\mathfrak{H}} = 1}} q_A(v) \qquad (5.44)$$

for any space $\mathcal{D} \subset Q(A)$ of dimension $d \geq k$. By taking a basis $(e_1, \ldots, e_d)$ of $\mathcal{D}$, we see that the term on the right of (5.44) is nothing but the k-th eigenvalue $\lambda_k(M_{\mathcal{D}})$ of the $d \times d$ matrix

$$(M_{\mathcal{D}})_{ij} := \varphi_A(e_i, e_j).$$

The inequality (5.44) therefore ensures that $\lambda_k(M_{\mathcal{D}})$ will always be an upper bound to the true k-th eigenvalue of A. In practice, we seek to increase the space $\mathcal{D}$ so that this eigenvalue converges towards that of A. By allowing the space $\mathcal{D}$ to vary, we can also express $\mu_k(A)$ in the form

$$\mu_k(A) = \inf_{\substack{\mathcal{D} \subset Q(A) \\ \dim(\mathcal{D}) \geq k}} \lambda_k(M_{\mathcal{D}}). \qquad (5.45)$$

Exercise 5.41 (Variational Principle for the Sum of Eigenvalues)   Justify (5.45) and then show that

$$\sum_{j=1}^{k} \mu_j(A) = \inf_{\substack{\mathcal{D} \subset Q(A) \\ \dim(\mathcal{D}) = k}} \mathrm{tr}\left(M_{\mathcal{D}}\right). \qquad (5.46)$$

This provides a characterization of the sum of the first k eigenvalues of an operator (when they exist), which is very useful for fermionic particles, as we will see later in Chap. 6. Formula (5.46) is usually attributed to Fan [Fan49].

The proof of Theorem 5.40 essentially relies on the fundamental property that any space of dimension k must intersect the orthogonal of a space of dimension $< k$.

Proof The equality of the two formulas on the right of (5.42) is shown using the density of $D(A)$ in $Q(A)$ for the associated norm. The numbers $\mu_k(A)$ form a non-decreasing sequence: $\mu_1(A) \le \mu_2(A) \le \cdots$. Let $\lambda_k(A)$ be the k-th eigenvalue of A below $\Sigma(A)$, counted with multiplicity, assuming it exists. Let W be the subspace generated by k eigenvectors v_j corresponding to the eigenvalues $\lambda_j(A)$ with $j \le k$. For any v in W, we have

$$\langle v, Av \rangle = \sum_{j=1}^{k} \lambda_j(A)|\langle v, v_j \rangle|^2 \le \lambda_k(A) \sum_{j=1}^{k} |\langle v, v_j \rangle|^2 = \lambda_k(A)\|v\|^2,$$

so that $\mu_k(A) \le \lambda_k(A)$. If A has less than k eigenvalues strictly below $\Sigma(A)$, we can use that $\mathbb{1}_{(-\infty, \Sigma(A)+\varepsilon]}(A)$ has infinite rank for all $\varepsilon > 0$ (as seen in the proof of Theorem 5.15). By taking W any subspace of dimension k in the range of the spectral projector $\mathbb{1}_{(-\infty, \Sigma(A)+\varepsilon]}(A)$, we have by the spectral theorem $\langle v, Av \rangle \le (\Sigma(A) + \varepsilon)\|v\|^2$ for all $v \in W$. By taking $\varepsilon \to 0$ we have therefore shown that $\mu_k(A) \le \Sigma(A)$ for all k.

It remains to prove the reverse inequality, which we do by induction on $k \ge 1$. For $k = 1$, the relation

$$\mu_1(A) = \inf_{\substack{v \in D(A) \\ \|v\|_{\mathfrak{H}}=1}} \langle v, Av \rangle = \min \sigma(A),$$

follows from the spectral theorem. Namely we use that for a multiplication operator M_a on a space $L^2(B, \mathrm{d}\mu)$,

$$\inf_{\int_B |v|^2 \mathrm{d}\mu=1} \int_B a(x)|v(x)|^2 \, \mathrm{d}\mu(x) = \inf a = \min \sigma(M_a).$$

The bottom of the spectrum is either equal to the first eigenvalue when it exists (with equality if and only if v is an associated eigenvector), or equal to the bottom of the essential spectrum. We then suppose that the assertion on $\mu_k(A)$ has been shown and prove it for $\mu_{k+1}(A)$. If $\mu_k(A) = \Sigma(A)$, there is nothing to prove because the sequence $\mu_j(A)$ is non-decreasing and less than or equal to $\Sigma(A)$, so $\mu_{k+1}(A) = \Sigma(A)$ as well. Thus, we can assume that $\mu_k(A) = \lambda_k(A) < \Sigma(A)$, which means that A has at least k eigenvalues (counted with multiplicity) strictly less than $\Sigma(A)$. Moreover, if $\lambda_{k+1}(A) = \lambda_k(A)$ (which is possible in case of degeneracy) then of course $\mu_{k+1}(A) \ge \mu_k(A) = \lambda_k(A) = \lambda_{k+1}(A)$. We are therefore left to deal with the situation where $\lambda_k(A) < \Sigma(A)$ and $\mathbb{1}_{(-\infty, \lambda_k(A)]}(A)$ is of rank exactly k. Let V_k be the space generated by the first k eigenvectors. If $W \subset D(A)$ is any subspace of dimension $k + 1$, then it must intersect $(V_k)^\perp$. But for all $v \in (V_k)^\perp \cap Q(A) = $

$\mathbb{1}_{(\lambda_k(A),+\infty)}(A)\mathfrak{H} \cap Q(A)$ such that $\|v\| = 1$, we have by functional calculus

$$q_A(v) \geq \min \sigma \left(A\mathbb{1}_{(\lambda_k(A),+\infty)}(A)_{|V_k^\perp} \right).$$

By the same argument as for $\lambda_1(A)$, the minimum on the right is $\lambda_{k+1}(A)$ (the first eigenvalue of $A\mathbb{1}_{(\lambda_k,+\infty)}(A)$ on $(V_k)^\perp$ if it exists) or $\Sigma(A)$ (if A has only k eigenvalues below $\Sigma(A)$). Thus, we have shown Formula (5.42).

Now, if in addition $\lambda_k(A) = \mu_k(A) < \Sigma(A)$, any subspace generated by k eigenvectors of eigenvalues $\leq \lambda_k(A)$ realizes the infimum on the right of (5.42). Let then j be such that $\lambda_k(A) = \lambda_{k+j}(A) < \mu_{k+j+1}(A)$ (which depends on the finite multiplicity of the eigenvalue $\lambda_k(A)$). A subspace $W \subset Q(A)$ of dimension k that intersects the range of the spectral projector $\mathbb{1}_{(\lambda_k(A),+\infty)}(A) = \mathbb{1}_{[\mu_{k+j+1}(A),+\infty)}(A)$ verifies for v in this space

$$q_A(v) = q_{A\mathbb{1}_{[\mu_{k+j+1}(A),+\infty)}(A)}(v) \geq \mu_{k+j+1}(A)\|v\|^2$$

so that

$$\max_{\substack{v\in W \\ \|v\|_{\mathfrak{H}}=1}} \langle v, Av \rangle \geq \mu_{k+j+1}(A) > \lambda_k(A).$$

Thus, we can only have equality if W is included in the image of the spectral projector $\mathbb{1}_{(-\infty,\lambda_k(A)]}(A)$. The proof for (5.43) is similar and left as an exercise. $\square$

A practical method to show that A has at least k eigenvalues below its essential spectrum follows immediately from the Courant-Fischer formula.

Corollary 5.42 (Existence of Eigenvalues Below $\Sigma(A)$) *Let A be a self-adjoint operator bounded from below on its domain $D(A) \subset \mathfrak{H}$ and $\Sigma(A) := \min \sigma_{\mathrm{ess}}(A) \in \mathbb{R}\cup\{+\infty\}$ the bottom of its essential spectrum. If there exists a subspace $W \subset Q(A)$ of dimension $\dim(W) = k$ such that*

$$\max_{\substack{v\in W \\ \|v\|=1}} q_A(v) < \Sigma(A),$$

then A has at least k eigenvalues (counted with multiplicity) strictly below $\Sigma(A)$.

The Courant-Fischer formula also allows us to compare the eigenvalues of operators (below the essential spectrum) by comparing their quadratic forms. The following result is also an immediate consequence of the Courant-Fischer formula.

Corollary 5.43 (Eigenvalues of Ordered Operators) *Let $(A, D(A))$ and $(B, D(B))$ be two bounded-below self-adjoint operators, such that $A \leq B$ in the sense of Definition 5.3. Then*

$$\mu_k(A) \leq \mu_k(B) \quad \text{for all } k \geq 1, \text{ and} \quad \Sigma(A) \leq \Sigma(B),$$

for the bottom of their essential spectra.

Example 5.44 (Eigenvalues of Dirichlet, Neumann and Robin) Let $\Omega \subset \mathbb{R}^d$ be a bounded open set whose boundary is piecewise Lipschitz. Let $(-\Delta)_{\text{Rob},\theta}$ be the Laplacian with Robin boundary conditions introduced in Sect. 3.3.4. Then the associated eigenvalues, denoted $\lambda_k(\theta)$, are all non-increasing functions of $\theta \in (0, 1)$. This follows from the Courant-Fischer formula, since the associated quadratic form

$$q_{(-\Delta)_{\text{Rob},\theta}}(u) = \int_\Omega |\nabla u(x)|^2 \, dx + \frac{1}{\tan(\pi\theta)} \int_{\partial\Omega} |u(x)|^2 \, dx,$$

$$\text{on} \quad Q((-\Delta)_{\text{Rob},\theta}) = H^1(\Omega). \qquad (5.47)$$

is a non-increasing function of θ. It is possible to show that $\theta \in (0, 1) \mapsto \lambda_k(\theta)$ is continuous and converges to the k-th eigenvalue of the Dirichlet Laplacian when $\theta \to 0^+$. In dimension $d = 1$ this is done in Exercise 5.56.

5.6.2 Discrete Spectrum of Schrödinger Operators

We can now discuss the existence or absence of negative eigenvalues for Schrödinger operators $-\Delta + V$, and their regularity with respect to V. The first result concerns the Lipschitz character of the Courant-Fischer levels μ_k defined in (5.42), which are the eigenvalues below the essential spectrum when they exist.

Theorem 5.45 (Eigenvalues Are Locally Lipschitz in V) *Let* $V \in L^p(\mathbb{R}^d, \mathbb{R}) + L^\infty_\varepsilon(\mathbb{R}^d, \mathbb{R})$ *with*

$$\begin{cases} p = 1 & \text{if } d = 1, \\ p > 1 & \text{if } d = 2, \\ p = \frac{d}{2} & \text{if } d \geq 3. \end{cases} \qquad (5.48)$$

There exists a constant $C = C(V)$ *(depending on V) such that*

$$\left| \mu_k(-\Delta + V) - \mu_k(-\Delta + V') \right| \leq C \left\| V - V' \right\|_{L^p(\mathbb{R}^d) + L^\infty(\mathbb{R}^d)} \qquad (5.49)$$

for all $k \geq 1$ and all $V' \in L^p(\mathbb{R}^d, \mathbb{R}) + L^\infty(\mathbb{R}^d, \mathbb{R})$ in a neighborhood of V for the norm of $L^p(\mathbb{R}^d) + L^\infty(\mathbb{R}^d)$ defined earlier in (1.34). Here $-\Delta + V$ and $-\Delta + V'$ are the Friedrichs self-adjoint realizations.

In the statement we have assumed that V is negligible at infinity because this is the situation that interests us, but not that V' is. It is therefore possible to consider perturbations in $L^\infty(\mathbb{R}^d)$.

Proof We start by writing $V = V_p + V_\infty$ and $V' = V'_p + V'_\infty$. Then, in the sense of quadratic forms,

$$-\Delta + V' = -\Delta + V'_p + V_\infty + V'_\infty - V_\infty$$

$$\geq -\Delta + V'_p + V_\infty - \left\| V'_\infty - V_\infty \right\|_{L^\infty(\mathbb{R}^d)}$$

$$\geq (1 - \varepsilon)(-\Delta + V) + \frac{\varepsilon}{2}(-\Delta + 2V)$$

$$+ \frac{\varepsilon}{2}\left(-\Delta + 2\frac{V'_p - V_p}{\varepsilon}\right) - \left\| V'_\infty - V_\infty \right\|_{L^\infty(\mathbb{R}^d)}. \qquad (5.50)$$

The terms have been grouped together so that a part of the Laplacian can be used to control the errors. The operator $-\Delta + 2V$ is bounded from below under our assumptions on V, so that

$$\frac{\varepsilon}{2}(-\Delta + 2V) \geq \frac{\varepsilon}{2}\mu_1(-\Delta + 2V).$$

Here $\mu_1(-\Delta + 2V) = \min \sigma(-\Delta + 2V)$ is the bottom of the spectrum of $-\Delta + 2V$, which is non-positive since V is negligible at infinity, hence $\Sigma(-\Delta + 2V) = 0$. Let us start with dimensions $d \geq 3$ where $p = d/2$. According to Proposition 1.16

$$-\Delta + v \geq 0, \qquad \text{for } \|v_-\|_{L^{d/2}(\mathbb{R}^d)} \leq (S_d)^{-1}$$

where S_d is the Sobolev constant. This suggests setting $\varepsilon = 2S_d \|V_p - V'_p\|_{L^p(\mathbb{R}^d)}$ so that $-\Delta + 2(V'_p - V_p)/\varepsilon \geq 0$. We obtain

$$-\Delta + V' \geq \left(1 - 2S_d \|V_p - V'_p\|_{L^p(\mathbb{R}^d)}\right)(-\Delta + V)$$

$$- S_d |\mu_1(-\Delta + 2V)| \|V'_p - V_p\|_{L^p(\mathbb{R}^d)} - \|V'_\infty - V_\infty\|_{L^\infty(\mathbb{R}^d)}.$$

This inequality is only interesting when $2S_d \|V_p - V'_p\|_{L^p(\mathbb{R}^d)} < 1$, which we assume in the following. Using the Courant-Fischer formula as in Corollary 5.43 and then optimizing with respect to V'_p and V'_∞, we get

$$\mu_k(-\Delta + V') \geq \mu_k(-\Delta + V) - \left(1 + S_d |\mu_1(-\Delta + 2V)|\right) \|V' - V\|_{L^p(\mathbb{R}^d) + L^\infty(\mathbb{R}^d)}$$

for $\|V - V'\|_{L^p(\mathbb{R}^d)+L^\infty(\mathbb{R}^d)} < (2S_d)^{-1}$. We have used that $\mu_k(-\Delta + V) \leq 0$ since V is negligible at infinity. With a similar reasoning (taking $\varepsilon < 0$ in (5.50)), we find

$$- \Delta + V' \leq \left(1 + 2S_d \|V - V'\|_{L^p(\mathbb{R}^d)+L^\infty(\mathbb{R}^d)}\right)(-\Delta + V)$$
$$+ S_d|\mu_1(-\Delta + 2V)|\|V'_p - V_p\|_{L^p(\mathbb{R}^d)} + \|V'_\infty - V_\infty\|_{L^\infty(\mathbb{R}^d)}.$$

which provides the desired inequality (5.49) with $C = 1 + S_d|\mu_1(-\Delta + 2V)|$.

In dimensions $d \in \{1, 2\}$, the operator $-\Delta + v$ is not necessarily positive when v is small, as we saw in Proposition 1.17. There is therefore an additional contribution coming from the minorization of $-\Delta + 2(V'_p - V_p)/\varepsilon$. This time we simply take $\varepsilon = 2\|V_p - V'_p\|_{L^p(\mathbb{R}^d)}$ which we assume ≤ 1. Let us start with dimension $d = 1$ where $p = 1$. Using Inequality (A.23) in Appendix A, we find

$$q_{-\Delta+v}(f) \geq \|f'\|^2_{L^2(\mathbb{R})} - 2\|f'\|_{L^2(\mathbb{R})}\|f\|_{L^2(\mathbb{R})}\|v_-\|_{L^1(\mathbb{R})} \geq -\|v_-\|^2_{L^1(\mathbb{R})}\|f\|^2_{L^2(\mathbb{R})}$$

for all $f \in L^2(\mathbb{R})$, which can also be stated in the form

$$- \Delta + v \geq -\|v_-\|^2_{L^1(\mathbb{R})}.$$

Thus, we have

$$- \Delta + 2\frac{V'_p - V_p}{\varepsilon} = -\Delta + \frac{V'_p - V_p}{\|V_p - V'_p\|_{L^1(\mathbb{R})}} \geq -1$$

and we obtain the desired inequality with $C = 2 + |\mu_1(-\Delta + 2V)|$. The argument is similar in dimension $d = 2$, using this time

$$- \Delta + v \geq -C_p \|v_-\|^{\frac{p}{p-1}}_{L^p(\mathbb{R}^d)}$$

which follows from the Gagliardo-Nirenberg inequality of Theorem A.15. □

If V' is negligible at infinity and $\mu_k(-\Delta + V) < 0$, the previous result therefore implies that $-\Delta + V'$ also has at least k strictly negative eigenvalues, for $\|V - V'\|_{L^p(\mathbb{R}^d)+L^\infty(\mathbb{R}^d)}$ small enough.

We now discuss the existence or absence of eigenvalues below the essential spectrum, that is, whether $\mu_k(-\Delta + V)$ is strictly negative or not. The intuition is that the potential V must be sufficiently negative somewhere in space. Under our assumptions on V, a deep local well will however only create finitely many negative eigenvalues (see Theorem 5.47 below and Problem B.4 in Appendix B). It is the behavior at infinity of the potential that matters to have infinitely many eigenvalues. In this direction we provide the following adaptation of Proposition 1.15.

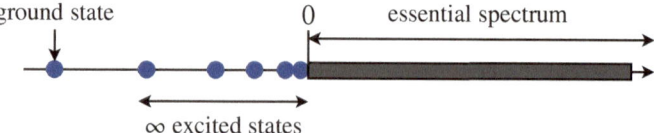

Theorem 5.46 (Infinity of Eigenvalues If V Decreases Slowly at Infinity) *Let* $V \in L^p(\mathbb{R}^d, \mathbb{R}) + L_\varepsilon^\infty(\mathbb{R}^d, \mathbb{R})$ *with p as in (5.48) and satisfying the upper estimate*

$$V(x) \leq -c|x|^{-\alpha} \tag{5.51}$$

for $|x|$ large enough with $c > 0$ and $0 < \alpha < 2$. Then the Friedrichs realization of the Schrödinger operator $-\Delta + V$ satisfies

$$\mu_k(-\Delta + V) < \Sigma(-\Delta + V) = 0$$

for all $k \geq 1$. It therefore has infinitely many strictly negative eigenvalues, which all have finite multiplicity and converge to 0.

This result, which applies for example to the hydrogen atom, means that the spectrum has the shape represented in Fig. 5.2 as soon as the potential V is negative at infinity and does not tend too quickly to 0, in the sense of (5.51). Let us recall that the negative eigenvalues explain the line spectrum obtained in spectroscopy experiments.

Proof The proof is essentially the same as that of Proposition 1.15. Consider any subspace W of $C_c^\infty(B_2 \setminus B_1)$ (functions with support in the annulus between the balls of radius 1 and 2), with $\dim(W) = k$. Then set

$$W_n = \left\{ \chi_n(x) = n^{-\frac{d}{2}} \chi(x/n), \quad \chi \in W \right\} = U_n W \subset C_c^\infty(B_{2n} \setminus B_n)$$

where $U_n v = n^{-d/2} v(\cdot/n)$ is the unitary consisting of dilating functions by a factor of $1/n$. The space W_n therefore has the same dimension as W. For all $\chi_n = U_n \chi \in W_n$ normalized in $L^2(\mathbb{R}^d)$, we have for n large enough

$$q_{-\Delta+V}(\chi_n) = \frac{1}{n^2} \int_{\mathbb{R}^d} |\nabla \chi(x)|^2 \, dx + \int_{B_2 \setminus B_1} V(nx)|\chi(x)|^2 \, dx$$

$$\leq \frac{1}{n^2} \int_{B_2 \setminus B_1} |\nabla \chi(x)|^2 \, dx - \frac{c}{n^\alpha} \int_{B_2 \setminus B_1} \frac{|\chi(x)|^2}{|x|^\alpha} \, dx$$

$$\leq \frac{1}{n^2} \max_{\substack{\chi \in W \\ \int_{\mathbb{R}^d} |\chi|^2 = 1}} \int_{B_2 \setminus B_1} |\nabla \chi(x)|^2 \, dx - \frac{c}{n^\alpha} \min_{\substack{\chi \in W \\ \int_{\mathbb{R}^d} |\chi|^2 = 1}} \int_{B_2 \setminus B_1} \frac{|\chi(x)|^2}{|x|^\alpha} \, dx.$$

Since W is of finite dimension, the minimum

$$\min_{\substack{\chi \in W \\ \int_{\mathbb{R}^d} |\chi|^2 = 1}} \int_{B_2 \setminus B_1} \frac{|\chi(x)|^2}{|x|^\alpha} \, dx$$

is achieved, therefore strictly positive. The Courant-Fischer formula (5.42) then implies

$$\mu_k(-\Delta + V) \leq \max_{\substack{\chi \in W \\ \int_{\mathbb{R}^d} |\chi|^2 = 1}} q_{-\Delta+V}(\chi_n)$$

which is strictly negative for large n, as the term $n^{-\alpha}$ is dominant. Thus, $\mu_k(-\Delta + V) < 0 = \Sigma(-\Delta + V)$ for all $k \geq 1$ and $-\Delta + V$ has infinitely many negative eigenvalues, which can only accumulate at 0 because the spectrum is bounded from below and $\sigma_{\mathrm{ess}}(-\Delta + V) = \mathbb{R}_+$ by Corollary 5.38. □

We will now see that the power $\alpha = 2$ is critical, in the sense that any potential that decays like $|x|^{-\alpha}$ at infinity with $\alpha > 2$ can only generate a finite number of eigenvalues. More precisely, we have already seen in Proposition 1.16 that if V is small in $L^{d/2}(\mathbb{R}^d)$, then there are no eigenvalues. The following result deals with the case of a potential of any size.

Theorem 5.47 (CLR Inequality) *Let us work in dimension $d \geq 3$. If $V \in L^{d/2}(\mathbb{R}^d) + L_\varepsilon^\infty(\mathbb{R}^d)$ and $V_- = \max(0, -V) \in L^{d/2}(\mathbb{R}^d)$, then the Friedrichs realization of the Schrödinger operator $-\Delta + V$ satisfies*

$$\mu_k(-\Delta + V) = \Sigma(-\Delta + V) = 0$$

for large k. More precisely, $-\Delta + V$ has only a finite number of non-positive eigenvalues, that is $\mathbb{1}_{(-\infty,0]}(-\Delta + V)$ has finite rank. There exists a universal constant $C_{\mathrm{CLR}}(d)$ depending only on the dimension d such that

$$\boxed{\operatorname{rank}\left(\mathbb{1}_{(-\infty,0]}(-\Delta + V)\right) \leq C_{\mathrm{CLR}}(d) \int_{\mathbb{R}^d} V(x)_-^{\frac{d}{2}} \, dx.} \tag{5.52}$$

An inequality like (5.52) cannot be valid in dimension $d = 1, 2$, even with other norms of V on the right. Indeed, we have seen in Proposition 1.17 that if $V \leq 0$ everywhere with $V \neq 0$, then $-\Delta + V$ always has a negative eigenvalue. This contradicts any inequality of the type (5.52) that one can imagine.

The inequality (5.52) is due to Cwikel [Cwi77], Lieb [Lie80] and Rozenblum [Roz72]. Its proof is beyond the scope of this book and we will not provide it. In Problem B.4 we show two weaker results. The first provides the finiteness of the discrete spectrum when V has compact support *in all dimensions $d \geq 1$*. The second shows the expected result that the discrete spectrum is finite, with the only assumption that $V_- \in L^{d/2}(\mathbb{R}^d)$ in dimension $d \geq 3$, but with a worse estimate than (5.52). We refer for example to [FLW22, Sec. 4.5] for the proof of (5.52).

The inequality (5.52) belongs to a very important class of estimates concerning negative eigenvalues of Schrödinger operators, commonly called **semi-classical inequalities** or **Lieb-Thirring inequalities**, named after their two inventors [LT75, LT76, LS10b, FLW22]. We will return to this later in Theorem 5.52. The semi-classical character comes from the fact that the term on the right of (5.52) is exact in the semi-classical limit, up to a multiplicative constant. More precisely, we will show in Sect. 5.7 that if we dilate the potential $V(x)$ into $V(\varepsilon x)$ with $\varepsilon \to 0$, so that it varies very slowly (this is the semi-classical limit), then

$$\lim_{\varepsilon \to 0} \varepsilon^d \, \text{rank} \left(\mathbb{1}_{(-\infty,0]}\big(- \Delta + V(\varepsilon x)\big) \right) = \frac{|\mathbb{S}^{d-1}|}{d(2\pi)^d} \int_{\mathbb{R}^d} V(x)_-^{\frac{d}{2}} \, dx. \qquad (5.53)$$

Thus, for a potential in the form $V(\varepsilon x)$, the number of eigenvalues behaves like ε^{-d} in the limit $\varepsilon \to 0$.

5.6.3 The Birman-Schwinger Principle

An important method for proving Theorem 5.47 and many other results concerning the spectrum of Schrödinger operators is the **Birman-Schwinger principle** [Bir61, Sch61], which we describe here in a somewhat informal way and which is then used in Problem B.4.

Let us start by describing this principle in finite dimension. Consider two Hermitian matrices A, B with A positive definite and B non-negative. Since A has its spectrum in $(0, +\infty)$, we wonder how large the matrix B must be for $A - B$ to have negative eigenvalues. We have that $\lambda = -E < 0$ is an eigenvalue of $A - B$ if and only if there exists $v \neq 0$ such that

$$(A + E)v = Bv.$$

If $v \in \ker(B)$ then we find $(A + E)v = 0$ which implies $v = 0$ because $A > 0$, and is absurd. So $Bv \neq 0$ and $w := \sqrt{B}v \neq 0$ because $\ker(B) = \ker(\sqrt{B})$. Using the fact that $-E \notin \sigma(A)$, we find using the relation $v = (A + E)^{-1}Bv$

$$w = \sqrt{B}v = \sqrt{B}(A + E)^{-1}Bv = \sqrt{B}(A + E)^{-1}\sqrt{B}\, w.$$

Thus, w is an eigenvector of the non-negative Hermitian matrix

$$K_E := \sqrt{B}(A + E)^{-1}\sqrt{B}$$

for the eigenvalue 1. Conversely, if we have a vector $w \neq 0$ such that $K_E w = w$ we can set $v = (A + E)^{-1}\sqrt{B}w$ which is such that $\sqrt{B}v = w$, therefore non-zero. By replacing w with $\sqrt{B}v$ in the expression for v, this provides $v = (A + E)^{-1}Bv$, that is $(A + E - B)v = 0$. So $-E$ is an eigenvalue of $A - B$. We deduce the Birman-Schwinger principle:

$$\boxed{\begin{array}{c} \lambda = -E < 0 \text{ is a negative eigenvalue of } A - B \\ \Longleftrightarrow 1 \text{ is an eigenvalue of } K_E = \sqrt{B}(A + E)^{-1}\sqrt{B}. \end{array}}$$

The previous argument even shows that $(A + E)^{-1}\sqrt{B}$ is a bijection from $\ker(K_E - 1)$ to $\ker(A - B + E)$. The multiplicities are therefore equal. As we have $(A + E_1)^{-1} \leq (A + E_2)^{-1}$ for $E_1 \geq E_2$, we see that the spectrum of K_E is composed of eigenvalues that are non-increasing as functions of $E \in [0, +\infty)$, by the Courant-Fischer formula. Moreover, they tend to 0 when $E \to +\infty$. In fact, these are Lipschitz functions. Thus, the picture is as in Fig. 5.3: the eigenvalues of $A - B$ correspond to the $\lambda_j = -E_j$ for which the spectrum of K_E crosses 1. In particular, by monotonicity and continuity, **the number of negative eigenvalues of** $A - B$ **is equal to the number of eigenvalues greater than 1 of** $K_0 = \sqrt{B}A^{-1}\sqrt{B}$ (counted with multiplicity). The latter can be estimated for example by:

$$\text{rank}\left(\mathbb{1}_{\mathbb{R}_-}(A - B)\right) = \text{rank}\left(\mathbb{1}_{[1,+\infty)}(K_0)\right) \leq \text{tr}\,(K_0)^m = \text{tr}\left(A^{-1}B\right)^m \quad (5.54)$$

for any integer $m \geq 1$ because

$$\#\{\lambda_n(K_0) \geq 1\} \leq \sum_n \lambda_n(K_0)^m.$$

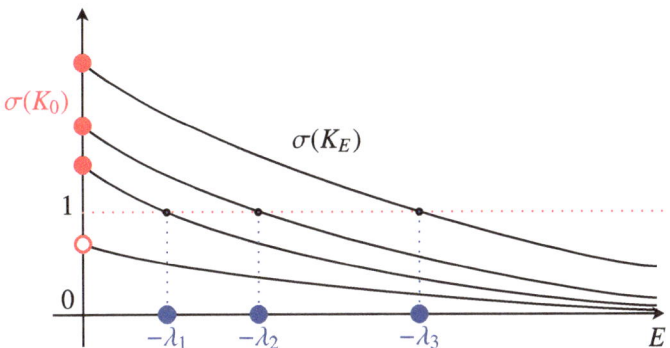

This discussion for matrices can be extended to Schrödinger operators [FLW22, Sec. 4.3.3]. That is, we can show that for a potential $0 \leq V \in L^p(\mathbb{R}^d, \mathbb{R}) + L_\varepsilon^\infty(\mathbb{R}^d, \mathbb{R})$ with p as in (5.48), we have that

$$\boxed{\begin{array}{l} \lambda = -E < 0 \text{ is a negative eigenvalue of } -\Delta - V \\ \qquad \Longleftrightarrow 1 \text{ is an eigenvalue of } K_E = \sqrt{V}(-\Delta + E)^{-1}\sqrt{V}. \end{array}}$$

Here the operator K_E is well defined and bounded when $E > 0$ using Theorem 5.24 and $-\Delta + V$ is the Friedrichs self-adjoint realization provided by Corollary 3.20. Understanding the number of negative eigenvalues of $-\Delta - V$ thus boils down to studying the operator

$$K_0 = \sqrt{V}(-\Delta)^{-1}\sqrt{V} = \lim_{E \to 0^+} \sqrt{V}(-\Delta + E)^{-1}\sqrt{V}.$$

The latter is more singular because of the inverse of the Laplacian. In fact, the operator K_0 is still compact in dimensions $d \geq 3$ by Remark 5.25. The proof of Theorem 5.47 then consists in estimating the number of its eigenvalues greater than or equal to 1 as a function of $\|V\|_{L^{d/2}(\mathbb{R}^d)}$. The details of an argument of this type based on Inequality (5.54) can be found in Problem B.4.

In dimensions $d \in \{1, 2\}$, the operator $(-\Delta)^{-1}$ is very singular because the function $k \mapsto |k|^{-2}$ is not integrable in a neighborhood of the origin. We can then show that K_0 is not bounded as soon as $V > 0$ and this is what always creates eigenvalues for $-\Delta - V$ in dimensions $d \in \{1, 2\}$, as we saw in Proposition 1.17. Read more about this in [RS78, Thm. XIII.11].

5.7 A Bit of Semi-classical Analysis*

The goal of this last section is to more precisely determine the behavior of the eigenvalues of the operator $-\Delta + V(\varepsilon x)$ in the limit $\varepsilon \to 0$. After a dilation of x by a factor of ε, this amounts to studying the operator $-\varepsilon^2 \Delta + V(x)$, that is, to make the constant $\hbar$ tend to 0, as discussed in Sect. 1.2.1. We therefore speak of a **semi-classical limit**. As we are working in a system of units where $\hbar^2 = 2m$, we prefer to change the spatial variations of the external potential V by the factor ε, rather than the physical constant $\hbar$. In the first section we start with the case where V is constant on a cube or on a bounded domain Ω, which will then be useful for the case of the whole space $\mathbb{R}^d$ with an arbitrary potential.

5.7.1 Weyl's Law on a Bounded Open Set

We have seen in Example 2.45 and Exercise 2.50 that the spectra of the Dirichlet and Neumann Laplacians on the unit cube $\Omega = (0, 1)^d$ are

$$\sigma\left((-\Delta)_{\text{Dir}}\right) = \pi^2 \left\{ \sum_{j=1}^{d} k_j^2, \ k_j \in \mathbb{N} \right\},$$

$$\sigma\left((-\Delta)_{\text{Neu}}\right) = \pi^2 \left\{ \sum_{j=1}^{d} k_j^2, \ k_j \in \mathbb{N} \cup \{0\} \right\}.$$

On any cube $C \subset \mathbb{R}^d$, the spectrum does not depend on the position of the cube. Moreover, by performing a dilation we can always reduce it to the case of the unit cube and we find that the spectrum is just divided by $|C|^{2/d}$. The difference between the two spectra becomes negligible when looking at a large set of eigenvalues, for example if we count the number of eigenvalues less than a level E and we take the limit $E \to +\infty$.

Lemma 5.48 (Weyl's Law for a Hypercube) *Let $C \subset \mathbb{R}^d$ be an arbitrary cube and $N_{\text{Dir/Neu}}(E, C)$ the number of eigenvalues of the Dirichlet/Neumann Laplacian, strictly less than E, counted with multiplicity. Then there exists a universal constant $K = K(d)$ (depending only on the dimension), such that*

$$\left| N_{\text{Dir/Neu}}(E, C) - \frac{|\mathbb{S}^{d-1}|}{d(2\pi)^d} E^{\frac{d}{2}} |C| \right| \leq K(d) \left(1 + |C|^{\frac{d-1}{d}} E^{\frac{d-1}{2}} \right). \tag{5.55}$$

Proof After dilation and translation, it suffices to show the result for the unit cube $C = (0, 1)^d$. The number of Dirichlet eigenvalues strictly less than E is equal to the number of vectors $k = (k_1, \ldots, k_d) \in \mathbb{N}^d$ such that $|k|^2 = \sum_{j=1}^{d} k_j^2 < E\pi^{-2}$. Let $\prod_{j=1}^{d}(k_j - 1, k_j)$ be the cube of which k is at the upper right corner. These cubes are all included in the intersection of the ball $B_{\sqrt{E}/\pi}$ of radius $\sqrt{E}/\pi$ with the quadrant $\{(x_1, \ldots, x_d) : x_i \geq 0\}$, see Fig. 5.4. As this region has volume

$$2^{-d} |B_{\sqrt{E}/\pi}| = \frac{|\mathbb{S}^{d-1}|}{d(2\pi)^d} E^{d/2},$$

we have

$$N_{\text{Dir}}(E, C) \leq \frac{|\mathbb{S}^{d-1}|}{d(2\pi)^d} E^{d/2}. \tag{5.56}$$

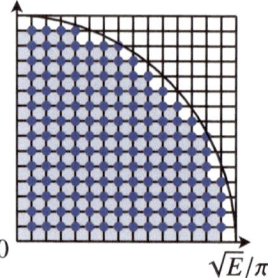 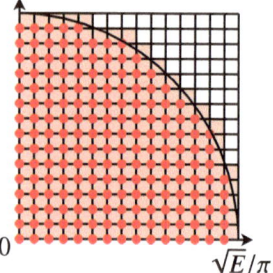

Fig. 5.4 The $k \in \mathbb{N}^2$ that provide all the eigenvalues $|k|^2$ of the Dirichlet (left) and Neumann (right) Laplacian inside the disk of radius E/π^2 and the cubes used in the proof of Lemma 5.48. © Mathieu Lewin 2021. All rights reserved

Similarly, for the Neumann Laplacian, it is natural to introduce the cubes $\prod_{j=1}^{d}(k_j, k_j + 1)$ where k is at the other end, and we find by the same argument

$$N_{\text{Neu}}(E, C) \geq \frac{|\mathbb{S}^{d-1}|}{d(2\pi)^d} E^{d/2}.$$

The difference between the number of eigenvalues of the two operators is equal to the number of $k \in (\mathbb{N} \cup \{0\})^d$ so that at least one of the components is zero, that is

$$N_{\text{Neu}}(E, C) - N_{\text{Dir}}(E, C)$$

$$= \# \left\{ (k_1, \ldots, k_d) \in (\mathbb{N} \cup \{0\})^d \ : \ \prod_{j=1}^{d} k_j = 0, \ |k|^2 < \frac{E}{\pi^2} \right\}.$$

According to the previous estimate (5.56) on the Dirichlet Laplacian in dimension ℓ instead of d, the number of vectors $k = (k_1, \ldots, k_d)$ that have exactly ℓ non-zero components is bounded by $K E^{\ell/2}$ for $K = \max_{\ell=1,\ldots,d} \ell^{-1}(2\pi)^{-\ell}|\mathbb{S}^{\ell-1}|$. So

$$N_{\text{Neu}}(E, C) - N_{\text{Dir}}(E, C) \leq K \sum_{\ell=0}^{d-1} E^{\frac{\ell}{2}} \leq dK \left(1 + E^{\frac{d-1}{2}} \right),$$

which concludes the proof.   □

Lemma 5.48 specifies that for any fixed cube C, the number of eigenvalues below E behaves like $E^{d/2}$ as $E \to \infty$. It turns out that this behavior is universal and is not specific to cubes. The following result, due to Weyl [Wey12], is a first step towards understanding the spectrum of Schrödinger operators with semi-classical methods.

Theorem 5.49 (Weyl's Law for Any Domain Ω) *Let $\Omega \subset \mathbb{R}^d$ be a bounded open set whose boundary has zero Lebesgue measure, $|\partial\Omega| = 0$. Let $N_{\text{Dir}}(E, \Omega)$ be the number of eigenvalues of the Dirichlet Laplacian on Ω, strictly less than E, counted* *with multiplicity. Then*

$$\lim_{E \to +\infty} \frac{N_{\text{Dir}}(E, \Omega)}{E^{d/2}} = \frac{|\mathbb{S}^{d-1}|}{d(2\pi)^d} |\Omega|. \tag{5.57}$$

With the change of variables $x' = \ell x$ we can see that $N_{\text{Dir}}(E, \ell\Omega) = N_{\text{Dir}}(E\ell^2, \Omega)$. We can therefore reformulate (5.57) in the form

$$\lim_{\ell \to +\infty} \frac{N_{\text{Dir}}(E, \ell\Omega)}{\ell^d} = \frac{|\mathbb{S}^{d-1}|}{d(2\pi)^d} E^{\frac{d}{2}} |\Omega|. \tag{5.58}$$

where now E is fixed and Ω is dilated by a factor $\ell \to +\infty$.

Proof The proof relies heavily on the Courant-Fischer principle and the comparisons of quadratic forms from Sect. 5.6.1. It can be read in [CH53, Sec. VI.4], in [RS78, Sec. XIII.15] or in [FLW22, Sec. 3.2]. Recall that the Dirichlet Laplacian is defined on any open set by its quadratic form $\int_\Omega |\nabla u|^2$ on $H_0^1(\Omega)$, as seen in Sect. 3.3.4. Without additional assumptions on the regularity of Ω, its operator domain is unknown.

Consider a tiling of $\mathbb{R}^d$ by cubes of side ε, that is

$$\varepsilon C_k = \prod_{j=1}^{d} [\varepsilon k_j, \varepsilon(k_j + 1)), \qquad k_j \in \mathbb{Z}.$$

In each of the cubes that are strictly inside Ω (Fig. 5.5), we can consider the eigenfunctions of the Dirichlet Laplacian inside this small cube, whose eigenvalues are strictly less than E. There are $N_{\text{Dir}}(E, \varepsilon C_0)$ such eigenvalues (the spectrum of the Dirichlet Laplacian is the same in each of these cubes, by translation invariance). Each of the eigenfunctions can be extended by 0 outside its cube and seen as a function of $H_0^1(\Omega)$. We then consider the space W spanned by all the vectors in

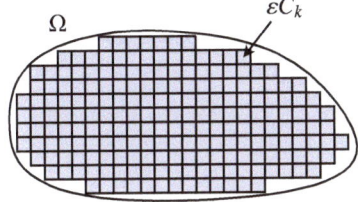 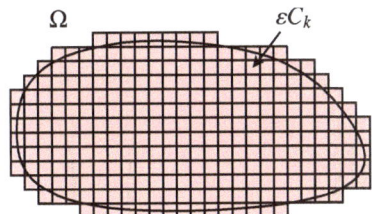

Fig. 5.5 Proof of Theorem 5.49: we use the eigenfunctions of all the small cubes εC_k included in the open set Ω as test functions for the Dirichlet Laplacian on Ω, which provides an upper bound on its eigenvalues (thus a lower bound on $N_{\text{Dir}}(E, \Omega)$). Then we use all the eigenfunctions of the cubes that intersect Ω with the Neumann condition, which provides a lower bound on the eigenvalues in the domain Ω (thus an upper bound on $N_{\text{Dir}}(E, \Omega)$). © Mathieu Lewin 2021. All rights reserved

each of the cubes, which has dimension

$$\dim(W) = N_{\text{Dir}}(E, \varepsilon C_0) \times \#\{k \in \mathbb{Z}^d \; : \; \varepsilon C_k \subset \Omega\}.$$

The restriction of the quadratic form of the Dirichlet Laplacian on Ω to the space W is associated with a matrix of size $\dim(W)$. This matrix is block diagonal, because two vectors in different cubes are orthogonal, since they have disjoint support. Each block is just the matrix of the quadratic form of the cube, whose eigenvalues are those of the Dirichlet Laplacian on the cube. We deduce that

$$q_{(-\Delta)}(v) = \int_{\Omega} |\nabla v|^2 < E \int_{\Omega} |v|^2, \qquad \forall v \in W.$$

By the Courant-Fischer Formula (5.42), this shows that $\mu_{\dim(W)}(-\Delta) < E$. The number of eigenvalues less than E in Ω is therefore at least equal to

$$N_{\text{Dir}}(E, \Omega) \geq \dim(W) = N_{\text{Dir}}(E, \varepsilon C_k) \times \#\{k \in \mathbb{Z}^d \; : \; \varepsilon C_k \subset \Omega\}.$$

The cubes that intersect Ω but are not inside are at most at distance $\varepsilon\sqrt{d}$ from the boundary, so that

$$|\Omega| \geq \varepsilon^d \#\{k \in \mathbb{Z}^d \; : \; \varepsilon C_k \subset \Omega\} \geq |\Omega| - |\partial\Omega_{\varepsilon\sqrt{d}}|,$$

where we have introduced the set

$$\partial\Omega_r := \{x \in \mathbb{R}^d \; : \; d(x, \partial\Omega) \leq r\}.$$

Using (5.55) we deduce that

$$N_{\text{Dir}}(E, \Omega) \geq \frac{|\mathbb{S}^{d-1}|}{d(2\pi)^d} E^{\frac{d}{2}} \left(|\Omega| - |\partial\Omega_{\varepsilon\sqrt{d}}| \right) + O\left(\varepsilon^{-1} E^{\frac{d-1}{2}} \right). \tag{5.59}$$

As Ω is assumed to be bounded, its boundary $\partial\Omega$ is compact. The sets $\partial\Omega_{\varepsilon\sqrt{d}}$ are therefore also compact and decrease towards $\partial\Omega$ when $\varepsilon \to 0^+$. We deduce by dominated convergence that

$$\lim_{\varepsilon \to 0^+} |\partial\Omega_{\varepsilon\sqrt{d}}| = \lim_{\varepsilon \to 0^+} \int_{\mathbb{R}^d} \mathbb{1}\left(d(x, \partial\Omega) \leq \varepsilon\sqrt{d} \right) dx = |\partial\Omega| = 0. \tag{5.60}$$

For a regular domain Ω, this term is even of order ε. The error in (5.59) is then of order $\varepsilon E^{d/2} + E^{(d-1)/2}\varepsilon^{-1}$ which suggests taking $\varepsilon = E^{-1/4}$. In the general case, it is not possible to exhibit a concrete ε, but we can always let $\varepsilon \to 0$ very slowly or, equivalently, first take $E \to \infty$ then $\varepsilon \to 0$. We find

$$\liminf_{E \to \infty} \frac{N_{\text{Dir}}(E, \Omega)}{E^{d/2}} \geq \frac{|\mathbb{S}^{d-1}|}{d(2\pi)^d} |\Omega|.$$

To obtain the opposite inequality, we now consider the set of all the small cubes εC_k that intersect Ω. As Ω is bounded, there are a finite number of them. In each of the εC_k, consider all the eigenfunctions of the Neumann Laplacian, whose corresponding eigenvalue is strictly less than E. These functions are in $H^1(\varepsilon C_k)$, hence in $L^2(\mathbb{R}^d)$ but not in $H^1(\Omega)$. Similarly as before, the space W' spanned by all these functions has dimension

$$\dim(W') = N_{\text{Neu}}(E, \varepsilon C_k) \times \#\{k \in \mathbb{Z}^d \;:\; \varepsilon C_k \cap \Omega \neq \emptyset\}$$

$$\leq \frac{|\mathbb{S}^{d-1}|}{d(2\pi)^d} E^{\frac{d}{2}} \left(|\Omega| + |\partial\Omega_{\varepsilon\sqrt{d}}|\right) + O\left(\varepsilon^{-d}E^{\frac{d-1}{2}}\right).$$

Now, for all $v \in H_0^1(\Omega)$, we have

$$\int_\Omega |\nabla v|^2 = \sum_{\varepsilon C_k \cap \Omega \neq \emptyset} \int_{\varepsilon C_k \cap \Omega} |\nabla v|^2.$$

If v is in $(W')^\perp \cap H_0^1(\Omega)$, we have by the spectral theorem

$$\int_{\varepsilon C_k \cap \Omega} |\nabla v|^2 \geq E \int_{\varepsilon C_k \cap \Omega} |v|^2. \tag{5.61}$$

for all k, because $v_{|\varepsilon C_k}$ is orthogonal to all the eigenvectors of the Neumann Laplacian of eigenvalue $< E$, in εC_k. Thus,

$$\int_\Omega |\nabla v|^2 \geq E \int_\Omega |v|^2, \qquad \forall v \in H_0^1(\Omega) \cap (W')^\perp$$

or again

$$\inf_{\substack{v \in H_0^1(\Omega) \cap (W')^\perp \\ \|v\|=1}} \int_\Omega |\nabla v|^2 \geq E.$$

By the second Courant-Fischer formula (5.43), this shows that $\mu_{\dim(W')+1}(-\Delta) \geq E$, because $\left(H_0^1(\Omega) \cap (W')^\perp\right)^\perp = W'$. Thus, $N_{\text{Dir}}(E, \Omega) \leq \dim(W')$, which concludes the proof of the theorem. □

From this proof we can remember that the Dirichlet conditions are very useful for obtaining upper bounds on the eigenvalues, since any function in $H_0^1(A)$ is in $H_0^1(B)$ when $A \subset B$. Conversely, the Neumann conditions are used to obtain lower bounds on the eigenvalues. The use of these two operators "in tandem", is called the **Dirichlet-Neumann bracketing** method.

The assumptions of Theorem 5.49 are too strong. For the Dirichlet Laplacian, the result remains true with the only assumption that Ω is of finite measure; it is not necessary to suppose that Ω is bounded, nor that $\partial\Omega$ has zero measure [Cie70, BS70, FLW22]. In the case of the Neumann Laplacian not treated here, the limit (5.57) is exactly the same but regularity assumptions on Ω are this time unavoidable [NS05].

For a domain Ω whose boundary is regular (for example C^1), the volume of (5.60) will be of order ε, in which case the above proof provides an estimate of the remainder in the form

$$N_{\mathrm{Dir}}(E, \Omega) = \frac{|\mathbb{S}^{d-1}|}{d(2\pi)^d}|\Omega| E^{\frac{d}{2}} + O\left(E^{\frac{d}{2}-\frac{1}{4}}\right).$$

This estimate is not optimal. Courant [Cou20] already obtained a better remainder in 1920, of order $O(E^{\frac{d-1}{2}}\log E)$ when Ω is sufficiently smooth. This was finally improved to $O(E^{\frac{d-1}{2}})$ in [See80]. In 1913, Weyl [Wey13] conjectured that

$$N_{\mathrm{Dir}}(E, \Omega) = \frac{|\mathbb{S}^{d-1}|}{d(2\pi)^d}|\Omega| E^{\frac{d}{2}} - \frac{|\mathbb{S}^{d-2}|}{4(d-1)(2\pi)^{d-1}}\mathrm{Per}(\partial\Omega)\, E^{\frac{d-1}{2}} + o\left(E^{\frac{d-1}{2}}\right),$$

$$(5.62)$$

where $\mathrm{Per}(\partial\Omega)$ denotes the surface of $\partial\Omega$. This question has since been the subject of many research works [Ivr16]. This asymptotic expansion was finally proved with more or less restrictive assumptions on the domain Ω. For $N_{\mathrm{Neu}}(E, \Omega)$ the expansion is the same but the sign of the second term is reversed.

Remark 5.50 (Can One Hear the Shape of a Drum?) From the behavior of the large eigenvalues of the Dirichlet Laplacian, we can find the volume $|\Omega|$ of the domain according to (5.57), or even the surface of its boundary with (5.62). In a now famous article [Kac66], Mark Kac asked in 1966 whether it was possible to find the exact shape of Ω by knowing all the eigenvalues of its Dirichlet Laplacian. The answer to this question turned out to be negative, in dimension $d \geq 2$ [GWW92].

5.7.2 Semi-classical Limit for $-\Delta + V$

Using arguments very similar to those of the proof of Theorem 5.49, we will now be able to show the semi-classical limit (5.53) announced previously.

Theorem 5.51 (Semi-classical Limit of Negative Eigenvalues) *We assume that $d \geq 1$. Let $V \in C_c^0(\mathbb{R}^d, \mathbb{R})$ (continuous with compact support) and f be a piecewise continuous function on $\mathbb{R}$, that vanishes on $(0, +\infty)$. Then we have*

$$\lim_{\varepsilon \to 0^+} \varepsilon^d \sum_j f\left(\lambda_j\left(-\Delta + V(\varepsilon x)\right)\right) = \frac{1}{(2\pi)^d} \iint_{\mathbb{R}^d \times \mathbb{R}^d} f\left(|p|^2 + V(x)\right) \mathrm{d}p\, \mathrm{d}x,$$

$$(5.63)$$

where the $\lambda_j(-\Delta + V(\varepsilon x))$ are all the non-positive eigenvalues of the self-adjoint operator $-\Delta + V(\varepsilon x)$ on $H^2(\mathbb{R}^d)$, repeated according to their multiplicity.

The integral on the right of (5.63) is finite because V is bounded and f is locally bounded, so that

$$\iint_{\mathbb{R}^d \times \mathbb{R}^d} \left| f\left(|p|^2 + V(x)\right) \right| dp\, dx \leq C \iint_{\mathbb{R}^d \times \mathbb{R}^d} \mathbb{1}\left(|p|^2 + V(x) \leq 0\right) dp\, dx$$

$$= C \int_{\mathbb{R}^d} \left(\int_{\mathbb{R}^d} \mathbb{1}\left(|p|^2 \leq V(x)_-\right) dp \right) dx$$

$$= C \frac{|\mathbb{S}^{d-1}|}{d} \int_{\mathbb{R}^d} V(x)_-^{\frac{d}{2}}\, dx,$$

where $C = \sup_{-\|V_-\|_{L^\infty} \leq x \leq 0} |f(x)|$.

The convergence (5.63) expresses the fact that when $\varepsilon \to 0$, the quantum observable $f(-\Delta + V(\varepsilon x))$ can be approximated by its classical counterpart $f(|p|^2 + V(\varepsilon x))$ on the phase space $\mathbb{R}^d \times \mathbb{R}^d$, see Sect. 1.5.5. The sum on the left of (5.63) is (formally) equal to the trace of the operator $\varepsilon^d f(-\Delta + V(\varepsilon x))$, so that (5.63) can be rewritten

$$\operatorname{tr} \left[f\left(-\Delta + V(\varepsilon x) \right) \right] \underset{\varepsilon \to 0}{\sim} \frac{1}{(2\pi)^d} \iint_{\mathbb{R}^d \times \mathbb{R}^d} f\left(|p|^2 + V(\varepsilon x)\right) dx\, dp$$

$$= \frac{1}{(2\pi\varepsilon)^d} \iint_{\mathbb{R}^d \times \mathbb{R}^d} f\left(|p|^2 + V(x)\right) dx\, dp$$

where $-i\nabla$ has been replaced by the classical variable p and the trace becomes an integral over the phase space, multiplied by $(2\pi)^{-d}$. In particular, if we take $f(x) = \mathbb{1}_{(-\infty,0]}(x)$, we obtain exactly the limit (5.53) concerning the number of non-positive eigenvalues.

The convergence (5.63) is true with weaker assumptions for the potential V and the function f. The condition that V is continuous is for example far too strong. The proof below works when V can be approached from below and above by a sequence of step functions, in $L^{d/2}(\mathbb{R}^d)$.

The intuition of the limit (5.63) is as follows. The potential $V(\varepsilon x)$ is very flat, since its derivative is of order ε (if it exists). It is therefore essentially constant on large cubes of size $\ell \gg 1$ provided that $\varepsilon\ell \ll 1$, by the mean value theorem. We therefore make a small controlled error by replacing V with a piecewise constant function on cubes of size $1 \ll \ell \ll 1/\varepsilon$ forming a tiling of $\mathbb{R}^d$. Then, by the **Dirichlet-Neumann bracketing** method used in the proof of Theorem 5.49, we can obtain an estimate on the eigenvalues of the operator in $\mathbb{R}^d$, in terms of the spectrum of the Dirichlet and Neumann Laplacians in each cube. Finally, as these cubes are very large, the semi-classical asymptotic (5.55) will lead to the result.

Proof Let us set $V_\varepsilon(x) = V(\varepsilon x)$. We write the proof first in the case where $f(x) = \mathbb{1}(x < 0)$. We must thus prove that the number $N(0, V_\varepsilon)$ of negative eigenvalues of the operator $-\Delta + V_\varepsilon$ behaves like

$$\frac{1}{(2\pi\varepsilon)^d} \iint_{\mathbb{R}^d \times \mathbb{R}^d} \mathbb{1}\left(|p|^2 \leq -V(x)\right) \mathrm{d}p\,\mathrm{d}x = \frac{|\mathbb{S}^{d-1}|}{d(2\pi\varepsilon)^d} \int_{\mathbb{R}^d} V(x)_-^{\frac{d}{2}}\,\mathrm{d}x,$$

as was already stated in (5.53). Since the proof is quite similar to that of Theorem 5.49, we only give the main ideas and leave the details as an exercise. Consider a tiling of $\mathbb{R}^d$ with cubes $\{\ell C_k\}_{k \in \mathbb{Z}^d}$ of volume ℓ^d, where $C_k = \prod_{j=1}^d [k_j, k_j + 1)$ and $1 \ll \ell \ll \varepsilon^{-1}$, for example $\ell = \varepsilon^{-1/2}$. We will replace $V(\varepsilon x)$ with a constant function in each of the cubes by taking either its maximum on the cube, or its minimum.

Consider all the cubes ℓC_k included in the support of V_ε (equivalently, $\varepsilon \ell C_k$ is in the support of V) and take the space generated by the eigenfunctions of the Dirichlet Laplacian on each cube ℓC_k, whose eigenvalues are strictly less than $-\max_{\ell C_k} V_\varepsilon = -\max_{\varepsilon \ell C_k} V$. These functions form a space on which the quadratic form associated with $-\Delta + V(\varepsilon x)$ is negative. This proves that the number of negative eigenvalues is at least equal to

$$\sum_{\varepsilon \ell C_k \subset \text{supp}(V)} N_{\text{Dir}}\left(-\max_{\varepsilon \ell C_k} V, \ell C_k\right)$$

$$\geq \frac{|\mathbb{S}^{d-1}|}{d(2\pi\varepsilon)^d} \sum_{\varepsilon \ell C_k \subset \text{supp}(V)} |\varepsilon \ell C_k| \left(-\max_{\varepsilon \ell C_k} V\right)_+^{\frac{d}{2}}$$

$$- K \sum_{\varepsilon \ell C_k \subset \text{supp}(V)} \left(1 + \ell^{d-1} \max_{\varepsilon \ell C_k} |V|^{\frac{d-1}{2}}\right).$$

We have used here the estimate (5.55) on $N_{\text{Dir}}(E, C)$. The first term is a Riemann sum that converges to $\int_{\mathbb{R}^d} V_-^{d/2}$ while the second is an $O(\varepsilon^{-d}\ell^{-1})$.

To have a bound in the other direction, we write as in the proof of Theorem 5.49

$$\int_{\mathbb{R}^d} |\nabla v|^2 + V|v|^2 \geq \sum_{\ell C_k \cap \text{supp}(V_\varepsilon) \neq \emptyset} \int_{\ell C_k} |\nabla v|^2 + V_\varepsilon |v|^2$$

$$\geq \sum_{\ell C_k \cap \text{supp}(V_\varepsilon) \neq \emptyset} \int_{\ell C_k} |\nabla v|^2 - \left(\min_{\varepsilon \ell C_k} V\right)_- \int_{\ell C_k} |v|^2.$$

The use of eigenfunctions with Neumann boundary conditions on each of the cubes ℓC_k, whose eigenvalues are less than $\left(\min_{\varepsilon \ell C_k} V\right)_-$ and the inequality (5.55) provides an upper bound on the number of negative eigenvalues, which allows us to conclude the proof of the theorem for $f(x) = \mathbb{1}(x < 0)$.

If we now take $f(x) = \mathbb{1}(x < a)$ with $a \leq 0$, we need to estimate the number of eigenvalues less than a of $-\Delta + V_\varepsilon$, which is equal to the number of negative eigenvalues of $-\Delta + V_\varepsilon - a$. The potential $V_\varepsilon - a$ does not tend to 0 at infinity but we can adapt the previous proof and obtain that the limit (5.63) is valid for $f(x) = \mathbb{1}(x < a)$ with $a \leq 0$.

By difference, we obtain the limit for $f = \mathbb{1}_{[a,b)}$ where $b \leq 0$. As $V \in L^\infty(\mathbb{R}^d)$, the operator $-\Delta + V_\varepsilon$ has its spectrum included in $[-\|V\|_{L^\infty(\mathbb{R}^d)}, +\infty)$. Only the restriction of f to $[-\|V\|_{L^\infty(\mathbb{R}^d)}, 0]$ therefore matters for the limit (5.63). However, any piecewise continuous function can be approximated from below and above by step functions, for which the limit has been shown. A density argument allows to conclude the proof. □

5.7.3 Lieb-Thirring Inequalities

The convergence (5.63) in the case where $f(x) = x_-$ concerns the sum of negative eigenvalues:

$$\lim_{\varepsilon \to 0^+} \varepsilon^d \sum_j \left|\lambda_j\left(-\Delta + V(\varepsilon x)\right)\right| = \frac{1}{(2\pi)^d} \iint_{\mathbb{R}^d \times \mathbb{R}^d} \left(|p|^2 + V(x)\right)_- \, dp \, dx$$

$$= \frac{2|\mathbb{S}^{d-1}|}{(2\pi)^d d(d+2)} \int_{\mathbb{R}^d} V(x)_-^{1+\frac{d}{2}} \, dx. \qquad (5.64)$$

For the second equality we just calculated the integral in p explicitly. It then seems natural to wonder if there exists a universal bound on the sum of eigenvalues, in terms of $\int_{\mathbb{R}^d} V(x)_-^{1+d/2} \, dx$, in the same way as we had the universal CLR bound (5.52) on the number of eigenvalues. This is the famous Lieb-Thirring inequality, which has played a fundamental role in the mathematical understanding of the behavior of fermionic matter in the limit of a large number of particles, that is, the stability of ordinary matter [LT75, LT76, Lie90, LS10b, FLW22, Fra23].

Theorem 5.52 (Lieb-Thirring Inequalities) *Let $V \in L^p(\mathbb{R}^d) + L^\infty_\varepsilon(\mathbb{R}^d)$ with p as in (5.48) such that $V_- = \max(0, -V) \in L^{\gamma+d/2}(\mathbb{R}^d, \mathbb{R})$, where*

$$\gamma \begin{cases} \geq 1/2 & \text{if } d = 1, \\ > 0 & \text{if } d = 2, \\ \geq 0 & \text{if } d \geq 3. \end{cases}$$

There exists a universal constant $C_{LT}(\gamma, d)$ such that the non-positive eigenvalues of the operator $-\Delta + V$ (repeated according to their multiplicity) satisfy the inequality

$$\boxed{\sum_j |\lambda_j(-\Delta + V)|^\gamma \leq C_{LT}(\gamma, d) \int_{\mathbb{R}^d} V(x)_-^{\gamma+\frac{d}{2}} \, dx.} \qquad (5.65)$$

Lieb and Thirring [LT75, LT76] actually only dealt with the cases where $\gamma > 1/2$ in dimension $d = 1$ and $\gamma > 0$ in dimensions $d \geq 2$. The case $\gamma = 0$ in dimensions $d \geq 3$ is the CLR inequality (5.52) already seen in Theorem 5.47. The case $\gamma = 1/2$ in dimension $d = 1$ was proven by Weidl in [Wei96]. We refer to the review article [Fra23] and the book [FLW22] for further comments and results on these inequalities.

It turns out that if we have shown the Lieb-Thirring inequality (5.65) for a γ, we can easily deduce the inequality for all $\gamma' \geq \gamma$. In dimensions $d \geq 3$, Theorem 5.52 therefore follows from the CLR inequality for $\gamma = 0$, which we have stated without proof in Theorem 5.47. We now explain this argument.

Proof (for $d \geq 3$ and $\gamma > 0$ Using Theorem 5.47) We have for $\gamma > 0$ and $x \in \mathbb{R}$

$$x_-^\gamma = \gamma \int_0^{x_-} \tau^{\gamma-1}\, d\tau = \gamma \int_0^\infty \mathbb{1}(x + \tau \leq 0)\, \tau^{\gamma-1}\, d\tau.$$

Using (5.52), this implies

$$\sum_j |\lambda_j(-\Delta + V)|^\gamma = \gamma \int_0^\infty \sum_j \mathbb{1}\big(\lambda_j(-\Delta + V) + \tau \leq 0\big)\, \tau^{\gamma-1}\, d\tau$$

$$= \gamma \int_0^\infty \#\big\{\lambda_j(-\Delta + V) + \tau \leq 0\big\}\, \tau^{\gamma-1}\, d\tau$$

$$\leq \gamma \int_0^\infty \#\big\{\lambda_j(-\Delta - (V + \tau)_-) \leq 0\big\}\, \tau^{\gamma-1}\, d\tau$$

$$\leq C_{\mathrm{CLR}}(d)\gamma \int_{\mathbb{R}^d} \int_0^\infty (V(x) + \tau)_-^{\frac{d}{2}}\, \tau^{\gamma-1}\, d\tau\, dx$$

$$= C' \int_{\mathbb{R}^d} V(x)_-^{\gamma + \frac{d}{2}}\, dx,$$

where $C' = C_{\mathrm{CLR}}(d)\gamma \int_0^1 (1 - \tau)^{\frac{d}{2}} \tau^{\gamma-1}\, d\tau$. For the first inequality we used

$$\lambda_j(-\Delta + V) + \tau = \lambda_j(-\Delta + V + \tau) \geq \lambda_j\big(-\Delta - (V + \tau)_-\big)$$

by the Courant-Fischer formula (5.42), since $V + \tau \geq -(V + \tau)_-$. This implies that the number of $\lambda_j(-\Delta + V)$ less than or equal to $-\tau$ is controlled by the number of non-positive eigenvalues of $-\Delta - (V + \tau)_-$. Note that $(V + \tau)_- = (V + \tau)\mathbb{1}(V \leq 0)\mathbb{1}(|V| \geq \tau)$ so that $(V + \tau)_-$ indeed belongs to $L^{d/2}(\mathbb{R}^d)$ for all $\tau > 0$.   □

Complementary Exercises

Exercise 5.53 (Discrete Laplacian) In the Hilbert space $\mathfrak{H} = \ell^2(\mathbb{Z}^d)$ of sequences $\mathbf{x} = (x(k))_{k \in \mathbb{Z}^d}$, we introduce the discrete derivative operator in the direction e_j by $(D_j \mathbf{x})(k) = x(k + e_j) - x(k)$.

1. Show that D_j is a bounded operator on $\mathfrak{H}$. What is its adjoint? Prove that the discrete Laplacian defined by $L = \sum_{j=1}^{d} D_j^* D_j$, can be expressed as

$$(L\mathbf{x})(k) = \sum_{j=1}^{d} 2x(k) - x(k + e_j) - x(k - e_j).$$

2. Show that L is self-adjoint and that its spectrum is included in $[0, +\infty)$.
3. We consider the isometry $U : \mathfrak{H} \to \mathfrak{K} = L^2((0, 1)^d)$ defined by $U\mathbf{x} = \sum_{k \in \mathbb{Z}^d} x(k)e^{2i\pi k \cdot x}$. Calculate $U D_j U^{-1}$ and $U L U^{-1}$ and deduce the spectra of D_j and L. Do they have eigenvalues?

We now show on an example, taken from [RS78, Sec. XIII.4], the importance of the condition that the operators are self-adjoint, in the Weyl theory on the stability of the essential spectrum by compact perturbations. We are in dimension $d = 1$ and we denote by $S = D_1 + 1$ the left shift. We consider the operator defined by $(B\mathbf{x})_n = \delta_{0n} x_1$, that is $B = |\delta_0\rangle\langle\delta_1|$.

4. Verify that B is compact.
5. Show that the essential spectrum of the operator $S - B$ is the entire closed unit disk.

Exercise 5.54 (Courant Principle) We work in $L^2(\mathbb{R})$, with a potential $V \in C_0^0(\mathbb{R}, \mathbb{R})$ (continuous and tending to 0 at infinity). Using the Courant-Fischer formula, show that any eigenfunction associated with the j-th eigenvalue $\lambda_j < 0$ of $-d^2/dx^2 + V(x)$ (counted with multiplicity) vanishes at most $k - 1$ times, where $k \geq j$ is the integer such that $\lambda_j = \lambda_k < \lambda_{k+1}$. Also show with the Cauchy-Lipschitz theorem that the eigenfunction changes sign at each of its zeros.

Exercise 5.55 (Beams and Compression) We consider a horizontal beam that is fixed at one end and can deform slightly in the vertical plane. We represent the beam by the segment $I = (0, 1)$ and we call $u(x)$ the vertical displacement of the beam at the point $x \in (0, 1)$. Due to the high rigidity of the system, Euler and Bernoulli showed that the energy of very small displacements involves a second derivative instead of a first derivative as for a vibrating string. We therefore introduce the quadratic form

$$q(u) := \int_0^1 |u''(x)|^2 \, dx \tag{5.66}$$

defined on the subspace $Q := \{u \in H^2(I) : u(0) = u'(0) = 0\}$ where the two constraints at 0 are due to the fixation of the beam [CH53, Sec. V.3.3]. Even though for a beam u is real-valued, we here allow u to take complex values and we work in the Hilbert space $\mathfrak{H} = L^2(I, \mathbb{C})$.

1. Show that $q + \| \cdot \|^2_{L^2(I)}$ is closed on Q and coercive.
2. Determine the unique self-adjoint operator A associated with q, in particular the associated boundary conditions.

Now we consider the situation where we apply a pressure P to the beam at 0 and 1, while maintaining its ends. If P is small enough, when we release this pressure the beam will return to its horizontal equilibrium position. But in case of too much pressure, the beam can bend and no longer find its initial shape (Fig. 5.6). The energy is this time described by the quadratic form

$$q_P(u) := \int_0^1 |u''(x)|^2 \, dx - P \int_0^1 |u'(x)|^2 \, dx \tag{5.67}$$

on the space $Q_0 := \{ u \in H^2(I) \ : \ u(0) = u(1) = 0 \}$, see [CH53, Sec. IV.12.13]. We define the critical pressure by

$$P_c := \inf_{\substack{0 \neq u \in Q_0 \\ u(0)=u(1)=0}} \frac{\int_0^1 |u''(x)|^2 \, dx}{\int_0^1 |u'(x)|^2 \, dx} = \inf_{\substack{u \in Q_0 \\ u(0)=u(1)=0 \\ \int_0^1 |u'(x)|^2 \, dx=1}} \int_0^1 |u''(x)|^2 \, dx. \tag{5.68}$$

3. Show that

$$\inf_{u \in Q_0} q_P(u) = \begin{cases} 0 & \text{if } P \leq P_c, \\ -\infty & \text{if } P > P_c. \end{cases}$$

4. Reformulate the minimization problem (5.68) in terms of the Dirichlet Laplacian on $I = (0, 1)$. Calculate P_c as well as the unique minimizer u of (5.68) up to phase.

Exercise 5.56 (Behavior of the Robin Laplacian on $I = (0, 1)$ in the Limit $\theta \to 1^-$) Let $I = (0, 1)$. We consider the self-adjoint realization of the Laplacian $A_{\text{Rob},\theta} u = -u''$ in the space $\mathfrak{H} = L^2(I)$, with the Robin boundary condition

$$D(A_{\text{Rob},\theta}) = \Big\{ u \in H^2(I) \ : \ \cos(\pi\theta)u(1) + \sin(\pi\theta)u'(1) = 0,$$

$$\cos(\pi\theta)u(0) - \sin(\pi\theta)u'(0) = 0 \Big\},$$

studied in Sects. 2.8.3 and 3.2.4. We call $\lambda_1(\theta) \leq \lambda_2(\theta) \leq \cdots$ the ordered eigenvalues of $A_{\text{Rob},\theta}$, repeated in case of multiplicity.

1. Using the Courant-Fischer formula, show that $\theta \mapsto \lambda_1(\theta)$ is continuous and strictly decreasing.
2. Show that $\lambda_1(1/2) = 0$ and deduce that $\lambda_1(\theta) > 0$ for all $\theta \in [0, 1/2)$, and that $\lambda_1(\theta) < 0$ for all $\theta \in (1/2, 1)$. State a Poincaré inequality for the Robin quadratic form when $\theta \in [0, 1/2)$.
3. Show that $\lambda_1(\theta) \to -\infty$ and $\lambda_2(\theta) \to -\infty$ when $\theta \to 1^-$.
4. Show that $\omega^2 \neq 0$ is an eigenvalue of the Robin Laplacian $A_{\text{Rob},\theta}$ (with eigenfunction $\alpha e^{i\omega x} + \beta e^{-i\omega x}$ with well-chosen α, β) if and only if

$$\Big(\cos(\pi\theta) + i\omega\sin(\pi\theta) \Big)^2 e^{i\omega} = \Big(\cos(\pi\theta) - i\omega\sin(\pi\theta) \Big)^2 e^{-i\omega}.$$

Show that 0 is an eigenvalue only for $\theta = 1/2$ and for $\theta = 1 - \pi^{-1}\arctan(1/2) \simeq 0, 85$. What is happening when $\theta \to 1^-$?

The curve in Fig. 5.7 represents the first three eigenvalues as a function of θ, while the one in Fig. 5.8 represents the third eigenfunction for various values of θ.

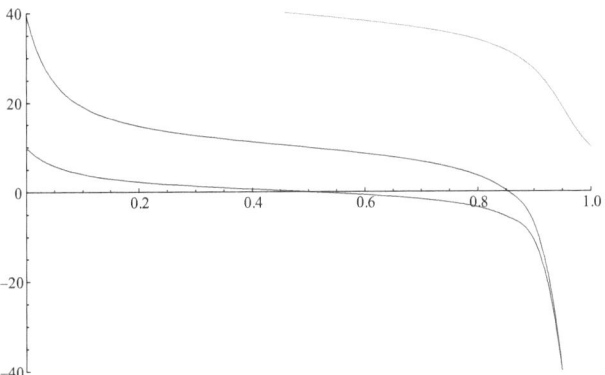

Fig. 5.7 Plot of the first three eigenvalues $\lambda_1(\theta) \leq \lambda_2(\theta) \leq \lambda_3(\theta)$ of the Robin Laplacian on the interval $I = (0, 1)$, as a function of the Robin parameter $\theta \in [0, 1)$. © Mathieu Lewin 2021. All rights reserved

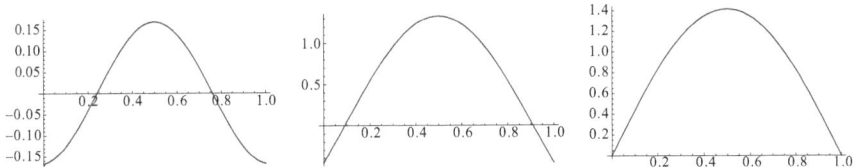

Fig. 5.8 Plot of the third eigenfunction of the Robin Laplacian on the interval $I = (0, 1)$ for $\theta = 0.7$, for $\theta = 0.97$ and at the limit $\theta \to 1^-$. © Mathieu Lewin 2021. All rights reserved

Chapter 6
N-particle Systems, Atoms, Molecules

This chapter is a small excursion into the world of Schrödinger operators describing N particles instead of just one particle. This is a vast subject, for which basic physical questions can lead to very difficult mathematical problems. We mention here some research results, old or recent, without always providing all the proofs. We have chosen to discuss in more detail the case of N electrons in a molecule whose nuclei are classical particles, which is the main system of interest for quantum chemists.

6.1 Hamiltonian for N Particles, Bosons and Fermions

We consider a system of N *identical* particles that evolve in $\mathbb{R}^d$, are subject to an external potential V and interact by pairs through a potential w. The classical energy of such a system (assuming their mass equals $m = 1/2$ for simplicity) is given by

$$E(x_1, p_1, \ldots, x_N, p_N) = \sum_{j=1}^{N} |p_j|^2 + V(x_j) + \sum_{1 \leq j < k \leq N} w(x_j - x_k)$$

where $x_j \in \mathbb{R}^d$ and $p_j \in \mathbb{R}^d$ are respectively the position and momentum of the j-th particle. We assume that w is even. We will discuss below the type of assumption we want to be able to cover for this function. The associated quantum system is posed on the Hilbert space

$$\mathfrak{H} = L^2\left((\mathbb{R}^d)^N, \mathbb{C}\right)$$

© The Editor(s) (if applicable) and The Author(s), under exclusive license
to Springer Nature Switzerland AG 2024
M. Lewin, *Spectral Theory and Quantum Mechanics*, Universitext,
https://doi.org/10.1007/978-3-031-66878-4_6

which includes wavefunctions $\Psi(x_1, \ldots, x_N)$ with the interpretation that

- $|\Psi(x_1, \ldots, x_N)|^2$ is the probability density that the first particle is at x_1, that the second is at x_2, etc.;
- $|\widehat{\Psi}(p_1, \ldots, p_N)|^2$ is the probability density that the first particle has a momentum p_1, that the second has a momentum p_2, etc.

As for the hydrogen atom in Chap. 1, the energy of this system in the state Ψ is therefore given by the formula

$$
\mathcal{E}^V(\Psi) = \sum_{j=1}^{N} \int_{\mathbb{R}^{dN}} |\nabla_{x_j} \Psi(x_1, \ldots, x_N)|^2 \, dx_1 \cdots dx_N
$$

$$
+ \sum_{j=1}^{N} \int_{\mathbb{R}^{dN}} V(x_j) |\Psi(x_1, \ldots, x_N)|^2 \, dx_1 \cdots dx_N
$$

$$
+ \sum_{1 \le j < k \le N} \int_{\mathbb{R}^{dN}} w(x_j - x_k) |\Psi(x_1, \ldots, x_N)|^2 \, dx_1 \cdots dx_N
$$

which is the quadratic form associated with the Hamiltonian operator

$$
H^V(N) = \sum_{j=1}^{N} -\Delta_{x_j} + \sum_{j=1}^{N} V(x_j) + \sum_{1 \le j < k \le N} w(x_j - x_k), \tag{6.1}
$$

the quantization of the classical energy E. In our notation $H^V(N)$ we have indicated the external potential V because it can typically be varied in experiments, while the interaction w is usually a characteristic of the particles that remains fixed in the study.

The interpretation we have given of $|\Psi(x_1, \ldots, x_N)|^2$ and $|\widehat{\Psi}(p_1, \ldots, p_N)|^2$ suggests that we can label the particles and know who is who at any given moment. In fact, if we observe the particles at two different times, it is impossible to know which particle went where, since they are exactly identical. Our modeling is thus not adequate and needs to be slightly modified. More precisely, our model must be invariant under the action of permutations of the labels. We should therefore work, not in $(\mathbb{R}^d)^N$, but rather in the quotient $(\mathbb{R}^d)^N / \mathfrak{S}_N$ corresponding to the action of the symmetric group $(x_1, \ldots, x_N) \mapsto (x_{\sigma(1)}, \ldots, x_{\sigma(N)})$ which permutes the indices. This quotient has topological properties that we will not mention and which play an important role in dimensions $d \in \{1, 2\}$ [LM77]. We will rather content ourselves with verifying that our modeling is indeed invariant under the action of the symmetric group.

We want $|\Psi(x_1, \ldots, x_N)|^2$ and $|\widehat{\Psi}(p_1, \ldots, p_N)|^2$ to be symmetric with respect to the exchanges of their variables, so that the labeling of the particles does not matter. But on the other hand, we must work in a vector space, due to the formalism of quantum mechanics described in Sect. 1.5. In Exercise 6.1 we show that there are only two possible linear constraints on the function Ψ:

- either we work with the assumption that Ψ is **symmetric** with respect to the exchanges of its variables, that is,

$$\Psi(x_{\sigma(1)}, \ldots, x_{\sigma(N)}) = \Psi(x_1, \ldots, x_N), \qquad \forall \sigma \in \mathfrak{S}_N; \qquad (6.2)$$

- or we work with the assumption that Ψ is **anti-symmetric** with respect to the exchanges of its variables, that is,

$$\Psi(x_{\sigma(1)}, \ldots, x_{\sigma(N)}) = \varepsilon(\sigma) \, \Psi(x_1, \ldots, x_N), \qquad \forall \sigma \in \mathfrak{S}_N \qquad (6.3)$$

where $\varepsilon(\sigma)$ is the signature of the permutation σ.

These two constraints imply that $|\Psi|^2$ and $|\widehat{\Psi}|^2$ are symmetric, as desired. Moreover, these are linear constraints that simply require working in the corresponding closed subspaces of $L^2((\mathbb{R}^d)^N, \mathbb{C})$, denoted in the following

$$L^2_s((\mathbb{R}^d)^N, \mathbb{C}) := \left\{ \Psi \in L^2((\mathbb{R}^d)^N, \mathbb{C}) \text{ satisfying (6.2) a.e.} \right\} \qquad (6.4)$$

and

$$L^2_a((\mathbb{R}^d)^N, \mathbb{C}) := \left\{ \Psi \in L^2((\mathbb{R}^d)^N, \mathbb{C}) \text{ satisfying (6.3) a.e.} \right\} \qquad (6.5)$$

and which are of course equipped with the usual norm of $L^2((\mathbb{R}^d)^N, \mathbb{C})$. Similarly, we can define the Sobolev spaces $H^k_{a/s}((\mathbb{R}^d)^N, \mathbb{C})$. The choice of symmetry or anti-symmetry condition depends on the type of particle being studied. Those that are modeled by a symmetric Ψ are called **bosons**, while those for which Ψ is anti-symmetric are called **fermions**. The *standard model* of particle physics states that all elementary particles building up matter are fermions (examples: electron, quark), while all the particles transporting energy are bosons (examples: photon, gluon, Higgs).

Depending on the scale at which we work, it is often convenient to describe **composite particles** (comprising several elementary particles) as a single entity. For example, in our study of the electron in the hydrogen atom, we assumed that the proton was a classical fixed particle. We can also describe it as a quantum particle, as in Example 1.22. However, we forget here that the proton actually contains three quarks, and also that there are isotopes of hydrogen that have neutrons, themselves composed of three quarks. The model can therefore be complexified depending on the scale at which we study the system. For a composite particle, the rule is that only the number of fermions matters to determine its type. It is a boson if the number of elementary fermions that compose it is even, and a fermion otherwise. For example, protons and neutrons, which are composed of three quarks, are fermions. Helium 4 (2 electrons, 2 protons and 2 neutrons) behaves like a boson while Helium 3 (2 electrons, 2 protons and 1 neutron), much rarer on earth, behaves like a fermion. This rule is quite intuitive. If we group the variables of a wavefunction and look at the sign appearing when we exchange these groups, we see that it only depends on the parity of the number of fermionic variables in each group.[1]

The Hamiltonian $H^V(N)$ often behaves very differently in the limit $N \to \infty$ depending on whether it is restricted to the subspaces of symmetric or anti-symmetric functions. Bosonic systems can be more unstable than fermionic systems as the number of particles grows. In the Coulomb case, only fermionic systems are stable in the limit $N \to \infty$ [LT75, Lie90, LS10b]. If we only take into account electrostatic forces, bosonic matter is in fact unstable [Lie79].

The intuitive explanation is that bosons are very social. They love being together, which can generate a high concentration of particles, a potential source of instability. We can put N bosons in the same state $u \in L^2(\mathbb{R}^d)$ with $\|u\|_{L^2(\mathbb{R}^d)} = 1$ by taking

$$\Psi_{\mathrm{BE}}(x_1, \dots, x_N) = u(x_1) \cdots u(x_N)$$

which is called a **Bose-Einstein condensate**. This is the quantum version of independent and identically distributed random variables in probability theory, since the spatial and momentum probability distributions are factorized:

$$|\Psi_{\mathrm{BE}}(x_1, \dots, x_N)|^2 = \prod_{j=1}^{N} |u(x_j)|^2, \qquad |\widehat{\Psi}_{\mathrm{BE}}(p_1, \dots, p_N)|^2 = \prod_{j=1}^{N} |\widehat{u}(p_j)|^2.$$

In this case the N bosons all do exactly the same thing. Fermions cannot adopt such a behavior. The anti-symmetry of Ψ is often called the **Pauli principle** and it implies a certain amount of correlations between the particles. For example, two fermions can never be in the same place since, when Ψ is a continuous function, the anti-symmetry implies

$$\Psi(x_1, \dots, x, \dots, x, \dots, x_N) = 0.$$

This natural small repulsion can stabilize the system in some situations. The question of understanding the mathematical implications of the Pauli principle on the stability of quantum systems in the limit $N \to \infty$ has been a source of hard problems in mathematical physics in the last decades.

Let us now discuss the interaction potential w. While electrons interact with the Coulomb potential (neglecting weak forces), composite particles have a complex interaction that can only be determined empirically, since it depends on their internal structure. The simplest is to describe it by a pair interaction potential w as in this

---

[1] For example, a system comprising four fermions is described by an anti-symmetric wavefunction $\Psi(x_1, x_2, x_3, x_4)$. Since $\Psi(x_1, x_2, x_3, x_4) = \Psi(x_3, x_4, x_1, x_2)$ we find that pairs of fermions behave like bosons.

Fig. 6.1 Typical form of the empirical interaction w for atoms. © Mathieu Lewin 2021. All rights reserved

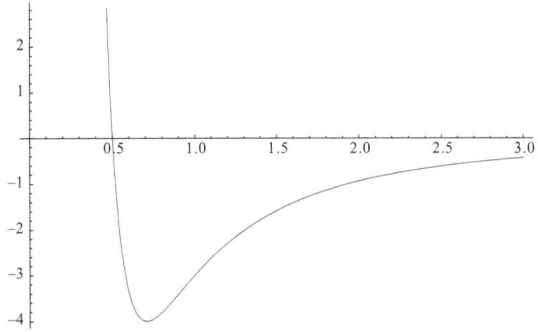

section. The latter is often assumed to have the form in Fig. 6.1, with a fairly strong repulsion at the origin and a rather fast decay at infinity. For noble gases, the attraction decreases like $-1/|x|^6$ at infinity (Van Der Waals attraction), but for general atoms it can decrease slower, for example if they can be polarized. It is therefore important that the mathematical theory be flexible enough in terms of assumptions about the potentials V and w, which are not always exactly known.

Atoms and Molecules We will study in more detail the case of electrons in atoms and molecules in Sect. 6.5. In the Born-Oppenheimer approximation, a molecule comprises N quantum electrons and M fixed classical nuclei, with charges $z_1, \ldots, z_M$ and located at $R_1, \ldots, R_M \in \mathbb{R}^3$. These M nuclei generate the Coulomb potential

$$V(x) = -\sum_{m=1}^{M} \frac{z_m}{|x - R_m|}$$

which is felt by the N electrons of the system. Moreover, the N electrons interact with each other through the Coulomb repulsion

$$\sum_{1 \leq j < k \leq N} \frac{1}{|x_j - x_k|}$$

which therefore corresponds to $w(x) = 1/|x|$. We are working here again in a system of units where $e^2/(4\pi\varepsilon_0) = \hbar^2/(2m_e) = 1$ where m_e is the electron mass. The Hamiltonian describing the N electrons is, therefore,

$$\boxed{H_{\text{mol}} = -\sum_{j=1}^{N} \Delta_{x_j} - \sum_{j=1}^{N} \sum_{m=1}^{M} \frac{z_m}{|x_j - R_m|} + \sum_{1 \leq j < k \leq N} \frac{1}{|x_j - x_k|}} \qquad (6.6)$$

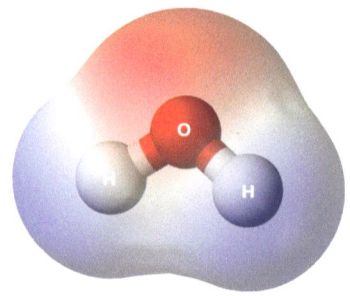

Fig. 6.2 For the water molecule, we have $z_1 = z_2 = 1$, $z_3 = 8$ and $N = 10$. The colors represent polarization

and it must be studied on $L_a^2((\mathbb{R}^3)^N, \mathbb{C})$ since electrons are fermions, as we explained. Here we have not taken into account the Coulomb repulsion between the nuclei

$$\sum_{1 \leq \ell < m \leq M} \frac{z_\ell z_m}{|R_\ell - R_m|}$$

because it is a constant, as long as they remain fixed. Of course, it must be added if we vary the positions of the nuclei and compare the energies obtained. We have also again neglected the spin of the electrons to simplify the writing of the model. It is quite surprising that the operator (6.6) describes all the atoms of the periodic table and all possible molecules, ranging from small objects like the water molecule H_2O (Fig. 6.2) to macro-molecules like DNA. This is one of the greatest achievements of quantum mechanics. The operator H_{mol}, whose formula fits on one line in (6.6), is supposed to describe the very different and complex physical behaviors of all these objects. However, we should temper our enthusiasm. Making concrete and precise predictions with the operator (6.6) turns out to be extraordinarily difficult, due to the very large dimension of the space $\mathbb{R}^{3N}$ in which this operator acts.

Exercise 6.1 (Bosons, Fermions and Representations of the Symmetric Group)
The permutation group $\mathfrak{S}_N$ acts on the space $L^2((\mathbb{R}^d)^N, \mathbb{C})$ by exchanging the variables of functions: $(U_\tau \Psi)(x_1, \ldots, x_N) = \Psi(x_{\tau^{-1}(1)}, \ldots, x_{\tau^{-1}(N)})$. As quantum states are always defined modulo phase (Sect. 1.5 of Chap. 1), we say that a normalized function $\Psi \in L^2((\mathbb{R}^d)^N, \mathbb{C})$ describes N indistinguishable particles when Ψ and $U_\tau \Psi$ are collinear for all $\tau \in \mathfrak{S}_N$:

$$\forall \tau \in \mathfrak{S}_N, \quad \exists \lambda_\tau \in \mathbb{C} : |\lambda_\tau| = 1, \ U_\tau \Psi = \lambda_\tau \Psi.$$

Then, the choice of a particle labeling has no influence on the physical state of the system. Verify that the U_τ's are unitary operators that form a group, $U_\tau U_{\tau'} = U_{\tau\tau'}$, with $U_{\mathrm{Id}} = 1$. Deduce that $\lambda_{\tau\tau'} = \lambda_\tau \lambda_{\tau'}$ for all $\tau, \tau' \in \mathfrak{S}_N$, then that Ψ is either symmetric ($\lambda_\tau = 1$), or anti-symmetric ($\lambda_\tau = \varepsilon(\tau)$, the signature of the permutation τ).

6.2 Self-adjointness

In this section we will show that the operator $H^V(N)$ defined in (6.1) is self-adjoint on $H^2_{a/s}((\mathbb{R}^d)^N, \mathbb{C})$, under physically reasonable conditions on the potentials V and w. We work in any dimension $d \geq 1$. Unfortunately, we cannot apply Theorem 3.4 in $\mathbb{R}^{dN}$. Indeed, the total potential

$$\sum_{j=1}^N V(x_j) + \sum_{1 \leq j < k \leq N} w(x_j - x_k)$$

cannot belong to any $L^p(\mathbb{R}^{dN})$ space for $p < \infty$, as it involves functions that depend on only one or two variables at a time. Moreover, the use of the Sobolev embedding in dimension dN would naturally lead us to use the space $L^{dN/2}(\mathbb{R}^{dN})$ which has a very bad dependence on N. Fortunately, the special form of the operator $H^V(N)$ allows us to show self-adjointness with the same assumptions on V and w as in $\mathbb{R}^d$, thus completely **independent of the number N of particles**.

Recall that the operator $H^V(1) = -\Delta + V(x)$ is self-adjoint on $D(H^V(1)) = H^2(\mathbb{R}^d)$ when $V \in L^p(\mathbb{R}^d) + L^\infty(\mathbb{R}^d)$, with

$$\begin{cases} p = 2 & \text{if } d \in \{1, 2, 3\}, \\ p > 2 & \text{if } d = 4, \\ p = \frac{d}{2} & \text{if } d \geq 5, \end{cases} \tag{6.7}$$

(Theorem 3.4). In this case V is infinitesimally $(-\Delta)$-bounded and we have the inequality

$$\|Vf\|^2_{L^2(\mathbb{R}^d)} \leq \varepsilon \|\Delta f\|^2_{L^2(\mathbb{R}^d)} + C_\varepsilon \|f\|^2_{L^2(\mathbb{R}^d)}, \qquad \forall f \in H^2(\mathbb{R}^d), \ \forall \varepsilon > 0. \tag{6.8}$$

The following result means that $H^V(N)$ is self-adjoint on $H^2_{a/s}((\mathbb{R}^d)^N)$ for all $N \geq 1$, with the same condition (6.7).

Theorem 6.2 (N-particle Hamiltonian: Self-adjointness) *We assume that V and w are in $L^p(\mathbb{R}^d, \mathbb{R}) + L^\infty(\mathbb{R}^d, \mathbb{R})$ with p satisfying Assumption (6.7). We also assume that w is even. Then, for all $N \geq 2$, the operator $H^V(N)$ is self-adjoint on*

$$D(H^V(N)) = \begin{cases} H^2((\mathbb{R}^d)^N, \mathbb{C}) \subset L^2((\mathbb{R}^d)^N, \mathbb{C}) & \text{(no symmetry)}, \\ H^2_s((\mathbb{R}^d)^N, \mathbb{C}) \subset L^2_s((\mathbb{R}^d)^N, \mathbb{C}) & \text{(bosons)}, \\ H^2_a((\mathbb{R}^d)^N, \mathbb{C}) \subset L^2_a((\mathbb{R}^d)^N, \mathbb{C}) & \text{(fermions)}, \end{cases}$$

and its spectrum is bounded from below in each of these three cases.

Proof The N-particle Laplace operator $\sum_{j=1}^{N} -\Delta_{x_j} = -\Delta$ is self-adjoint on $H^2((\mathbb{R}^d)^N)$, as we saw in Theorem 2.36. As the subspaces $H^2_s((\mathbb{R}^d)^N)$ and $H^2_a((\mathbb{R}^d)^N)$ are closed in $H^2((\mathbb{R}^d)^N)$ and they are invariant by $-\Delta$ in the sense of Exercise 4.45, the operator $-\Delta$ remains self-adjoint when it is restricted to these two subspaces. Simply, the equation $(1 - \Delta)\Psi = \Phi$ which admits a unique solution Ψ for all $\Phi \in L^2((\mathbb{R}^d)^N)$ verifies $\Psi \in H^2_{s/a}((\mathbb{R}^d)^N)$ as soon as $\Phi \in L^2_{s/a}((\mathbb{R}^d)^N)$ (write it in Fourier). According to Theorem 2.28, this shows the self-adjointness of $-\Delta$ on the two subspaces $H^2_{s/a}((\mathbb{R}^d)^N)$. The spectrum equals $\sigma(-\Delta) = \mathbb{R}_+$ (Exercise 6.3). By the Rellich-Kato Theorem 3.1, it suffices to show that each of the terms appearing in the definition of the total potential is infinitesimally $(-\Delta)$-bounded. We start for example with the function $V(x_1)$ and calculate

$$\|V(x_1)\Psi\|^2_{L^2} = \int_{\mathbb{R}^d} \cdots \left(\int_{\mathbb{R}^d} V(x_1)^2 |\Psi(x_1, \ldots, x_N)|^2 \, dx_1 \right) \cdots dx_N$$

$$\leq \varepsilon \int_{\mathbb{R}^d} \cdots \int_{\mathbb{R}^d} |\Delta_{x_1}\Psi(x_1, \ldots, x_N)|^2 \, dx_1 \cdots dx_N$$

$$+ C_\varepsilon \int_{\mathbb{R}^d} \cdots \int_{\mathbb{R}^d} |\Psi(x_1, \ldots, x_N)|^2 \, dx_1 \cdots dx_N.$$

Here we have used Inequality (6.8) in the variable x_1, fixing all the other variables $x_2, \ldots, x_N$, which is allowed by Fubini. As

$$\int_{(\mathbb{R}^d)^N} |\Delta_{x_1}\Psi|^2 = \int_{(\mathbb{R}^d)^N} |k_1|^4 |\widehat{\Psi}|^2 \leq \int_{(\mathbb{R}^d)^N} \left(\sum_{j=1}^{d} |k_j|^2 \right)^2 |\widehat{\Psi}|^2 = \int_{(\mathbb{R}^d)^N} |\Delta\Psi|^2,$$

this shows that $V(x_1)$ is infinitesimally $(-\Delta)$-bounded. The argument is exactly the same for $V(x_j)$.

To deal with the interaction, we first notice that in (6.8) the term on the right is invariant by translations while the one on the left is not. By replacing f by $f(\cdot + R)$, we therefore obtain

$$\|V(\cdot - R)f\|^2_{L^2(\mathbb{R}^d)} \leq \varepsilon \|f\|^2_{H^2(\mathbb{R}^d)} + C_\varepsilon \|f\|^2_{L^2(\mathbb{R}^d)},$$

$$\forall f \in H^2(\mathbb{R}^d), \ \forall R \in \mathbb{R}^d, \ \forall \varepsilon > 0. \qquad (6.9)$$

We used the same trick in the proof of Kato's inequality in Corollary 1.4. Thus, we can now write

$$\|w(x_1 - x_2)\Psi\|^2_{L^2} = \int_{\mathbb{R}^d} \cdots \left(\int_{\mathbb{R}^d} w(x_1 - x_2)^2 |\Psi(x_1, \ldots, x_N)|^2 \, dx_1 \right) \cdots dx_N$$

$$\leq \varepsilon \int_{\mathbb{R}^d} \cdots \int_{\mathbb{R}^d} |\Delta_{x_1}\Psi(x_1, \ldots, x_N)|^2 \, dx_1 \cdots dx_N$$

$$+ C_\varepsilon \int_{\mathbb{R}^d} \cdots \int_{\mathbb{R}^d} |\Psi(x_1, \ldots, x_N)|^2 \, dx_1 \cdots dx_N$$

where we used (6.9) with $R = x_2$, fixing again $x_2, \ldots, x_N$ by Fubini's theorem. This shows that all the terms of the potential are infinitesimally $(-\Delta)$-bounded, so the total potential is. The behavior in N of our estimates is quite bad, but this is not important for self-adjointness since we can take ε as small as we like.

Since the N-particle Laplacian $-\Delta$ is non-negative, the operator $H^V(N)$ is bounded from below, by the last part of the statement of the Rellich-Kato Theorem 3.1. But the previous arguments can also be adapted to $\int_{(\mathbb{R}^d)^N} V(x_j)|\Psi|^2$ and $\int_{(\mathbb{R}^d)^N} w(x_j - x_k)|\Psi|^2$ to infer the energy bound

$$\mathcal{E}^V(\Psi) \geq (1 - \varepsilon) \int_{(\mathbb{R}^d)^N} |\nabla \Psi|^2 - C_\varepsilon \int_{(\mathbb{R}^d)^N} |\Psi|^2 \tag{6.10}$$

This also implies that $H^V(N)$ is bounded from below, this time by Theorem 2.33.

$\square$

Exercise 6.3 (Laplacian: Bosons and Fermions) Show that the spectrum of the self-adjoint operator $-\Delta$ defined on $H_s^2((\mathbb{R}^d)^N)$ and $H_a^2((\mathbb{R}^d)^N)$ is still equal to $\mathbb{R}_+$.

Exercise 6.4 (Atoms and Molecules) Using Kato's inequality (1.30), find an estimate on the constant C_ε as a function of ε, N, M (the number of nuclei) and $\max(|z_m|)$ (the maximum charge of the nuclei), for the operator (6.6).

With the weaker assumption $V \in L^p(\mathbb{R}^d, \mathbb{R}) + L^\infty(\mathbb{R}^d, \mathbb{R})$ where

$$\begin{cases} p = 1 & \text{if } d = 1, \\ p > 1 & \text{if } d = 2, \\ p = \frac{d}{2} & \text{if } d \geq 3, \end{cases} \tag{6.11}$$

we were able to construct in Corollary 3.20 the **Friedrichs self-adjoint realization** of the operator $H^V(1) = -\Delta + V$, whose domain is given by

$$D(-\Delta + V) = \left\{ u \in H^1(\mathbb{R}^d) : (-\Delta + V)u \in L^2(\mathbb{R}^d) \right\}.$$

The same proof as that of Theorem 6.2 allows us to deduce the following result.

Theorem 6.5 (N-particle Hamiltonian: Self-adjointness II) *We assume that V and w are in $L^p(\mathbb{R}^d, \mathbb{R}) + L^\infty(\mathbb{R}^d, \mathbb{R})$ with p satisfying Assumption (6.11). We also assume that w is even. Then, for all $N \geq 2$, the operator $H^V(N)$ is self-adjoint on*

$$D(H^V(N)) = \left\{ \Psi \in Q(H^V(N)) : H^V(N)\Psi \in L^2((\mathbb{R}^d)^N) \right\}$$

where

$$Q(H^V(N)) = \begin{cases} H^1((\mathbb{R}^d)^N, \mathbb{C}) \subset L^2((\mathbb{R}^d)^N, \mathbb{C}) & \textit{(no symmetry)}, \\ H^1_s((\mathbb{R}^d)^N, \mathbb{C}) \subset L^2_s((\mathbb{R}^d)^N, \mathbb{C}) & \textit{(bosons)}, \\ H^1_a((\mathbb{R}^d)^N, \mathbb{C}) \subset L^2_a((\mathbb{R}^d)^N, \mathbb{C}) & \textit{(fermions)}, \end{cases}$$

and its spectrum is bounded from below in each of these three cases.

Of course, there are similar results for potentials that diverge at infinity or satisfy the assumptions of Sect. 3.3.

6.3   Essential Spectrum: HVZ Theorem

We saw in Corollary 5.38 that the essential spectrum of a Schrödinger operator $-\Delta + V$ is

$$\sigma_{\mathrm{ess}}(-\Delta + V) = [0, +\infty)$$

when V is negligible at infinity. This corresponds to the energies of a particle that has escaped to infinity and therefore does not feel anymore the potential V.

The situation is much more complicated for the essential spectrum of $H^V(N)$ describing N particles. Indeed, any number $k \in \{1, \dots, N\}$ of them can now escape to infinity. If those that leave remain close to each other, they will continue to interact through the potential w. To state the theorem describing this situation, we call

$$E^V_{a/s}(N) := \min \sigma\left(H^V(N)\right), \qquad \Sigma^V_{a/s}(N) := \min \sigma_{\mathrm{ess}}\left(H^V(N)\right)$$

the bottom of the spectrum (which may be an eigenvalue or not), and the bottom of the essential spectrum, when the operator $H^V(N)$ is considered on $L^2_{a/s}((\mathbb{R}^d)^N, \mathbb{C})$.

Theorem 6.6 (HVZ) *Suppose that* $V, w \in L^p(\mathbb{R}^d, \mathbb{R}) + L^\infty_\varepsilon(\mathbb{R}^d, \mathbb{R})$, *with* p *satisfying* (6.11) *and that* w *is even. Then the essential spectrum is a half-line*

$$\sigma_{\mathrm{ess}}\left(H^V(N)\right) = \left[\Sigma^V_{a/s}(N), +\infty\right) \tag{6.12}$$

with

$$\boxed{\Sigma^V_{a/s}(N) = \min\left\{E^V_{a/s}(N-k) + E^0_{a/s}(k), \; k = 1, \dots, N\right\}.} \tag{6.13}$$

If V ≡ 0, we have

$$E^0_{a/s}(N) = \Sigma^0_{a/s}(N) = \min\left\{E^0_{a/s}(N-k) + E^0_{a/s}(k),\ k = 1, \ldots, N-1\right\}.$$

$$(6.14)$$

There is a similar result when $H^V(N)$ is considered on the whole space $L^2((\mathbb{R}^d)^N)$. In fact, in this case $E^V(N) = E^V_s(N)$ and $\Sigma^V(N) = \Sigma^V_s(N)$, that is, the bottom of the spectrum and of the essential spectrum are the same as in the symmetric case.

Formula (6.13) was proven by Zhislin [Zhi60], Van Winter [Van64] and Hunziker [Hun66] in the 1960s and we will not provide the proof here. It means that the essential spectrum begins when k particles have escaped, while $N - k$ remain in a neighborhood of the support of V. The minimum energy of such a system is the sum of that of the $N - k$ remaining particles $E^V(N - k)$ and the energy $E^0(k)$ of the particles that have escaped and no longer feel the potential V (the latter being negligible at infinity by hypothesis). We then need to find the optimal number k of particles to send to infinity so that the energy obtained is as small as possible, hence the minimum over $k = 1, \ldots, N$. When $V \equiv 0$ we need to remove the case $k = N$ as stated in (6.14) because otherwise the formula would be a tautology.

If the result is quite intuitive, the proof of Theorem 6.6 is not so simple since the three numbers $\Sigma^V_{a/s}(N)$, $E^V_{a/s}(N - k)$ and $E^0_{a/s}(k)$ that we must compare concern operators defined on the three different spaces $L^2_{a/s}((\mathbb{R}^d)^N)$, $L^2_{a/s}((\mathbb{R}^d)^{N-k})$ and $L^2_{a/s}((\mathbb{R}^d)^k)$. The tensor product structure then plays a fundamental role. For example, for (6.13) the idea is that any Weyl sequence (Ψ_n) associated with $\Sigma^V_{a/s}(N)$, that is such that $\|\Psi_n\| = 1$, $\Psi_n \rightharpoonup 0$ and

$$\left(H^V(N) - \Sigma^V_{a/s}(N)\right)\Psi_n \to 0,$$

should behave like

$$\Psi_n(x_1, \ldots, x_N) \simeq \Phi(x_1, \ldots, x_{N-k})\Phi'_n(x_{N-k+1}, \ldots, x_N) \qquad (6.15)$$

(which must in addition be symmetrized or anti-symmetrized), where Φ is the first eigenfunction of $H^V(N - k)$, assuming it exists, and $\Phi'_n \rightharpoonup 0$ is a Weyl sequence associated with $E^0_{a/s}(k) = \Sigma^0_{a/s}(k)$. We note that such a tensor product (6.15) converges weakly to 0, even when Φ is fixed. In the article [Lew11], Theorem 6.6 is proved using a special weak topology, different from that of $L^2_{a/s}((\mathbb{R}^d)^N)$, capable of detecting a factorization of the type (6.15).

The case where $w \geq 0$ is simpler and is the one that will concern us most when we study electrons, for which $w(x) = 1/|x|$ in dimension $d = 3$.

Corollary 6.7 (HVZ for Repulsive Systems) *Suppose that $V, w \in L^p(\mathbb{R}^d, \mathbb{R}) + L^\infty_\varepsilon(\mathbb{R}^d, \mathbb{R})$ with p satisfying (6.11) and that $w \geq 0$ is an even function. Then we have $E^0_{a/s}(N) = 0$ for all $N \geq 1$ and*

$$\boxed{\Sigma^V_{a/s}(N) = E^V_{a/s}(N-1).} \tag{6.16}$$

The result specifies that for a non-negative interaction w, the essential spectrum starts when one particle has escaped to infinity. Pulling out more particles would only increase the energy.

Proof When there is no possible confusion, we remove the index a/s. When $w \geq 0$, the associated quadratic form $\mathcal{E}^0$ is non-negative, so $E^0(N) \geq 0$. Taking a test function in the form $\Psi_n(x_1, \ldots, x_N) = n^{-Nd/2}\Psi(x_1/n, \ldots, x_N/n)$ where $\Psi \in C^\infty_c$, we find $\mathcal{E}^0(\Psi_n) \to 0$, which therefore shows that $E^0(N) = 0$. By the HVZ Theorem 6.6, we have $E^V(N) \leq \Sigma^V(N) \leq E^V(N-1)$ because $E^0(1) = 0$ so $N \mapsto E^V(N)$ is non-increasing. Thus, $E^V(N-k) + E^0(k) = E^V(N-k) \geq E^V(N-1)$ for all $k = 1, \ldots, N$ and $\Sigma^V(N) = \min\{E^V(N-k) + E^0(k), k = 1, \ldots, N\} = E^V(N-1)$. $\qquad\square$

6.4   Particles Without Interaction

Before studying systems with interaction, it is useful to start with the much simpler case where

$$\boxed{w \equiv 0.}$$

We state here a result for the operator $-\Delta + V$ which is in fact general and is proved in a similar way for any operator in the form $\sum_{j=1}^N A_j$ on a symmetric or anti-symmetric tensor product, in any Hilbert space [RS72, Sec. VIII.10]. One way to realize these operators is, by the spectral theorem, to consider a multiplication operator in the form

$$\sum_{j=1}^N a(x_j), \qquad \text{on} \qquad L^2_{a/s}(B^N, d\mu^{\otimes N}), \tag{6.17}$$

for example with $B = \sigma(A) \times \mathbb{N}$ and $a(s, n) = s$. We will leave as an exercise the extension of the following result to this more general framework.

Theorem 6.8 (Spectrum of $H^V(N)$ When $w \equiv 0$) *We assume that V satisfies the assumptions of Theorem 6.6 and we call $\mu_k(-\Delta + V) \leq 0$ the k-th Courant-Fischer level of the operator $-\Delta + V$, which is equal to the k-th eigenvalue counted with* *multiplicity or to 0 if the operator has less than k negative eigenvalues. Then for* $w \equiv 0$ *we have*

$$E_s^V(N) = N\mu_1(-\Delta + V) \quad \text{and} \quad \Sigma_s^V(N) = (N-1)\mu_1(-\Delta + V) \qquad (6.18)$$

in the bosonic case, and

$$E_a^V(N) = \sum_{j=1}^{N} \mu_j(-\Delta + V) \quad \text{and} \quad \Sigma_a^V(N) = \sum_{j=1}^{N-1} \mu_j(-\Delta + V) \qquad (6.19)$$

in the fermionic case. Moreover, the eigenvalues of $H^V(N)$ *are given by*

$$\sigma_{\text{pt}}\big(H^V(N)\big) = \Big\{ \lambda'_{j_1}(-\Delta + V) + \cdots + \lambda'_{j_N}(-\Delta + V), \quad 1 \le j_1 \le \cdots \le j_N \Big\} \tag{6.20}$$

in the bosonic case, and

$$\sigma_{\text{pt}}\big(H^V(N)\big) = \Big\{ \lambda'_{j_1}(-\Delta + V) + \cdots + \lambda'_{j_N}(-\Delta + V), \quad 1 \le j_1 < \cdots < j_N \Big\} \tag{6.21}$$

in the fermionic case, where the $\lambda'_j(-\Delta + V)$ *are* all *the eigenvalues of the operator* $-\Delta + V$ *(including positive eigenvalues when they exist), repeated according to their multiplicity.*

Here we see a striking difference between bosons and fermions, as represented in Fig. 6.3. The ground state energy of a bosonic system without interactions is obtained by putting all the particles in the same lowest energy state, that is, by taking the Bose-Einstein condensate

$$\Psi_{\text{BE}}(x_1, \ldots, x_N) = u_1(x_1) \cdots u_1(x_N) = (u_1)^{\otimes N}(x_1, \ldots, x_N), \qquad (6.22)$$

which is the associated eigenfunction of $H^V(N)$. Here u_1 is the first eigenfunction of $-\Delta + V$, assuming it exists (in this case it is always unique up to a phase by Theorem 1.18). In contrast, the anti-symmetry of Ψ implies that two fermions cannot be in the same state, which forces the fermions to occupy the N smallest eigenvalues (counted with multiplicity). The corresponding eigenfunction is the **Slater determinant**

$$\Psi_{\text{Slat}}(x_1, \ldots, x_N) = \frac{1}{\sqrt{N!}} \sum_{\sigma \in \mathfrak{S}_N} \varepsilon(\sigma) \, u_1(x_{\sigma(1)}) \cdots u_N(x_{\sigma(N)}) = \frac{\det(u_i(x_j))}{\sqrt{N!}} \tag{6.23}$$

consisting in anti-symmetrizing the tensor product $u_1 \otimes \cdots \otimes u_N$. The fermionic problem does not have a minimizer if $-\Delta + V$ has less than N non-positive

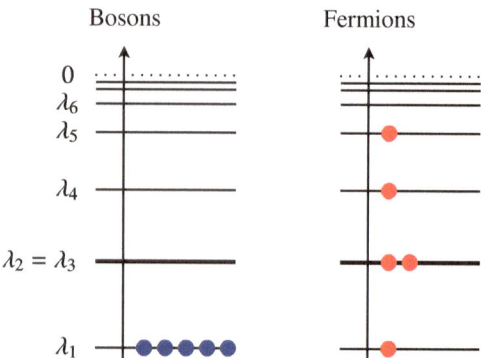

Fig. 6.3 Computation of the first eigenvalue of the operator $H^V(N)$ with $N = 5$ when $w \equiv 0$, as stated in Theorem 6.8. In the bosonic case, $E_s^V(N)$ is obtained by putting all particles in the ground state of the operator $-\Delta + V$, which forms a Bose-Einstein condensate. For fermions, $E_a^V(N)$ is obtained by filling the N lowest eigenstates of $-\Delta + V$ starting from the lowest, without redundancy (except in case of multiplicity). © Mathieu Lewin 2021. All rights reserved

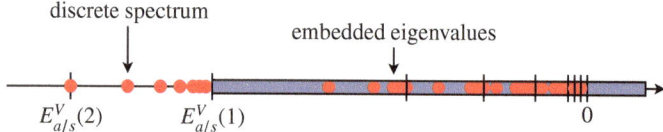

Fig. 6.4 Spectrum of the operator $H^V(2)$ when $w \equiv 0$, as stated in Theorem 6.8. © Mathieu Lewin 2021. All rights reserved

eigenvalues, while the bosonic problem always has a minimizer if $-\Delta + V$ has at least one.

Let us finally note that the spectrum of $H^V(N)$ contains many eigenvalues embedded in the essential spectrum. If the operator $-\Delta + V$ has infinitely many eigenvalues, then $H^V(N)$ has eigenvalues that have many accumulation points. For example, if $N = 2$, the spectrum contains eigenvalues that accumulate at all the $\mu_j(-\Delta + V)$, as shown in Fig. 6.4. These embedded eigenvalues are very unstable and they will generically all disappear when an interaction is turned on.

Proof (of Theorem 6.8) Let us start with (6.18). By taking a function $\Psi = u^{\otimes N}$ where $u \in H^2(\mathbb{R}^d)$ is normalized, we find that

$$E_s^V(N) \leq \left\langle \Psi, H^V(N)\Psi \right\rangle = N \left(\int_{\mathbb{R}^d} |\nabla u|^2 + \int_{\mathbb{R}^d} V|u|^2 \right).$$

By the Courant-Fischer principle, this shows that

$$E_s^V(N) \leq N \inf_{\substack{u \in H^2(\mathbb{R}^d) \\ \|u\|_{L^2}=1}} \langle u, (-\Delta + V)u \rangle = N\mu_1(-\Delta + V) = NE^V(1).$$

Conversely, we have by symmetry of Ψ and Fubini

$$\left\langle \Psi, H^V(N)\Psi \right\rangle = N \int_{\mathbb{R}^{dN}} \left(|\nabla_{x_1}\Psi|^2 + V(x_1)|\Psi|^2 \right)$$

$$= N \int_{(\mathbb{R}^d)^{N-1}} \left(\underbrace{\int_{\mathbb{R}^d} \left(|\nabla_{x_1}\Psi|^2 + V(x_1)|\Psi|^2 \right) dx_1}_{\geq E^V(1) \int_{\mathbb{R}^d} |\Psi(x_1,x_2,\dots)|^2\, dx_1} \right) dx_2 \cdots dx_N$$

$$\geq N E^V(1) \|\Psi\|^2.$$

This indeed shows that $E_s^V(N) = N E^V(1)$. The rest follows from the HVZ Theorem 6.6.

The argument is more difficult for fermions. By taking a function in the form (6.23) where $u_1, \dots, u_N \in H^2(\mathbb{R}^d)$ is any orthonormal system in $L^2(\mathbb{R}^d)$, a tedious but completely elementary calculation gives

$$\|\Psi\| = 1 \quad \text{and} \quad \left\langle \Psi, H^V(N)\Psi \right\rangle = \sum_{j=1}^N \left\langle u_j, (-\Delta + V)u_j \right\rangle.$$

By minimizing with respect to the orthonormal systems $u_1, \dots, u_N$ we deduce from the variational principle for sums of eigenvalues stated in (5.46) that

$$E_a^V(N) \leq \mu_1(-\Delta + V) + \cdots + \mu_N(-\Delta + V).$$

The proof of the lower bound is more elaborate. We want to show that

$$\mathcal{E}^V(\Psi) \geq \mu_1(-\Delta + V) + \cdots + \mu_N(-\Delta + V) \tag{6.24}$$

for all $\Psi \in H_a^1((\mathbb{R}^d)^N)$ such that $\|\Psi\| = 1$. Let (u_j) be any Hilbert basis of $L^2(\mathbb{R}^d)$, with $u_j \in D(-\Delta + V)$ for all j. The anti-symmetrized tensor products

$$u_{j_1} \wedge \cdots \wedge u_{j_N}(x_1, \dots, x_N) := \frac{1}{\sqrt{N!}} \sum_{\sigma \in \mathfrak{S}^N} \varepsilon(\sigma)\, u_{j_1}(x_{\sigma(1)}) \cdots u_{j_N}(x_{\sigma(N)}),$$

with $1 \leq j_1 < \cdots < j_N$ form a Hilbert basis of $L_a^2((\mathbb{R}^d)^N)$ (left as an exercise). We can therefore show (6.24) for any finite linear combination of these functions, the general case being deduced by density. A simple way to truncate the series is to fix an integer $K \geq 1$ and look at the space

$$\mathcal{V}_K := \text{span}\left(u_{j_1} \wedge \cdots \wedge u_{j_N} \right)_{1 \leq j_1 < \cdots < j_N \leq K},$$

which is of dimension $\binom{K}{N}$. We can verify that this space is unchanged if we choose another basis of $\mathrm{span}(u_1, \ldots, u_K)$. Consider then the $K \times K$ matrix given by

$$M_{ij} = \langle u_i, (-\Delta + V)u_j \rangle,$$

which represents the projection of $-\Delta + V$ on $\mathrm{span}(u_1, \ldots, u_K)$. As M is a Hermitian matrix, it is diagonalizable in an orthonormal basis, that is, there exists a unitary $U \in U(K)$ such that

$$\langle v_i, (-\Delta + V)v_j \rangle = \langle v_i, (-\Delta + V)v_i \rangle \delta_{ij} \quad \text{with the vectors} \quad v_j = \sum_{k=1}^{K} U_{kj} u_k.$$

Any function $\Psi \in \mathcal{V}_K$ can therefore be written in the form

$$\Psi = \sum_{1 \le j_1 < \cdots < j_N \le K} c_{j_1, \ldots, j_N} \, v_{j_1} \wedge \cdots \wedge v_{j_N}, \qquad \sum_{1 \le j_1 < \cdots < j_N \le K} |c_{j_1, \ldots, j_N}|^2 = 1.$$

Another simple but tedious calculation shows that the matrix of $H^V(N)$ in $\mathcal{V}_K$ is diagonal in the basis that diagonalizes the projection of $-\Delta + V$, which provides

$$\mathcal{E}^V(\Psi) = \sum_{1 \le j_1 < \cdots < j_N \le K} |c_{j_1, \ldots, j_N}|^2 \langle v_{j_1} \wedge \cdots \wedge v_{j_N}, H^V(N) v_{j_1} \wedge \cdots \wedge v_{j_N} \rangle$$

$$= \sum_{1 \le j_1 < \cdots < j_N \le K} |c_{j_1, \ldots, j_N}|^2 \sum_{k=1}^{N} \langle v_{j_k}, (-\Delta + V)v_{j_k} \rangle$$

$$\ge \big(\mu_1(-\Delta + V) + \cdots + \mu_N(-\Delta + V)\big) \underbrace{\sum_{1 \le j_1 < \cdots < j_N \le K} |c_{j_1, \ldots, j_N}|^2}_{=1}.$$

In the last line, we have again used the variational principle for sums of eigenvalues in (5.46). This concludes the proof of (6.19).

While it is easy to see that for $u_{j_1}, \ldots, u_{j_N}$ eigenfunctions of $-\Delta + V$, $u_{j_1} \otimes_s \cdots \otimes_s u_{j_N}$ and $u_{j_1} \wedge \cdots \wedge u_{j_N}$ are eigenfunctions of $H^V(N)$, with eigenvalue $\lambda_{j_1} + \cdots + \lambda_{j_N}$, it is more difficult to verify that these are the only possible eigenvalues, as stated in (6.20) and (6.21). For this, it is probably simpler to apply the spectral theorem and show it for operators in the form (6.17). Indeed, by induction on N, we have $\mu^{\otimes N}(\{s_1 + \cdots + s_N = \lambda\}) > 0$ if and only if $\lambda = \lambda_1 + \cdots + \lambda_N$ with $\mu(\{\lambda\} \times \mathbb{N}) > 0$. This is a consequence of Fubini, since

$$\mu^{\otimes N}(\{s_1 + \cdots + s_N = \lambda\})$$

$$= \int_{(\sigma(A) \times \mathbb{N})^N} \mathbb{1}(s_1 + \cdots + s_N = \lambda) \, d\mu(s_1, n_1) \cdots d\mu(s_N, n_N)$$

$$= \int_{(\sigma(A)\times\mathbb{N})^{N-1}} \left(\underbrace{\int_{\sigma(A)\times\mathbb{N}} \mathbb{1}(s_N = \lambda - s_1 - \cdots - s_{N-1}) d\mu(s_1, n_1)}_{=0 \text{ except if } \lambda - s_1 - \cdots - s_{N-1} \in \sigma(A)} \right)$$

$$\cdots d\mu(s_N, n_N).$$

This concludes the proof. □

6.5 Atoms and Molecules*

In this section, we present some results concerning the spectrum of the Hamiltonian $H^V(N)$ describing the N quantum electrons of a molecule, which also contains M classical nuclei, of charges $z_1, \ldots, z_M \in (0, +\infty)$ and located at $R_1, \ldots, R_M \in \mathbb{R}^3$. As explained previously, this amounts to choosing the potentials

$$V(x) = - \sum_{m=1}^{M} \frac{z_m}{|x - R_m|}, \qquad w(x) = \frac{1}{|x|}, \tag{6.25}$$

in dimension $d = 3$. All the previous results apply, since $V, w \in L^2(\mathbb{R}^3) + L^\infty(\mathbb{R}^3)$ and $V, w \to 0$ at infinity. In particular, we obtain

Corollary 6.9 (HVZ for Atoms and Molecules) *For V, w given by (6.25) in dimension $d = 3$, we have $\Sigma_{a/s}^V(N) = E_{a/s}^V(N-1)$.*

Even though electrons are fermions, it is interesting to study in detail the bosonic case, to understand the differences with the fermionic case.

6.5.1 Existence of Eigenvalues, Ionization Conjecture

An important question is that of the existence or non-existence of eigenvalues below the essential spectrum, which represent stationary states of the system, between which the electrons can navigate when they are excited, and which explain the line spectrum observed in a spectroscopy experiment. Intuitively, the M nuclei will not be able to maintain a large number of electrons close to them. While the electrons are indeed attracted to the nuclei (the potential V is negative), they also repel each other (w is positive), which makes the presence of too many electrons energetically unfavorable. The following theorem provides a complete description of the number of eigenvalues below $\Sigma^V(N)$, in terms of the total charge of the system.

Theorem 6.10 (Existence or Non-existence of Bound States for Molecules) *We
assume that V and w are given by (6.25) in dimension $d = 3$, with $z_m > 0$, and we
call*

$$Z := \sum_{m=1}^{M} z_m$$

the total nuclear charge.

* (Neutral or positively charged molecules [Zhi60, ZS65]). *If $N < Z + 1$, then
  $H^V(N)$ has **infinitely many eigenvalues** below its essential spectrum, that is*

$$\mu_k\big(H^V(N)\big) < \Sigma_{a/s}^V(N),$$

  for all $k \geq 1$.
* (Negatively charged molecules [Zhi71, Yaf76, VZ77, Sig82]). *If $N \geq Z+1$, then
  $H^V(N)$ has **at most a finite number of eigenvalues** below its essential spectrum,
  that is we have*

$$\mu_k\big(H^V(N)\big) = \Sigma_{a/s}^V(N),$$

  for k large enough.
* (Non-existence if N is large [Rus82, Sig82, Sig84]). *There exists a critical
  number $N_{a/s}(V)$ such that $H^V(N)$ possesses **no eigenvalue** below its essential
  spectrum for all $N > N_{a/s}(V)$, that is we have*

$$E_{a/s}^V(N) = E_{a/s}^V(N-1) = \Sigma_{a/s}^V(N).$$

* (Estimate on $N_{a/s}(V)$ [Lie84]). *We have*

$$N_{a/s}(V) < 2Z + M. \tag{6.26}$$

The statement is illustrated in Fig. 6.5. The theorem states that the total charge
of the system is the right criterion to determine whether there is an infinity or not
of eigenvalues below the essential spectrum. A neutral or positively charged system
(with more protons than electrons) always has infinitely many excited states, while
a negatively charged system only has a finite number, or even none if there are too
many electrons. It should be noted that if the z_m's are all integers (as is the case in
application), then the condition becomes $N \leq Z$ or $N > Z$. However, the theorem
is valid when the z_m's are any positive real numbers.

The idea of the proof of the existence of an infinity of eigenvalues below $\Sigma_{a/s}^V(N)$
is quite simple and it is very similar to that used for Theorem 5.46. In fact, the first
part of Theorem 6.10 is valid in any dimension d if we replace $1/|x|$ by a potential
behaving like $1/|x|^\alpha$ at infinity, with $0 < \alpha < 2$, for both V and w. The argument

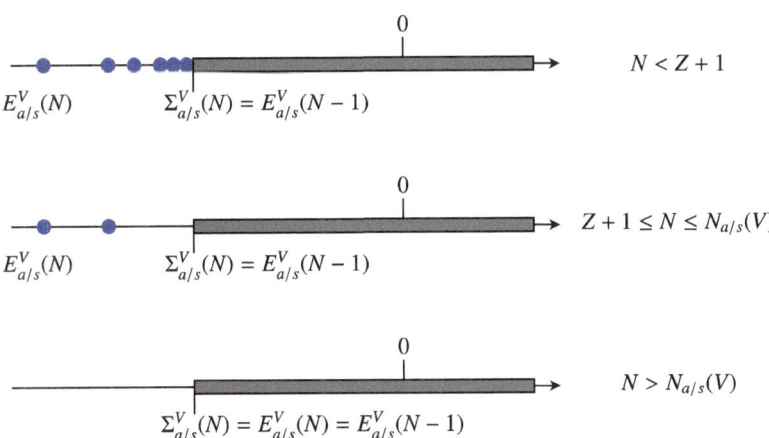

Fig. 6.5 Illustration of the results of Theorem 6.10. © Mathieu Lewin 2021. All rights reserved

consists in showing that it is not energetically favorable for one electron to escape to infinity. It would indeed feel the Coulomb potential generated by all the other particles, which form a system of total charge $Z - (N - 1) = Z + 1 - N > 0$, therefore attractive.

An important question is also to determine from which value of N the molecule cannot accept more electrons, that is, what is the maximum degree of negative ionization of a molecule. This amounts to estimating the constant $N_{a/s}(V)$ of Theorem 6.10. In nature, we do not observe very negatively charged atoms, which suggests that, at least for $M = 1$ (atoms), we should have

$$N_a(V) \overset{?}{\le} Z + C, \qquad \text{for } M = 1, \tag{6.27}$$

where C is a universal constant probably equal to 1 or 2. This assertion is called the **ionization conjecture** and it is a famous unsolved problem [FLLS22, Chap. 34]. For the case of molecules, the estimate probably takes the form

$$N_a(V) \overset{?}{\le} Z + CM, \qquad \text{for } M \ge 1. \tag{6.28}$$

For $M = 1$, the best estimate known to date is due to Nam [Nam12]:

$$N_a(V) < 1.22Z + 3Z^{1/3}, \qquad \text{for } M = 1.$$

Even if electrons are fermions, Theorem 6.10 holds the same for bosons. On the contrary, the ionization conjecture (6.27) is known to be **false for the bosonic** $N_s(V)$, as we will see a bit later. Bosons love each other so much that they can form very negative stable ions with many more electrons than protons, in spite of their strong Coulomb repulsion. The proof of the conjecture (6.27) should therefore use the fermionic nature of electrons in a crucial way.

Most of the research works have been focused on the existence or absence of eigenvalues below the essential spectrum, as these are easily accessible by the Courant-Fischer formula in Theorem 5.40. While it is physically reasonable to imagine that the absence of eigenvalues below $\Sigma_{a/s}^V(N)$ implies the total absence of eigenvalue in the entire spectrum, this fact is not known mathematically. However, it has been proven in [LL13] with a method based on the time-dependent Schrödinger equation that $H^V(N)$ has no eigenvalue at all for $N \geq 4Z + 1$, when $M = 1$.

6.5.2   The Limit $N \sim \kappa Z \to \infty$ for Atoms

Let us conclude this chapter with a walk through the periodic table. Let us recall that the latter contains all the neutral atoms, thus with $M = 1$ and $N = Z$. In theory this table should stop at $Z = 137$ due to relativistic effects that have been neglected in this book (read about this Remark B.1 in Appendix B), but for non-relativistic particles it continues indefinitely. Even if Schrödinger's equation becomes increasingly complicated and involves wavefunctions Ψ depending on an increasing number of variables, it turns out that great simplifications appear in the limit $N = Z \to \infty$. As we will see, the leading order is given by a **nonlinear model of mean field type**, posed on $\mathbb{R}^3$.

To learn more about the behavior of quantum matter, we will allow ourselves to visit generalized periodic tables, possibly without any physical reality. We will assume that $N \sim \kappa Z$ with $\kappa > 0$ instead of just $N = Z$, to study the stability of highly ionized atoms. We will also allow electrons to be bosons. In this case, we will see that the ionization conjecture is false ("bosonic" atoms are stable even when they are negatively ionized with $\kappa > 1$) and that the system collapses completely. It is therefore the Pauli principle that ensures the stability of atoms, in addition to preventing their too strong negative ionization.

To state the main theorem, it is useful to introduce the electronic density

$$\rho_\Psi^{(1)}(x) := N \int_{(\mathbb{R}^3)^{N-1}} |\Psi(x, x_2, \ldots, x_N)|^2 \, dx_2 \cdots dx_N, \qquad (6.29)$$

which gives the local average number of electrons. More precisely,

$$\int_\Omega \rho_\Psi^{(1)} = \int_{(\mathbb{R}^3)^N} \left(\sum_{j=1}^N \mathbb{1}_\Omega(x_j) \right) |\Psi(x_1, \ldots, x_N)|^2 \, dx_1 \cdots dx_N$$

is the average number of electrons in the domain $\Omega \subset \mathbb{R}^3$. The following result is a summary of several research works obtained during the years 1980s–1990s. It was possible to show that the average behavior of the electrons in a very heavy atom is given by a nonlinear problem in $\mathbb{R}^3$, this latter being different for bosons and fermions.

Theorem 6.11 (Atoms with $N \sim \kappa Z \to \infty$) *We take $V(x) = -Z/|x|$ and $w(x) = 1/|x|$ in dimension $d = 3$. Let $\kappa > 0$ be fixed.*

- (Bosons [BL83, Sol90, Bac91, BLLS93]). *We have*

$$
\lim_{\substack{N \to \infty \\ N/Z \to \kappa}} \frac{E_s^V(N)}{N^3} = \inf_{\substack{u \in H^1(\mathbb{R}^3) \\ \int_{\mathbb{R}^3} |u|^2 = 1}} \left\{ \int_{\mathbb{R}^3} |\nabla u(x)|^2 \, \mathrm{d}x - \int_{\mathbb{R}^3} \frac{|u(x)|^2}{\kappa |x|} \, \mathrm{d}x \right.
$$
$$
\left. + \frac{1}{2} \iint_{\mathbb{R}^3 \times \mathbb{R}^3} \frac{|u(x)|^2 |u(y)|^2}{|x - y|} \, \mathrm{d}x \, \mathrm{d}y \right\}. \qquad (6.30)
$$

The problem on the right admits a unique solution u_κ up to a phase, when $\kappa \leq \kappa_c \simeq 1.21$, and no minimizer for $\kappa > \kappa_c$. We have

$$
\boxed{\lim_{Z \to \infty} \frac{N_s(V)}{Z} = \kappa_c \simeq 1.21.} \qquad (6.31)
$$

Assume now $0 < \kappa \leq \kappa_c$. The function u_κ is radial-decreasing and strictly positive. The electronic density (6.29) satisfies

$$
\frac{\rho_{\Psi_N}^{(1)}(x)}{N} - N^3 |u_\kappa(Nx)|^2 \to 0 \qquad (6.32)
$$

strongly in $L^1(\mathbb{R}^3) \cap L^3(\mathbb{R}^3)$, for any eigenvector Ψ_N associated with the eigenvalue $E_s^V(N)$ of $H^V(N)$.

- (Fermions [LS77, Lie81, LSST88]). *We have*

$$
\lim_{\substack{N \to \infty \\ N/Z \to \kappa}} \frac{E_a^V(N)}{N^{7/3}} = \inf_{\substack{\rho \in L^1 \cap L^{5/3}(\mathbb{R}^3, \mathbb{R}^+) \\ \int_{\mathbb{R}^3} \rho = 1}} \left\{ \frac{3}{5} c_{\mathrm{TF}} \int_{\mathbb{R}^3} \rho(x)^{\frac{5}{3}} \, \mathrm{d}x - \int_{\mathbb{R}^3} \frac{\rho(x)}{\kappa |x|} \, \mathrm{d}x \right.
$$
$$
\left. + \frac{1}{2} \iint_{\mathbb{R}^3 \times \mathbb{R}^3} \frac{\rho(x)\rho(y)}{|x - y|} \, \mathrm{d}x \, \mathrm{d}y \right\} \qquad (6.33)
$$

where $c_{\mathrm{TF}} = \pi^{4/3} 2^{2/3} 3^{2/3}$. The problem on the right admits a unique solution ρ_κ when $\kappa \leq 1$, and no minimizer for $\kappa > 1$. We have

$$
\boxed{\lim_{Z \to \infty} \frac{N_a(V)}{Z} = 1.} \qquad (6.34)
$$

Assume now $0 < \kappa \le 1$. *The function* ρ_κ *is radial-decreasing, with compact support if* $\kappa < 1$ *and strictly positive for* $\kappa = 1$ *with*

$$\rho_1(x) \underset{|x| \to \infty}{\sim} \frac{(3c_{TF})^3}{\pi^3 |x|^6}. \tag{6.35}$$

The electronic density satisfies

$$\frac{\rho_{\Psi_N}^{(1)}(x)}{N} - N\rho_\kappa(N^{\frac{1}{3}}x) \to 0 \tag{6.36}$$

strongly in $L^1(\mathbb{R}^3) \cap L^{5/3}(\mathbb{R}^3)$, *for any eigenvector* Ψ_N *associated with the eigenvalue* $E_a^V(N)$ *of* $H^V(N)$.

Let us start by commenting on this result in the fermionic case. The limit (6.36) seems to suggest that

$$\rho_{\Psi_N}^{(1)}(x) \approx N^2 \rho_\kappa(N^{\frac{1}{3}}x), \tag{6.37}$$

when $N \to \infty$ and $N/Z \to \kappa$, where ρ_κ is the unique solution of the minimization problem to the right of (6.33), called the Thomas-Fermi model [Tho27, Fer27]. Even if the limit (6.37) is unfortunately not true pointwise (because we have multiplied by N), it is believed that the term on the right provides a good representation of the different scales at play in the electronic density $\rho_{\Psi_N}^{(1)}$, that is, the term on the left is either equivalent, or at least comparable to the term on the right, depending on the region where x is located [Lie81, Lie90]. The general shape of the term on the right (and thus possibly of $\rho_{\Psi_N}^{(1)}$) is provided in Fig. 6.6.

Let us start by placing ourselves near the nucleus, at a distance $|x| \sim N^{-1/3}$, where the **core electrons** live. By making the change of variable $x = N^{-1/3}y$, the limit (6.36) of the theorem implies

$$\frac{\rho_{\Psi_N}^{(1)}(N^{-\frac{1}{3}}y)}{N^2} \underset{N \to \infty}{\longrightarrow} \rho_\kappa(y)$$

in $L^1(\mathbb{R}^3)$ and almost everywhere. As $\int_{\mathbb{R}^3} \rho_\kappa = 1$, this region therefore contains the majority of the N electrons in the system. They seek to compensate for the large charge Z of the nucleus, with a very high density of order N^2 whose profile is given by the function ρ_κ.

The minimization problem of (6.33) is nonlinear and it only has a solution ρ_κ under the assumption that $\kappa \le 1$, where κ is the limiting proportion of the number of electrons relative to the nuclear charge. It is this property of the effective nonlinear model that can be used to show the limit (6.34) concerning the maximum number of electrons $N_a(V)$ that can be bound by the nucleus. If this is a good indication that

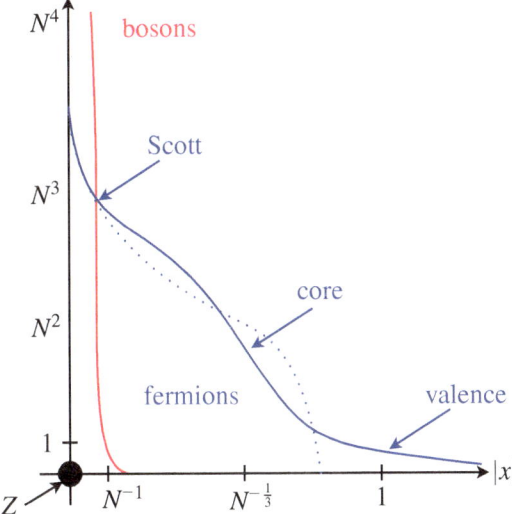

Fig. 6.6 Expected behavior of the electron density $\rho^{(1)}_{\psi_N}$ of a neutral atom at the limit $N \to \infty$, following [Lie81, Lie90] (second part of Theorem 6.11). Most of the electrons gather into a small ball of radius $\sim N^{-1/3}$, where the density is of order N^2. At a distance of order N^{-1} from the nucleus, the density is higher, of order N^3. This correction due to Scott comes from the singularity of the Coulomb potential and generates a correction of order N^2 in the energy expansion. At distances of order one from the nucleus, the density is of order 1 and this is where chemistry takes place. If we take $N/Z \sim \kappa < 1$, that is, we put too few electrons, there are no valence electrons (dotted curve). If the electrons were bosons, the whole system would be very concentrated at the scale $\sim N^{-1}$ where the density would be of order N^4 (first part of Theorem 6.11). © Mathieu Lewin 2021. All rights reserved

the ionization conjecture (6.27) might be true, there is still a long way to go between $N_a(V) = Z + o(Z)$ and $N_a(V) = Z + O(1)$.

Let us now move away from the nucleus, at a distance $|x| \gg N^{-1/3}$, for example $|x|$ of order one. When $\kappa = 1$, the Thomas-Fermi density has the behavior (6.35) in $|x|^{-6}$ at infinity. The powers of N cancel out exactly, so that the term on the right of (6.37) behaves like

$$N^2 \rho_1(N^{\frac{1}{3}}x) \xrightarrow[N\to\infty]{} \frac{(3c_{TF})^3}{\pi^3 |x|^6} \tag{6.38}$$

for any fixed x. A famous conjecture states that the true density $\rho^{(1)}_{\psi_N}$ is also bounded and strictly positive at a finite distance from the nucleus [Lie81, Lie90], but not necessarily equal to the term on the right of (6.38). This would reflect the existence of a finite number of **valence electrons**, which are subject to a finite electrostatic potential since the high charge of the nucleus Z is compensated by the core electrons. These valence electrons participate in all chemical phenomena.

When $\kappa < 1$ the function ρ_κ has compact support, so there should be no valence electrons at all, as the charge of the nucleus is not well screened.

If it is true that the density $\rho_{\Psi_N}^{(1)}$ has the expected behavior at a finite distance from the nucleus, this would have various interesting consequences. For example, the radius of an atom, defined as that of the ball where $N-1$ electrons are found [Sol16], should remain finite as $N \to \infty$. This is an experimental reality [PA09, Fig. 2] that atoms in the periodic table do not grow much with the number of electrons.

If we move even further from the nucleus, at a distance $|x| \gg 1$, the prediction (6.37) is no longer valid. We indeed expect an exponential decay of $\rho_{\Psi_N}^{(1)}$ for $\kappa = 1$, and not a heavy tail in $|x|^{-6}$ as in (6.38) [Lie81, Lie90].

Finally, let us return very close to the nucleus, at a distance $|x| \ll N^{-1/3}$. It is known that $\rho_\kappa(y) \sim (c_{TF}\kappa|y|)^{-3/2}$ when $y \to 0$, which this time provides

$$ N^2 \rho_\kappa(N^{\frac{1}{3}}x) \quad \underset{|x|N^{\frac{1}{3}} \to 0}{\sim} \quad \left(\frac{N}{c_{TF}\kappa|x|} \right)^{\frac{3}{2}}. $$

This now suggests that the density rises sharply and is of order N^3 at a distance $|x| \sim N^{-1}$. This has been proved and is called the **Scott correction** [Sco52, SW87, ILS96].

According to (6.33), the energy of atoms behaves like $-CN^{7/3}$ for large $N = Z$ for a constant C. If we use this crude asymptotics for real atoms where $N = Z$ is not so large, the energy obtained is only about 15% lower than the predictions of other much more elaborate models such as Hartree-Fock [Eng88]. This is quite surprising, considering the simplicity of the Thomas-Fermi model. Various works have been devoted to the rigorous determination of the following terms in the power expansion of N, or to the inclusion of other effects (for example, relativistic). The next terms, in N^2 and $N^{5/3}$, are known explicitly,[2] which improves the predictions on the ground state energy of atoms. Unfortunately, chemistry mainly takes place at the $O(1)$ scale, which corresponds to processes involving the few valence electrons. We are therefore very far from understanding the richness of the behavior of atoms in the periodic table.

A "bosonic" atom behaves completely differently. According to (6.32) it is much more concentrated, at the scale N^{-1}, and completely collapses. No valence electron remains, even for $\kappa = 1$ of $\kappa = \kappa_c$, because u_κ decreases very quickly (like $e^{-|x|}$ if $\kappa < \kappa_c$ and $e^{-\sqrt{|x|}}$ for $\kappa = \kappa_c$). The ground state energy is also much lower (of order $-N^3$) than that of fermions (of order $-N^{7/3}$). As $\kappa_c \simeq 1.21 > 1$, bosonic atoms can be negatively ionized according to (6.31), which does not correspond to any physical reality. In particular, the ionization conjecture (6.27) is false for bosons.

---

[2] The $N^{5/3}$ term was first derived by Fefferman and Seco in the beginning of the 1990s, in an impressive series of papers which spans over more than 800 pages [FS90].

It may seem surprising that the **linear** equation associated with the first eigenvalue $E_{a/s}^V(N)$ of the Schrödinger Hamiltonian simplifies as in the limits (6.30) and (6.33), and furthermore leads to **nonlinear** problems. This is a very common phenomenon for very dense (classical or quantum) systems, where a large number of particles occupy a small space. The general idea is that each particle is subject to a large number of collisions because it has many neighbors. By the law of large numbers, the interaction is then replaced by an **average interaction**, seen by all the particles of the system, which is at the origin of the nonlinearity of the limiting problem. This is called a **mean field regime**.

It turns out that the two results of Theorem 6.11 (for, respectively, bosons and fermions) are particular cases of more general theorems concerning the mean field limit with any potentials V and w. These can be found in [LNR14, Rou16, Lew23] for bosons and [FLS18] for fermions.

In this section we have provided a selection of some results concerning a very particular N-particle model, namely describing the electrons in an atom or a molecule. There are many other quantum systems that pose interesting mathematical challenges. The interested reader can read more in [RS78, LS10b, LSSY05], for instance.

Chapter 7
Periodic Schrödinger Operators, Electronic Properties of Materials

This chapter is an introduction to the study of Schrödinger operators $-\Delta + V$ when V is a periodic function on $\mathbb{R}^d$. The latter is used to describe quantum particles evolving in an infinite ordered medium like electrons in a crystal. We present Bloch-Floquet theory, which is the main mathematical tool used to explain the electrical behavior of solids. We will omit some technical details that can be read in [RS78, Sec. XIII.16].

7.1 Self-adjointness

Throughout this chapter we consider a discrete lattice of $\mathbb{R}^d$

$$\mathscr{L} := \left\{ \sum_{j=1}^{d} z_j v_j, \qquad z_1, \ldots, z_d \in \mathbb{Z} \right\}$$

where the vectors $v_1, \ldots, v_d$ form any basis of $\mathbb{R}^d$ (not necessarily orthonormal). A **unit cell of** $\mathscr{L}$ is by definition an open and bounded set $C \subset \mathbb{R}^d$, such that the $(C + \ell)_{\ell \in \mathscr{L}}$ form a tiling of $\mathbb{R}^d$ (they are pairwise disjoint and their closures cover the entire space $\mathbb{R}^d$). We can take for instance the **Voronoi cell**, also called **Wigner-Seitz cell**,

$$C = \left\{ x \in \mathbb{R}^d \ : \ |x - \ell| > |x|, \ \forall \ell \in \mathscr{L} \setminus \{0\} \right\} \tag{7.1}$$

which contains all the points x that are closer to the origin than to any other point of the lattice $\mathscr{L}$. For example, for the canonical orthonormal basis of $\mathbb{R}^d$ we obtain the cubic lattice $\mathscr{L} = \mathbb{Z}^d$ with the unit cell $C = (-1/2, 1/2)^d$. In dimension $d = 2$,

© The Editor(s) (if applicable) and The Author(s), under exclusive license to Springer Nature Switzerland AG 2024
M. Lewin, *Spectral Theory and Quantum Mechanics*, Universitext,
https://doi.org/10.1007/978-3-031-66878-4_7

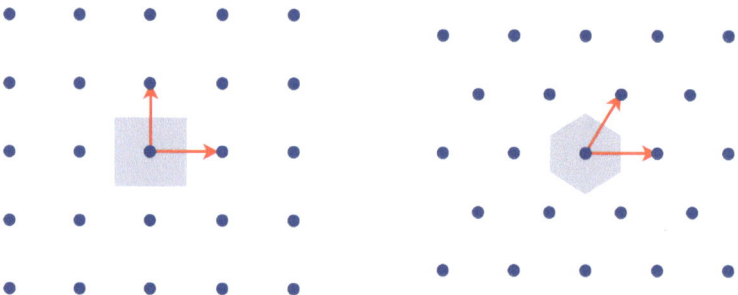

Fig. 7.1 Example of square (left) and triangular (right) lattices in $\mathbb{R}^2$, with two vectors v_1, v_2 generating the lattice and with the Wigner-Seitz unit cell C emphasized in color. © Mathieu Lewin 2021.

the basis

$$v_1 = \begin{pmatrix} 1 \\ 0 \end{pmatrix}, \qquad v_2 = \frac{1}{\sqrt{2}} \begin{pmatrix} 1 \\ 1 \end{pmatrix}$$

provides the triangular lattice and C is in this case a hexagon (Fig. 7.1).

A measurable function V is called $\mathcal{L}$-**periodic** when it satisfies $V(x+\ell) = V(x)$ for all $\ell \in \mathcal{L}$ and almost every $x \in \mathbb{R}^d$. In this case, V is almost everywhere determined by its values on the bounded set C. If V belongs to $L^p(C)$ for a $1 \le p < \infty$ then it belongs to $L^p(\Omega)$ for any bounded Ω, since the latter can be covered by a finite number of copies of C. If $p = +\infty$ then $V \in L^\infty(\mathbb{R}^d)$. We now study operators in the form $-\Delta + V(x)$ where $V(x)$ is an $\mathcal{L}$-periodic function, starting with the question of self-adjointness. The following result is a generalization of Corollaries 3.5 and 3.20 to the periodic setting. Since $L^\infty(C) \subset L^p(C)$ for all $1 \le p \le \infty$ because C is bounded, we do not use a sum of L^p spaces as we did in $\mathbb{R}^d$ and work with the simple assumption that $V \in L^p(C, \mathbb{R})$ with p as small as possible.

Theorem 7.1 (Self-adjointness of Periodic Operators) *Let V be a real-valued $\mathcal{L}$-periodic function on $\mathbb{R}^d$, such that $V \in L^p(C, \mathbb{R})$ for*

$$\begin{cases} p = 1 & \text{if } d = 1, \\ p > 1 & \text{if } d = 2, \\ p = \frac{d}{2} & \text{if } d \ge 3. \end{cases} \qquad (7.2)$$

Then the quadratic form

$$\mathcal{E}^V(u) := \int_{\mathbb{R}^d} |\nabla u(x)|^2 \, dx + \int_{\mathbb{R}^d} V(x)|u(x)|^2 \, dx$$

is well defined on $H^1(\mathbb{R}^d)$ and is bounded from below. Moreover, $\mathcal{E}^V + C\|\cdot\|^2_{L^2(\mathbb{R}^d)}$ is equivalent to the square of the norm of $H^1(\mathbb{R}^d)$ for a sufficiently large constant C. The quadratic form $\mathcal{E}^V$ is associated with a unique self-adjoint operator $H = -\Delta + V$ defined on

$$D(H) = \left\{ u \in H^1(\mathbb{R}^d) \; : \; (-\Delta + V)u \in L^2(\mathbb{R}^d) \right\}$$

by $Hu := (-\Delta + V)u$. If in addition

$$\begin{cases} p = 2 & \text{if } d \in \{1, 2, 3\}, \\ p > 2 & \text{if } d = 4, \end{cases} \tag{7.3}$$

then $D(H) = H^2(\mathbb{R}^d)$ and H is the closure of the minimal operator $H^{\min} = -\Delta + V$ defined on $D(H^{\min}) = C_c^\infty(\mathbb{R}^d)$.

Proof In order to apply the results of Chap. 3, we need to show that V is infinitesimally $(-\Delta)$-bounded in the sense of forms under condition (7.2) and in the sense of operators for (7.3), that is, for every $\varepsilon > 0$ there exists a constant C_ε such that

$$\left| \int_{\mathbb{R}^d} V(x)|u(x)|^2 \, dx \right| \leq \varepsilon \int_{\mathbb{R}^d} |\nabla u(x)|^2 \, dx + C_\varepsilon \int_{\mathbb{R}^d} |u(x)|^2 \, dx, \quad \forall u \in H^1(\mathbb{R}^d) \tag{7.4}$$

and

$$\int_{\mathbb{R}^d} V(x)^2 |u(x)|^2 \, dx \leq \varepsilon \int_{\mathbb{R}^d} |\Delta u(x)|^2 \, dx + C_\varepsilon \int_{\mathbb{R}^d} |u(x)|^2 \, dx, \quad \forall u \in H^2(\mathbb{R}^d), \tag{7.5}$$

respectively. Let χ be a function with compact support that equals 1 on $(-1/2, 1/2)^d$ and vanishes outside $(-1, 1)^d$. We introduce the cube $Q_z := z + (-1/2, 1/2)^d$, the enlarged cube $Q'_z = z + (-1, 1)^d$, and $\chi_z(x) = \chi(x - z)$ for $z \in \mathbb{Z}^d$. Even though $\mathscr{L}$ is not necessarily a cubic lattice, we use such a lattice for our proof. The function V is also in $L^p(Q_0)$, since Q_0 can be covered by a finite number of copies of the unit cell C of the lattice $\mathscr{L}$. In fact Q_z can be covered by a finite number of copies of C, this number being uniformly bounded with respect to $z \in \mathbb{R}^d$. The same is true for Q'_z. Using that $V\mathbb{1}_{Q'_z} \in L^p(Q'_z)$, Lemma 1.10 provides

$$\int_{Q_z} |V||u|^2 \leq \int_{Q'_z} |V||\chi_z u|^2 \leq \varepsilon \int_{\mathbb{R}^d} |\nabla(\chi_z u)|^2 + C_\varepsilon \int_{\mathbb{R}^d} \chi_z^2 |u|^2,$$

where the constant C_ε is independent of z. More precisely, the proof of Lemma 1.10 informs us that C_ε only depends on M and ε, when M is chosen so that $\|V\mathbb{1}(|V| \geq$

$M)\|_{L^p(Q'_z)} \leq C\varepsilon$ for a universal constant C related to the Sobolev or Gagliardo-Nirenberg inequality depending on the dimension d. But by periodicity we have

$$\|V\mathbb{1}(|V| \geq M)\|_{L^p(Q'_z)} \leq N^{1/p}\|V\mathbb{1}(|V| \geq M)\|_{L^p(C)}$$

where N is the number of copies of C needed to cover Q_z, which is uniformly bounded. This allows us to choose M, hence C_ε, independently of z. By expanding the gradient and using the inequality $|a + b|^2 \leq 2a^2 + 2b^2$, we obtain

$$\int_{Q_z} |V||u|^2 \leq 2\varepsilon \int_{Q'_z} |\nabla u|^2 + \left(C_\varepsilon + 2\varepsilon\|\nabla\chi\|_{L^\infty}^2\right)\int_{Q'_z} |u|^2.$$

It remains to sum over z and to use that each small cube Q_z is counted 2^d times in the sum of integrals involving the larger cube Q'_z, which provides

$$\int_{\mathbb{R}^d} |V||u|^2 \leq 2^{d+1}\varepsilon\int_{\mathbb{R}^d}|\nabla u|^2 + 2^d\left(C_\varepsilon + 2\varepsilon\|\nabla\chi\|_{L^\infty}^2\right)\int_{\mathbb{R}^d}|u|^2$$

and shows (7.4). The proof for (7.5) is similar and left as an exercise.   □

7.2   Bloch-Floquet Theory

In order to determine the spectrum of $-\Delta + V$ when V is $\mathscr{L}$-periodic, we will replace the Fourier transform (which diagonalizes the Laplacian) by a different unitary map, better adapted to the lattice $\mathscr{L}$. It is called the **Bloch transform** or the **Bloch-Floquet transform** [Flo83, Blo29]. For this we need to introduce the **dual lattice** $\mathscr{L}^*$ of $\mathscr{L}$. Let M be the $d \times d$ matrix whose columns are the coordinates of the v_i's in the canonical basis of $\mathbb{R}^d$. Then $\mathscr{L}^*$ is by definition the lattice associated with the basis of the rows of the matrix $2\pi M^{-1}$. It is also the largest subgroup of $\mathbb{R}^d$ such that

$$k \cdot \ell \in 2\pi\mathbb{Z}, \qquad \forall \ell \in \mathscr{L}, \ k \in \mathscr{L}^*. \tag{7.6}$$

Its Voronoi unit cell B defined as in (7.1) is called the **Brillouin zone**. For example, for the cubic lattice $\mathscr{L} = \mathbb{Z}^d$ we have $\mathscr{L}^* = 2\pi\,\mathbb{Z}^d$ and $B = (-\pi, \pi)^d$. The dual $\mathscr{L}^*$ of the triangular lattice in Fig. 7.1 is a triangular lattice rotated by $90°$ and dilated by $2\pi\sqrt{2}$.

Next we introduce the Bloch transform. The idea is to rewrite the family of plane waves $(e^{ip\cdot x})_{p\in\mathbb{R}^d}$ in the form of a two-parameter family

$$\left(e^{i(k+\xi)\cdot x}\right)_{\substack{k\in\mathscr{L}^* \\ \xi\in B}} \tag{7.7}$$

where we notice that $p = k + \xi$ covers all of $\mathbb{R}^d$ when $k \in \mathcal{L}^*$ and $\xi \in \overline{B}$. Any function f of the Schwartz class $\mathcal{S}(\mathbb{R}^d)$ decomposes on this family, starting from the usual Fourier transform:

$$f(x) = \frac{1}{(2\pi)^{d/2}} \int_{\mathbb{R}^d} \widehat{f}(p) e^{ip \cdot x} \, dp = \frac{1}{(2\pi)^{d/2}} \sum_{k \in \mathcal{L}^*} \int_B \widehat{f}(k + \xi) e^{i(k+\xi) \cdot x} \, d\xi.$$

This suggests introducing the function of two parameters

$$f_\xi(x) := \frac{1}{|C|^{1/2}} \sum_{k \in \mathcal{L}^*} \widehat{f}(k + \xi) e^{i(k+\xi) \cdot x} \tag{7.8}$$

which is called the **Bloch transform** of f. The constant

$$\frac{1}{|C|^{1/2}} = \frac{|B|^{1/2}}{(2\pi)^{d/2}} \tag{7.9}$$

is chosen so that $e^{i(k+\xi) \cdot x} |C|^{-1/2}$ is normalized in $L^2(C)$. The function f_ξ defined in (7.8) satisfies the Born-von Kármán condition

$$f_\xi(x + \ell) = f_\xi(x) e^{i\xi \cdot \ell}, \qquad \forall \ell \in \mathcal{L}, \tag{7.10}$$

with respect to x (re-read on this subject Sect. 2.8.1 in dimension $d = 1$ and Exercise 2.50 for $d \geq 2$) and is $\mathcal{L}^*$-periodic with respect to ξ,

$$f_{k+\xi}(x) = f_\xi(x), \qquad \forall k \in \mathcal{L}^*.$$

We can therefore always restrict it to $\xi \in B$ and $x \in C$. We have the reconstruction formula

$$\boxed{f(x) = \frac{1}{|B|^{1/2}} \int_B f_\xi(x) \, d\xi}$$

which is similar to the one that gives $f(x)$ in terms of $\widehat{f}(p) e^{ip \cdot x}$. While the Fourier transform allows one to decompose f as an average of arbitrary plane waves, the Bloch transform only involves functions $f_\xi(x)$ satisfying the Born-von Kármán property (7.10) with a fixed lattice $\mathcal{L}$. The price to pay is that $f_\xi(x)$ is not proportional to a simple function like $e^{ip \cdot x}$.

We can also look for the function whose Fourier coefficients are $|C|^{-1/2} \widehat{f}(k+\xi)$, which provides another expression for f_ξ, similar to Poisson's summation formula:

$$\boxed{f_\xi(x) := \frac{1}{|C|^{1/2}} \sum_{k \in \mathcal{L}^*} \widehat{f}(k + \xi) e^{i(k+\xi) \cdot x} = \frac{1}{|B|^{1/2}} \sum_{\ell \in \mathcal{L}} f(\ell + x) e^{-i\xi \cdot \ell}.}$$

$$\tag{7.11}$$

Indeed, we have

$$
\begin{aligned}
\widehat{f}(k+\xi) &= \frac{1}{(2\pi)^{d/2}} \int_{\mathbb{R}^d} f(x) e^{-i(k+\xi)\cdot x}\, dx \\
&= \frac{1}{(2\pi)^{d/2}} \sum_{\ell \in \mathscr{L}} \int_C f(\ell + y) e^{-i(k+\xi)\cdot(\ell+y)}\, dy \\
&= \frac{1}{|B|^{1/2}} \int_C \left(\sum_{\ell \in \mathscr{L}} f(\ell + y) e^{-i\xi\cdot\ell} \right) \frac{e^{-i(k+\xi)\cdot y}}{|C|^{1/2}}\, dy.
\end{aligned}
$$

Like for the Fourier transform, we will now extend the Bloch transform to all of $L^2(\mathbb{R}^d)$.

Theorem 7.2 (Plancherel) *For all f in the Schwartz class $\mathcal{S}(\mathbb{R}^d)$, we have Parseval's formula*

$$
\int_{\mathbb{R}^d} |f(x)|^2\, dx = \int_B \left(\int_C |f_\xi(y)|^2\, dy \right)\, d\xi, \tag{7.12}
$$

which allows us to extend the Bloch transform into a unique isomorphism of Hilbert spaces

$$
\mathcal{B} : f \in L^2(\mathbb{R}^d) \mapsto f_\xi(x) \in L^2\big(B, L^2(C)\big) = L^2(B \times C).
$$

Proof For $f \in \mathcal{S}(\mathbb{R}^d)$ we write

$$
\begin{aligned}
\int_B \left(\int_C |f_\xi(y)|^2\, dy \right) d\xi &= \frac{1}{|B|} \sum_{\ell,\ell' \in \mathscr{L}} \int_B e^{i\xi\cdot(\ell'-\ell)}\, d\xi \int_C \overline{f(\ell'+y)} f(\ell+y)\, dy \\
&= \sum_{\ell \in \mathscr{L}} \int_C |f(\ell+y)|^2\, dy = \int_{\mathbb{R}^d} |f(x)|^2\, dx.
\end{aligned}
$$

The extension to all of $L^2(\mathbb{R}^d)$ follows.   □

We mentioned that $f_\xi(x)$ was $\mathscr{L}^*$-periodic in ξ and satisfied the Born-von Kármán condition (7.10) in x, for $f \in \mathcal{S}(\mathbb{R}^d)$. It is sometimes convenient to swap these two properties. This is achieved through the transformation

$$
\widetilde{f}_\xi(x) := f_\xi(x) e^{-i\xi\cdot x}.
$$

We then have

$$\widetilde{f_\xi}(x) := \frac{1}{|C|^{1/2}} \sum_{k\in\mathscr{L}^*} \widehat{f}(k+\xi)e^{ik\cdot x} = \frac{1}{|B|^{1/2}} \sum_{\ell\in\mathscr{L}} f(\ell+x)e^{-i\xi\cdot(\ell+x)}$$

and the new reconstruction formula

$$f(x) = \frac{1}{|B|^{1/2}} \int_B e^{-i\xi\cdot x}\,\widetilde{f_\xi}(x)\,\mathrm{d}\xi.$$

The function $(\xi, x) \mapsto \widetilde{f_\xi}(x)$ is now $\mathscr{L}$-periodic in x and satisfies the Born-von Kármán condition in ξ. The corresponding isomorphism on $L^2(\mathbb{R}^d)$ will be denoted $\widetilde{\mathcal{B}}$. The name "Bloch transform" is used interchangeably for $\mathcal{B}$ or $\widetilde{\mathcal{B}}$ and it is easy to switch from one to the other by multiplying by the phase $e^{\pm i\xi\cdot x}$.

Let us now move on to the study of the action of $\mathcal{B}$ and $\widetilde{\mathcal{B}}$ on spatial derivatives and to the identification of the two self-adjoint operators $\mathcal{B}(-\Delta)\mathcal{B}^{-1}$ and $\widetilde{\mathcal{B}}(-\Delta)\widetilde{\mathcal{B}}^{-1}$ in the Hilbert space $L^2(B \times C)$. For all $f \in S(\mathbb{R}^d)$, the second formula of (7.11) shows that spatial derivatives commute with the Bloch transform $\mathcal{B}$:

$$(\partial^\alpha f)_\xi = \partial_x^\alpha f_\xi.$$

For the second Bloch transform $\widetilde{\mathcal{B}}$, we rather get

$$\widetilde{(\partial^\alpha f)_\xi} = (\partial_x + i\xi)^\alpha \widetilde{f_\xi}.$$

Using the Parseval relation (7.12), this provides an expression of the kinetic energy in Bloch variables:

$$\int_{\mathbb{R}^d} |\nabla f(x)|^2\,\mathrm{d}x = \int_B \int_C |\nabla_x f_\xi(x)|^2\,\mathrm{d}x\,\mathrm{d}\xi = \int_B \int_C \left|(-i\nabla_x + \xi)\widetilde{f_\xi}(x)\right|^2\,\mathrm{d}x\,\mathrm{d}\xi,$$

$$(7.13)$$

for all $f \in S(\mathbb{R}^d)$. As $S(\mathbb{R}^d)$ is dense in $H^1(\mathbb{R}^d)$, this allows us to identify the image of $H^1(\mathbb{R}^d)$ by the two isomorphisms $\mathcal{B}$ and $\widetilde{\mathcal{B}}$. We mentioned that f_ξ satisfies the Born-von Kármán condition (7.10) in x, that is, belongs to the space $H^1_{\mathrm{per},\xi}(C)$, while $\widetilde{f_\xi} \in H^1_{\mathrm{per}}(C)$ is simply periodic. These properties survive after closure and we find

$$\mathcal{B}H^1(\mathbb{R}^d) = \left\{f \in L^2(B \times C) \text{ such that } f_\xi \in H^1_{\mathrm{per},\xi}(C)\right.$$

$$\left.\text{for almost all } \xi \in B \text{ with } \int_B \|f_\xi\|^2_{H^1(C)}\,\mathrm{d}\xi < \infty\right\}$$

$$=: L^2\big(B, H^1_{\mathrm{per},\xi}(C)\big).$$

As the two quadratic forms

$$\int_C |(-i\nabla + \xi)u(x)|^2\, dx + \int_C |u(x)|^2\, dx \quad \text{and} \quad \int_C |\nabla u(x)|^2\, dx + \int_C |u(x)|^2\, dx$$

are equivalent on $H_{\text{per}}^1(C)$, uniformly with respect to ξ in the compact $\overline{B}$, we find similarly that

$$\mathcal{B}H^1(\mathbb{R}^d) = \left\{ f \in L^2(B \times C) \text{ such that } \widetilde{f}_\xi \in H_{\text{per}}^1(C) \right.$$

$$\left. \text{for almost all } \xi \in B \text{ with } \int_B \|\widetilde{f}_\xi\|_{H^1(C)}^2\, d\xi < \infty \right\}$$

$$=: L^2\big(B, H_{\text{per}}^1(C)\big).$$

With similar definitions, we also have

$$\mathcal{B}H^2(\mathbb{R}^d) = L^2\big(B, H_{\text{per},\xi}^2(C)\big), \qquad \widetilde{\mathcal{B}}H^2(\mathbb{R}^d) = L^2\big(B, H_{\text{per}}^2(C)\big).$$

This allows us to identify the two self-adjoint operators $\mathcal{B}(-\Delta)\mathcal{B}^{-1}$ and $\widetilde{\mathcal{B}}(-\Delta)\widetilde{\mathcal{B}}^{-1}$ which are

$$\begin{cases} \mathcal{B}(-\Delta)\mathcal{B}^{-1} = (-\Delta_{\text{per},\xi})_x, & D\big(\mathcal{B}(-\Delta)\mathcal{B}^{-1}\big) = L^2\big(B, H_{\text{per},\xi}^2(C)\big), \\ \widetilde{\mathcal{B}}(-\Delta)\widetilde{\mathcal{B}}^{-1} = |(P_{\text{per}})_x + \xi|^2, & D\big(\widetilde{\mathcal{B}}(-\Delta)\widetilde{\mathcal{B}}^{-1}\big) = L^2\big(B, H_{\text{per}}^2(C)\big). \end{cases}$$

$$(7.14)$$

Our notation means that the two operators $-\Delta_{\text{per},\xi}$ and P_{per} act on the variable x for almost every $\xi \in B$. Recall that these operators were defined in Sects. 2.8.1 and 2.8.3 in dimension $d = 1$ and in Exercise 2.50 for $d \geq 2$. For a fixed ξ, the spectrum is

$$\sigma\big((-\Delta)_{\text{per},\xi}\big) = \sigma\big(|P_{\text{per}} + \xi|^2\big) = \big\{|k + \xi|^2\big\}_{k \in \mathscr{L}^*},$$

with the corresponding eigenfunctions $e^{i(k+\xi)\cdot x}$ and $e^{ik\cdot x}$ and $k \in \mathscr{L}^*$. Like for multiplication operators (Theorem 4.3), the total spectrum in $L^2(B \times C)$ is the closure of the union of the spectra for each ξ, which provides the spectrum $[0, +\infty)$ of the Laplacian when (ξ, k) runs through all $B \times \mathscr{L}^*$ (Fig. 7.2).

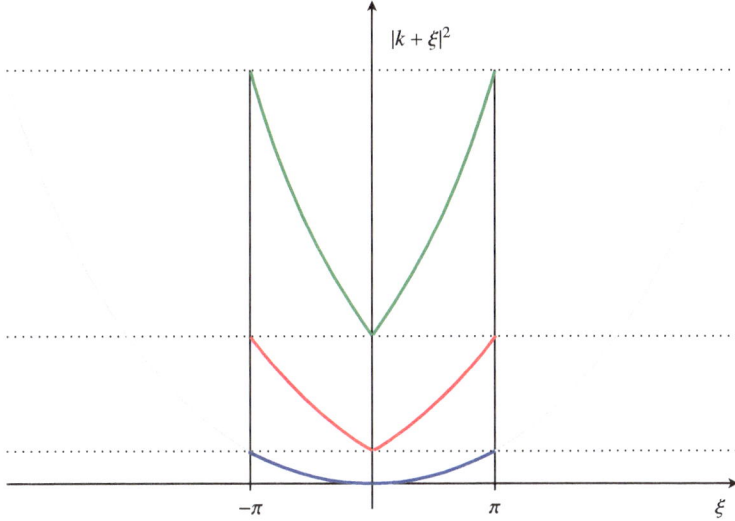

Fig. 7.2 Effect of the Bloch transform on the spectrum of the Laplacian, in dimension 1 with the lattice $\mathscr{L} = \mathbb{Z}$. The parabola $p \mapsto |p|^2$ is cut into pieces according to $|p|^2 = |k + \xi|^2$ where $\xi \in B = (-\pi, \pi)$ and $k \in \mathscr{L}^* = 2\pi\mathbb{Z}$. Each fiber $\{|k + \xi|^2\}_{k \in 2\pi\mathbb{Z}}$ at a fixed ξ corresponds to the spectrum of the Laplacian with the Born-von Kármán condition (7.10) at the edge of $C = (-1/2, 1/2)$ defined in Sect. 2.8.3 or, equivalently, to the operator $|-i\frac{d}{dx} + \xi|^2$ with periodic condition. © Mathieu Lewin 2021.

7.3   Diagonalization of Periodic Schrödinger Operators

When V is $\mathscr{L}$-periodic for a given lattice $\mathscr{L}$, it is natural to involve the Bloch transform associated with the lattice $\mathscr{L}$ to study the Schrödinger operator $-\Delta + V$. The most important property is that such an operator commutes with this transform, that is, we formally have

$$\left((-\Delta + V)f\right)_\xi (x) = -\Delta_x f_\xi(x) + V(x) f_\xi(x). \tag{7.15}$$

More precisely, our operators $\mathcal{B}(-\Delta + V)\mathcal{B}^{-1}$ and $\widetilde{\mathcal{B}}(-\Delta + V)\widetilde{\mathcal{B}}^{-1}$ are defined by the Friedrichs method using the two expressions of the quadratic form in Bloch variables

$$\int_{\mathbb{R}^d} |\nabla f|^2 + V|f|^2 = \int_B \left(\int_C |\nabla_x f_\xi(x)|^2 \, dx + \int_C V(x) |f_\xi(x)|^2 \, dx \right) d\xi$$

$$= \int_B \left(\int_C |(-i\nabla_x + \xi)\widetilde{f}_\xi(x)|^2 \, dx + \int_C V(x) |\widetilde{f}_\xi(x)|^2 \, dx \right) d\xi. \tag{7.16}$$

It is more convenient to use the second Bloch transform $\widetilde{\mathcal{B}}$, even if the argument is very similar for $\mathcal{B}$. For each $\xi \in B$, we therefore consider the self-adjoint operator on $L^2(C)$

$$\widetilde{H}_\xi := |-i\nabla_x + \xi|^2 + V(x), \qquad D(\widetilde{H}_\xi) = \left\{ f \in H^1_{\text{per}}(C) \; : \; \widetilde{H}_\xi f \in L^2(C) \right\}$$

(7.17)

obtained by the Friedrichs method, where $\widetilde{H}_\xi f$ is interpreted in the sense of distributions on the open set C. As

$$|-i\nabla_x + \xi|^2 = -\Delta_x - 2i\xi \cdot \nabla_x + |\xi|^2$$

and the domain is by definition included in $H^1_{\text{per}}(C)$, we always have $-2i\xi \cdot \nabla f \in L^2(C)$ in the domain, so that the latter is in fact independent of ξ:

$$D(\widetilde{H}_\xi) = D(\widetilde{H}_0) = \left\{ f \in H^1_{\text{per}}(C) \; : \; (-\Delta + V)f \in L^2(C) \right\}.$$

For each $\xi \in B$, the operator $\widetilde{H}_\xi$ has a compact resolvent (Sect. 5.4). We call $\lambda_n(\xi)$ its eigenvalues ordered increasingly and repeated according to their multiplicity. They tend to $+\infty$. The following theorem specifies that the spectrum of $-\Delta + V$ is the union of the images of these eigenvalues, which are continuous functions in the variable ξ. The spectrum of $-\Delta + V$ is therefore a countable union of compact intervals, called the **Bloch bands**.

Theorem 7.3 (Diagonalization of Periodic Schrödinger Operators) *Let V be a real-valued $\mathcal{L}$-periodic function on $\mathbb{R}^d$, such that $V \in L^p(C)$ with p satisfying (7.2). We call $\widetilde{H}_\xi = |-i\nabla_x + \xi|^2 + V(x)$ the Friedrichs realization (7.17) on $L^2(C)$ with the periodic boundary condition and $\lambda_n(\xi)$ its eigenvalues ordered increasingly and repeated according to their multiplicity.*

(i) For each $n \geq 1$, $\xi \mapsto \lambda_n(\xi)$ is a Lipschitz function on $\overline{B}$.
(ii) The spectrum of the operator $-\Delta + V(x)$ defined in Theorem 7.1 is

$$\boxed{\sigma(-\Delta + V) = \bigcup_{n \geq 1} \lambda_n(\overline{B}),}$$

(7.18)

where the band $\lambda_n(\overline{B}) = [\min_{\overline{B}} \lambda_n, \max_{\overline{B}} \lambda_n]$ is the range of λ_n on the compact set $\overline{B}$.
(iii) The spectrum of $-\Delta + V(x)$ contains no eigenvalues.

The spectrum of the operator $-\Delta + V$ defined on the whole space $\mathbb{R}^d$ can thus be obtained from the eigenvalues of the operators $\widetilde{H}_\xi = |-i\nabla + \xi|^2 + V(x)$ defined on the bounded open set C with the periodic boundary condition, by varying ξ in the Brillouin zone B. More precisely, the spectrum is the union of the bands which

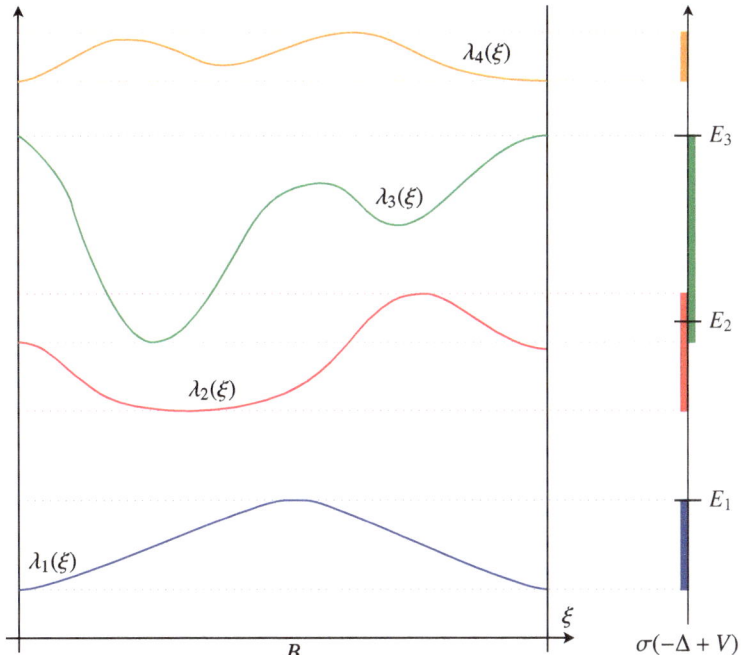

Fig. 7.3 Schematic representation of the eigenvalues $\lambda_n(\xi)$ of the operator $\widetilde{H}_\xi = |-i\nabla_x + \xi|^2 + V(x)$ on $L^2(C)$ with the periodic boundary condition, and link with the spectrum of the operator $-\Delta + V$ on $\mathbb{R}^d$ (on the right) when ξ goes through the Brillouin zone B. The energies E_k are discussed later in Sect. 7.4.2.

are the images of the eigenvalues $\lambda_n(\xi)$ when ξ varies in B, as shown in Fig. 7.3. This echoes the case of the Laplacian seen in Fig. 7.2, except that the bands can now partially overlap or not touch at all, depending on the shape of the periodic potential V.

Proof The regularity of $\lambda_n(\xi)$ follows from the methods developed in Sects. 5.1.3 and 5.6.2.[1] As seen in Exercise 2.50, it follows from Theorem 2.34 that the eigenvalues of the periodic Laplacian $-\Delta_{\text{per}}$ on C are $\{|k|^2\}_{k \in \mathscr{L}^*}$ and that the eigenfunctions are given by the Fourier basis $(e^{ik \cdot x})_{k \in \mathscr{L}^*}$. We have

$$\frac{|k|^2}{2} - |\xi|^2 \le |k + \xi|^2 \le 2|k|^2 + 2|\xi|^2$$

for all $k \in \mathscr{L}^*$. Since B is bounded, this implies by functional calculus

$$c_1(1 - \Delta_{\text{per}}) \le |P_{\text{per}} + \xi|^2 + C \le c_2(1 - \Delta_{\text{per}})$$

---

[1] As long as the $\lambda_n(\xi)$ do not cross, they are even real analytic functions according to Theorem 5.7.

for constants $C, c_1, c_2 > 0$ that can be chosen independent of $\xi \in B$. With our assumption (7.2) on p, the potential V is infinitesimally $(-\Delta_{per})$-bounded in the sense of quadratic forms, and we therefore have a similar inequality

$$c_1(1 - \Delta_{per}) \leq \tilde{H}_\xi + C \leq c_2(1 - \Delta_{per}), \tag{7.19}$$

with other constants denoted the same for convenience. By a counting argument similar to Lemma 5.48 we have

$$c_1 \, n^{\frac{2}{d}} \leq \lambda_n(-\Delta_{per}) + C \leq c_2 \, n^{\frac{2}{d}}$$

and the Courant-Fischer formula then provides

$$c_1 \, n^{\frac{2}{d}} \leq \lambda_n(\xi) + C \leq c_2 \, n^{\frac{2}{d}} \tag{7.20}$$

for other constants $c_1, c_2 > 0$. In other words, the λ_n's tend to infinity uniformly with respect to ξ. Finally, since

$$\tilde{H}_\xi = \tilde{H}_\zeta - 2i(\xi - \zeta) \cdot \nabla + |\xi|^2 - |\zeta|^2$$

and $\pm 2\xi \cdot (-i\nabla) \leq |\xi|(1 - \Delta_{per})$, Inequality (7.19) implies

$$\left(1 - \frac{|\zeta - \xi|}{c_2} - \frac{|\zeta^2 - \xi^2|}{c_2}\right)(\tilde{H}_\xi + C)$$

$$\leq \tilde{H}_\zeta + C \leq \left(1 + \frac{|\zeta - \xi|}{c_1} + \frac{|\zeta^2 - \xi^2|}{c_1}\right)(\tilde{H}_\xi + C).$$

By Theorem 5.6, this shows that each λ_n is in fact a Lipschitz function on $\overline{B}$, as in the proof of Theorem 5.45.

Next we show the main assertion *(ii)* of the theorem that $\sigma(-\Delta + V) = \bigcup_{n \geq 1} \lambda_n(\overline{B})$. We start with the inclusion $\supset$ and therefore consider $n \geq 1$ and $\xi_0 \in \overline{B}$, as well as a normalized eigenvector $f_{\xi_0} \in H^1_{per}(C)$ of eigenvalue $\lambda_n(\xi_0)$ for the operator $\tilde{H}_{\xi_0}$,

$$|-i\nabla + \xi_0|^2 f_{\xi_0}(x) + V(x) f_{\xi_0}(x) = \lambda_n(\xi_0) f_{\xi_0}(x).$$

This equation holds in the sense of distributions on C. We can extend f_{ξ_0} to a periodic function on all $\mathbb{R}^d$ (still denoted f_{ξ_0}) and which then belongs to $H^1_{loc}(\mathbb{R}^d)$. The equation is in this case valid in the sense of distributions on all $\mathbb{R}^d$. We now introduce the test function

$$g_\varepsilon(x) = \varepsilon^{d/2} e^{i\xi_0 \cdot x} f_{\xi_0}(x) \chi(\varepsilon x) \in H^1(\mathbb{R}^d)$$

where χ is a C^∞ positive function with compact support, such that $\int_{\mathbb{R}^d} \chi^2 = 1$. The norm of g_ε is

$$\int_{\mathbb{R}^d} |g_\varepsilon(x)|^2 \, dx = \varepsilon^d \int_{\mathbb{R}^d} |f_{\xi_0}(x)|^2 |\chi(\varepsilon x)|^2 \, dx.$$

It is classical that this integral converges to

$$\lim_{\varepsilon \to 0} \varepsilon^d \int_{\mathbb{R}^d} |f_{\xi_0}(x)|^2 \chi(\varepsilon x)^2 \, dx = \left(\frac{1}{|C|} \int_C |f_{\xi_0}(x)|^2 \, dx \right) \left(\int_{\mathbb{R}^d} \chi(x)^2 \, dx \right) = \frac{1}{|C|}.$$

Indeed, for g a periodic and locally integrable function and $\varphi \in \mathcal{S}(\mathbb{R}^d)$, we have

$$\varepsilon^d \int_{\mathbb{R}^d} g(x) \varphi(\varepsilon x) \, dx = \frac{(2\pi)^{d/2}}{|C|^{1/2}} \sum_{k \in \mathcal{L}^*} c_k(g) \, \overline{\widehat{\varphi}(k/\varepsilon)}$$

$$\xrightarrow[\varepsilon \to 0]{} \frac{(2\pi)^{d/2}}{|C|^{1/2}} c_0(g) \, \widehat{\varphi}(0) = \frac{1}{|C|} \left(\int_C g(x) \, dx \right) \left(\int_{\mathbb{R}^d} \varphi(x) \, dx \right), \qquad (7.21)$$

since the Fourier transform of a periodic function is a Dirac comb. Thus, modulo a factor $|C|$, our function g_ε is normalized in the limit. We can now calculate, in the sense of distributions on $\mathbb{R}^d$,

$$\left(-\Delta + V - \lambda_n(\xi_0) \right) g_\varepsilon = -2\varepsilon^{1+\frac{d}{2}} \nabla f_{\xi_0}(x) \cdot \nabla \chi(\varepsilon x) - \varepsilon^{2+\frac{d}{2}} f_{\xi_0}(x) \Delta \chi(\varepsilon x).$$

Since $f_{\xi_0} \in H^1_{\text{loc}}(\mathbb{R}^d)$, the function on the right belongs to $L^2(\mathbb{R}^d)$ and this proves that g_ε belongs to the domain of $-\Delta + V$. Furthermore, we have

$$\left\| \left(-\Delta + V - \lambda_n(\xi_0) \right) g_\varepsilon \right\|_{L^2(\mathbb{R}^d)} \leq 2\varepsilon^{1+d/2} \left\| \nabla f_{\xi_0}(x) \cdot \nabla \chi(\varepsilon x) \right\|_{L^2(\mathbb{R}^d)}$$

$$+ \varepsilon^{2+d/2} \left\| f_{\xi_0}(x) \Delta \chi(\varepsilon x) \right\|_{L^2(\mathbb{R}^d)}.$$

The two terms are respectively of order ε and ε^2 since, according to (7.21),

$$\lim_{\varepsilon \to 0} \left\| \varepsilon^{\frac{d}{2}} \nabla f_{\xi_0} \cdot \nabla \chi(\varepsilon \cdot) \right\|_{L^2(\mathbb{R}^d)}^2 = \frac{1}{|C|} \left(\int_C |\nabla f_{\xi_0}|^2 \right) \left(\int_{\mathbb{R}^d} |\nabla \chi|^2 \right)$$

and

$$\lim_{\varepsilon \to 0} \left\| \varepsilon^{\frac{d}{2}} f_{\xi_0} \Delta \chi(\varepsilon \cdot) \right\|_{L^2(\mathbb{R}^d)}^2 = \frac{1}{|C|} \left(\int_C |f_{\xi_0}|^2 \right) \left(\int_{\mathbb{R}^d} |\Delta \chi|^2 \right).$$

The convergence to 0 of $\left\| \left(-\Delta + V - \lambda_n(\xi_0) \right) g_\varepsilon \right\|_{L^2(\mathbb{R}^d)}$ and that of $\|g_\varepsilon\|_{L^2(\mathbb{R}^d)}$ to $|C|^{-1}$ imply that $\lambda_n(\xi_0)$ belongs to the spectrum of $-\Delta + V$ by the Weyl characterization of Theorem 2.30.

It remains to prove the converse inclusion. For this, we can for example use that

$$\|(-\Delta + V - \mu)f\|^2_{L^2(\mathbb{R}^d)} = \int_B \int_C \left| (|-i\nabla + \xi|^2 + V)\widetilde{f_\xi} - \mu \widetilde{f_\xi} \right|^2 \, dx \, d\xi$$

for all $f \in D(-\Delta+V)$. This is a formula similar to Parseval's identity (7.12) whose proof (left as an exercise) is slightly complicated by the fact that $-\Delta \widetilde{f_\xi}$ and $V \widetilde{f_\xi}$ are not necessarily both in $L^2(C)$. If μ does not belong to the union of $\lambda_n(\overline{B})$, then there exists an $\eta > 0$ such that $|\lambda_n(\xi) - \mu| \geq \eta$ for all $\xi \in \overline{B}$ and all $n \geq 1$, because the λ_n are continuous and tend to infinity when $n \to \infty$ according to (7.20). By the spectral theorem, we then have

$$\left\| (|-i\nabla + \xi|^2 + V)f_\xi - \mu f_\xi \right\|^2_{L^2(C)} \geq \eta^2 \|f_\xi\|^2_{L^2(C)}$$

and we therefore conclude that for $\mu \notin \cup \lambda_n(\overline{B})$, we have

$$\|(-\Delta + V - \mu)f\|^2_{L^2(\mathbb{R}^d)} \geq \eta^2 \int_B \int_C |\widetilde{f_\xi}(x)|^2 \, dx \, d\xi = \eta^2 \, \|f\|^2_{L^2(\mathbb{R}^d)} \, .$$

By Theorem 2.30, this shows that $\mu \notin \sigma(-\Delta + V)$, which completes the proof of the formula (7.18) for the spectrum.

The absence of eigenvalues in *(iii)* follows from the regularity of the functions $\lambda_n(\xi)$ and it is a difficult statement with our minimal assumptions on p. The first result of this type is due to Thomas [Tho73] in the case $p = 2$ in dimension $d = 3$ and it was not until the late 1990s [BS98, BS99, She01] that a proof covering Assumption (7.2) was found. We refer to these references for the proof of (iii). □

7.4 Infinite Fermionic Systems and Electronic Properties of Materials*

In condensed matter physics, operators of the type $-\Delta + V$ with V a periodic function describe infinite pure crystalline materials (without defect), in which electrons can move. While it is interesting to study a single electron evolving in this landscape, it is often more physically relevant to consider **infinitely many electrons**, also periodically distributed. Think of table salt, which is a lattice of sodium atoms Na^+ (comprising 10 electrons and a nucleus of charge $Z = 11$) and chlorine Cl^- (comprising 18 electrons and a nucleus of charge $Z = 17$). When the interactions between the electrons are neglected, we will see in this section that such an infinite system is naturally modeled by the **spectral projector**

$$\mathbb{1}_{(-\infty, E]}(-\Delta + V)$$

introduced in Sect. 4.5 of Chap. 4, where the energy level E is determined so that the number of electrons per unit volume is equal to a desired constant $\rho > 0$. It is then the position of this constant E in the spectrum of $-\Delta + V$ that explains some of the conduction properties of materials.

7.4.1 Thermodynamic Limit, Density of States

The usual method to describe an infinite system is to first consider a finite piece of it, restricted to a bounded domain $\Omega \subset \mathbb{R}^d$. Then, we let Ω grow so that it covers all space and we study the limit of some important observables of the system. We speak of a **thermodynamic limit** [Rue99, CLL98]. Most of these observables will diverge, but we can for example study their value per unit volume (we divide by the volume of $|\Omega|$) or differences of well-chosen observables.

For Ω a bounded open set of $\mathbb{R}^d$, we denote by $(-\Delta + V)_{|\Omega}$ the Friedrichs realization with Dirichlet condition at the boundary of Ω. As in Sect. 3.3.4, it is the unique self-adjoint operator whose quadratic form equals $\int_\Omega (|\nabla u|^2 + V|u|^2)$ on the form domain $H_0^1(\Omega)$. This quadratic form is closed, because it is equivalent to the square of the norm of $H^1(\Omega)$ under our assumptions on V, according to Theorem 7.1. By the same argument as in Sect. 5.4.3, the operator $(-\Delta + V)_{|\Omega}$ has a compact resolvent and we call $\lambda_{\Omega,i}$ its eigenvalues (ordered increasingly and repeated according to their multiplicity). We simply denote by $-\Delta + V$ the Friedrichs realization on all $\mathbb{R}^d$, provided by Theorem 7.1, and by $\lambda_n(\xi)$ its Bloch-Floquet eigenvalues as in Theorem 7.3.

The following result provides some information about the limit where Ω tends towards the whole space $\mathbb{R}^d$, which will then be used to describe the behavior of our infinity of electrons. Of course, we expect the operator $(-\Delta + V)_{|\Omega}$ to converge in some sense towards $-\Delta + V$, but the exact notion of convergence is more subtle than it may seem. The eigenvalues will form an increasingly dense set that will tend "on average" towards the continuous spectrum of $-\Delta + V$, but may also include outliers that do not converge at all (see Remark 7.5 and Fig. 7.4 below).

Theorem 7.4 (Thermodynamic Limit for a Periodic Operator) *Let $\Omega \subset \mathbb{R}^d$ be a bounded open set containing 0, whose boundary has zero Lebesgue measure, $|\partial\Omega| = 0$. Let V be a real-valued $\mathscr{L}$-periodic function on $\mathbb{R}^d$, such that $V \in L^p(C)$ with p satisfying (7.2). Then, for any piecewise continuous function f with support included in $(-\infty, E]$ for an $E \in \mathbb{R}$, we have*

$$\lim_{\ell\to\infty} \frac{1}{|\Omega|\ell^d} \sum_i f\left(\lambda_{\ell\Omega,i}\right) = \frac{1}{(2\pi)^d} \sum_{n\geq 1} \int_B f\left(\lambda_n(\xi)\right) d\xi. \tag{7.22}$$

Moreover, for any $g \in L^2(\mathbb{R}^d)$, we have the convergence

$$f\left((-\Delta + V)_{|\ell\Omega}\right)\mathbb{1}_{\ell\Omega}g \xrightarrow[\ell\to\infty]{} f(-\Delta + V)g \qquad in\ L^2(\mathbb{R}^d) \tag{7.23}$$

where the function on the left is by convention extended by 0 outside of $\ell\Omega$.

The proof of Theorem 7.4 is provided at the end of this chapter in Sect. 7.4.3. We have considered the Dirichlet condition for simplicity, but the statement is exactly the same for the Neumann boundary condition, when Ω is smooth enough. The statement includes two parts. The limit (7.23) is a kind of convergence of the truncated model on $\ell\Omega$ towards that on all $\mathbb{R}^d$. This limit is not at all related to the fact that V is a periodic function. The reader can verify that the proof provided in Sect. 7.4.3 remains valid for any reasonable potential V, assuming however that f is continuous if $-\Delta + V$ has eigenvalues.

The limit (7.22) is more specific to the periodic case and concerns the precise way in which the eigenvalues accumulate to form the continuous spectrum of $-\Delta + V$. Taking $f = \mathbb{1}_I$ for some finite interval $I = [a, b]$, we deduce from (7.22) that there are always of the order of ℓ^d eigenvalues inside the spectrum, that is, when $I \subset \sigma(-\Delta + V)$. Outside the spectrum, that is, between the bands, there are $o(\ell^d)$ eigenvalues. For $I = (-\infty, E]$, we find that the number of eigenvalues per unit volume located below a level E converges

$$\lim_{\ell \to \infty} \frac{\#\{\lambda_{\ell\Omega,i} \leq E\}}{\ell^d |\Omega|} = \mathrm{n}(E), \tag{7.24}$$

where the limit is

$$\mathrm{n}(E) := \frac{1}{(2\pi)^d} \sum_{n \geq 1} \int_B \mathbb{1}_{(-\infty,E]}\big(\lambda_n(\xi)\big) \, \mathrm{d}\xi. \tag{7.25}$$

This is a non-decreasing function whose derivative $\mathrm{n}'(E)$ is called the **density of states** because it provides the average limit density of eigenvalues in the vicinity of an energy E. The function $\mathrm{n}(E)$ is called the **integrated density of states**. Note that the sum on the right of (7.25) is finite since the $\lambda_n(\xi)$'s tend to infinity according to (7.20). In principle, the derivative n' is a positive measure (since n is increasing) but in fact it is a locally integrable function because it has been shown in [BS98, BS99, She01] that the λ_n's cannot be constant on sets of non-zero measure. One says that the operator $-\Delta + V$ has a **purely absolutely continuous** spectrum. The limit (7.22) can also be rewritten in the form

$$\lim_{\ell \to \infty} \frac{1}{|\Omega|\ell^d} \sum_i f\big(\lambda_{\ell\Omega,i}\big) = \int_{\mathbb{R}} f(E)\, \mathrm{n}'(E)\, \mathrm{d}E$$

or again, with a weak local convergence in the sense of measures,

$$\frac{1}{|\Omega|\ell^d} \sum_i \delta_{\lambda_{\ell\Omega,i}} \xrightarrow[\ell \to \infty]{} \mathrm{n}'.$$

The sum on the left is called the **empirical measure** associated with the eigenvalues. It simply returns the local number of eigenvalues, divided by the total volume.

When $V \equiv 0$ the eigenvalues of the Bloch-Floquet transform are $|k + \xi|^2$ with $k \in \mathscr{L}^*$ and $\xi \in B$, as shown in Fig. 7.2. We then have

$$\mathrm{n}(E) = \frac{1}{(2\pi)^d} \int_{\mathbb{R}^d} \mathbb{1}_{(-\infty, E]}(|p|^2) \, \mathrm{d}p = \frac{|\mathbb{S}^{d-1}|}{d(2\pi)^d} E^{\frac{d}{2}}$$

and we find exactly the Weyl asymptotic of Theorem 5.49 concerning the number of eigenvalues of the Dirichlet Laplacian in a large domain.

Remark 7.5 (Spectral Pollution) The convergence (7.22) does not prevents from the presence of $o(\ell^d)$ eigenvalues outside the spectrum of $-\Delta + V$, called **spurious eigenvalues**. The presence of such outliers is called **spectral pollution** [DP04, LS10a] and it is sometimes a consequence of topological properties [Gon20, Sec. III]. Using the Courant-Fischer principle, the reader can show as an exercise that $\lambda_{\ell\Omega, 1}$ does converge towards the bottom of the spectrum of $-\Delta + V$, so that spectral pollution can only occur in spectral gaps (between the Bloch bands) and never below the spectrum. For a numerical example, see Fig. 7.4.

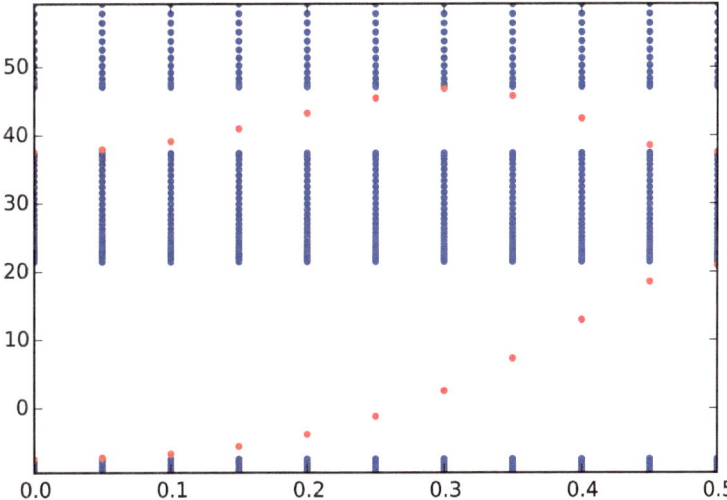

Fig. 7.4 Numerical calculation of the bottom of the spectrum of the operator $-\frac{\mathrm{d}^2}{\mathrm{d}x^2} + 30 \cos(2\pi x)$, restricted to the interval $\Omega = [t, 30 + t]$ with Dirichlet boundary condition, and plotted vertically for various values of t in the horizontal axis. As predicted by Theorem 7.4, the eigenvalues form a very dense set and we observe the appearance of the first three Bloch bands. However, there exists also a unique eigenvalue in each spectral gap (in red on the figure), which varies quite a bit with the parameter t and would persist even if the size of the interval Ω were increased. This is referred to as spectral pollution [DP04, LS10a]. The corresponding eigenfunction is localized at the edge of the interval [CEM12] and its existence is due to topological constraints [Gon20, Sec. III]. For similar calculations see [BL07, Gon20].

7.4.2 Fermi Sea, Insulators, Metals

Let us now see what Theorem 7.4 teaches us about our infinite system of electrons. We fix an average density $\rho > 0$ of electrons per unit volume and find out how the N electrons of the system are arranged in the thermodynamic limit $N \to \infty$ for this density. We therefore take a sufficiently smooth bounded open set Ω of volume $|\Omega| = 1$ and dilate it by a factor $\ell = (N/\rho)^{1/d}$ chosen so that the number of electrons per unit volume is exactly equal to the given ρ:

$$\frac{N}{|\ell\Omega|} = \frac{N}{\ell^d} = \rho.$$

As we neglect the interactions between the electrons, the latter are described by the N-particle Hamiltonian seen in Sect. 6.4 of Chap. 6

$$H^V(N) = \sum_{j=1}^{N} -\Delta_{x_j} + V(x_j)$$

in the anti-symmetric space $L_a^2((\ell\Omega)^N)$, with the Dirichlet condition at the boundary of $\ell\Omega$. According to Theorem 6.8 and its proof, we then know that the smallest eigenvalue $\lambda_1(H^V(N))$ is obtained by placing the N electrons in the first N eigenfunctions of the operator $(-\Delta + V)_{|\ell\Omega}$ on $\ell\Omega$. Their total energy is, therefore,

$$\lambda_1\big(H^V(N)\big) = \sum_{i=1}^{N} \lambda_{\ell\Omega,i} \tag{7.26}$$

and a corresponding eigenfunction is the Slater determinant

$$\Psi_N(x_1, \dots, x_N) = \frac{\det\big(u_{N,i}(x_j)\big)}{\sqrt{N!}}, \tag{7.27}$$

where of course the $u_{N,i}$'s are the eigenfunctions associated with the eigenvalues $\lambda_{\ell\Omega,i}$ of the operator $(-\Delta + V)_{|\ell\Omega}$. We want to understand the behavior of these two objects in the limit $N \to \infty$.

Let us start with the ground state energy of the system, that is, the first eigenvalue (7.26). This is not exactly in the form studied in Theorem 7.4. Indeed, if $\lambda_{\ell\Omega,N} < \lambda_{\ell\Omega,N+1}$ we can write

$$\sum_{i=1}^{N} \lambda_{\ell\Omega,i} = \sum_{1} f_N\big(\lambda_{\ell\Omega,i}\big)$$

but $f_N(x) = x\mathbb{1}_{(-\infty,\lambda_N(\ell\Omega)]}(x)$ unfortunately depends on N. If $\lambda_{\ell\Omega,N}$ converges to a real number E, we expect the limit to coincide with the one where f_N is replaced by $f(x) = x\mathbb{1}_{(-\infty,E]}(x)$. Even if the convergence of $\lambda_{\ell\Omega,N}$ is not ensured in all cases due to Remark 7.5, the few outliers that do not converge do not play a role at the order of interest and we can prove the following result.

Corollary 7.6 (Convergence of the Energy Per Unit Volume) *Let $\rho > 0$ and ε_ρ be a real number such that* $n(\varepsilon_\rho) = \rho$ *where* n *is the integrated density of states defined in (7.25). Then we have*

$$\lim_{\ell\to\infty} \frac{1}{\ell^d} \sum_{i=1}^{N} \lambda_{\ell\Omega,i} = \int_{-\infty}^{\varepsilon_\rho} E\,n'(E)\,dE$$

$$= \frac{1}{(2\pi)^d} \sum_{n\geq 1} \int_B \lambda_n(\xi)\mathbb{1}_{(-\infty,\varepsilon_\rho]}\big(\lambda_n(\xi)\big)\,d\xi, \qquad (7.28)$$

where we recall that $\ell = (N/\rho)^{1/d}$ and $|\Omega| = 1$.

The real number ε_ρ is called the **Fermi level**. According to (7.25), the function n is strictly increasing on the spectrum and constant in the spectral gaps between the Bloch bands. Thus, ε_ρ is unique when n takes the value ρ on the spectrum. Otherwise, ε_ρ can be taken equal to any real number in the spectral gap where n is constant equal to ρ. The chosen value does not matter because the $\lambda_n(\xi)$'s do not penetrate into the spectral gap.

Proof We call

$$\varepsilon_\rho^- = \min\{E \;:\; n(E) = \rho\}, \qquad \varepsilon_\rho^+ = \max\{E \;:\; n(E) = \rho\}$$

the minimum and maximum possible values of ε_ρ. When n takes the value ρ on the spectrum of $-\Delta + V$ we simply have $\varepsilon_\rho^- = \varepsilon_\rho^+ = \varepsilon_\rho$. Let $\eta > 0$. According to Theorem 7.4, the number of eigenvalues below $\varepsilon_\rho^\pm \pm \eta$ behaves like

$$\lim_{\ell\to\infty} \frac{\#\{\lambda_{\ell\Omega,i} \leq \varepsilon_\rho^\pm \pm \eta\}}{\ell^d} = n(\varepsilon_\rho^\pm \pm \eta).$$

However, by definition of $\varepsilon_\rho^\pm$, we have $n(\varepsilon_\rho^- - \eta) < \rho$ and $n(\varepsilon_\rho^+ + \eta) > \rho$. Since $\ell^d = N/\rho$, this proves that there are $N\,n(\varepsilon_\rho^\pm \pm \eta)/\rho + o(N) < N$ eigenvalues below $\varepsilon_\rho^- - \eta$ and $N\,n(\varepsilon_\rho^+ + \eta)/\rho + o(N) > N$ eigenvalues below $\varepsilon_\rho^+ + \eta$. There are also $N\,(n(\varepsilon_\rho^+ + \eta) - n(\varepsilon_\rho^+ + \eta))/\rho + o(N)$ eigenvalues in the interval $[\varepsilon_\rho^- - \eta, \varepsilon_\rho^+ + \eta]$. We introduce the functions

$$f_\pm(x) = x\mathbb{1}_{(-\infty,\varepsilon_\rho^\pm \pm \eta]}(x), \qquad \varphi(x) = \mathbb{1}_{[\varepsilon_\rho^- - \eta,\varepsilon_\rho^+ + \eta]}(x). \qquad (7.29)$$

For N large enough, we have $\varepsilon_\rho^- - \eta < \lambda_{\ell\Omega,N} < \varepsilon_\rho^+ + \eta$ and therefore

$$\sum_{i=1}^{N} \lambda_{\ell\Omega,i} \geq \sum_{i\geq 1} f_-\left(\lambda_{\ell\Omega,i}\right) + (\varepsilon_\rho^- - \eta) \sum_{i\geq 1} \varphi\left(\lambda_{\ell\Omega,i}\right).$$

After taking the limit, we find from (7.22) in Theorem 7.4

$$\liminf_{N\to\infty} \frac{1}{\ell^d} \sum_{i=1}^{N} \lambda_{\ell\Omega,i} \geq \int_{-\infty}^{\varepsilon_\rho^- - \eta} E\mathrm{n}'(E)\,\mathrm{d}E + (\varepsilon_\rho^- - \eta)\big(\mathrm{n}(\varepsilon_\rho^+ + \eta) - \mathrm{n}(\varepsilon_\rho^- - \eta)\big).$$

The same argument shows that

$$\limsup_{N\to\infty} \frac{1}{\ell^d} \sum_{i=1}^{N} \lambda_{\ell\Omega,i} \leq \int_{-\infty}^{\varepsilon_\rho^+ + \eta} E\mathrm{n}'(E)\,\mathrm{d}E + (\varepsilon_\rho^+ + \eta)\big(\mathrm{n}(\varepsilon_\rho^+ + \eta) - \mathrm{n}(\varepsilon_\rho^- - \eta)\big).$$

The result (7.28) follows by taking $\eta \to 0^+$ and using that n is a continuous function, constant equal to ρ on $[\varepsilon_\rho^-, \varepsilon_\rho^+]$. $\qquad\qquad\square$

Remark 7.7 (Convergence of the Last Eigenvalue) If $\varepsilon_\rho^+ = \varepsilon_\rho^- = \varepsilon_\rho$, the proof above shows that $\lambda_{\ell\Omega,N}$ converges to ε_ρ. If $\varepsilon_\rho^+ > \varepsilon_\rho^-$ we only know that the accumulation points of $\lambda_{\ell\Omega,N}$ will all be in $[\varepsilon_\rho^-, \varepsilon_\rho^+]$ and we cannot hope for anything better according to Remark 7.5 and Fig. 7.4.

Now that we have found the limit of the energy, we wish to understand that of the quantum state of the electrons, described by the wavefunction Ψ_N of (7.27). The latter has a number of variables tending to infinity and it is not very clear how to study its limit. Fortunately, Slater determinants are very special functions that can be described by simpler mathematical objects, which will allow us to make some progress.

The crucial property that we will use is that a Slater determinant actually only depends on the vector space generated by the orthonormal functions that compose it, up to a phase. Indeed, another orthonormal basis $(v_{N,1}, \ldots, v_{N,N})$ of $\mathrm{span}(u_{N,i})$ can be written $v_{N,i} = \sum_{j=1}^{N} U_{ij} u_{N,j}$ where the matrix $U \in \mathrm{U}(N)$ is unitary. A calculation shows that the new Slater determinant $\Psi_N' := \det(v_{N,i}(x_j))/\sqrt{N!}$ equals $\Psi_N' = \det(U)\Psi_N$. Since $|\det(U)| = 1$ and two collinear quantum states describe the same physical state (Sect. 1.5), we deduce that the state of the system only depends on the space $\mathrm{span}(u_{N,i})$ of dimension N, and not on the individual functions. We therefore need to understand the limit of this sequence of subspaces of $L^2(\mathbb{R}^d)$, of increasing dimension.

It is equivalent and simpler to work with the orthogonal projection on this space, which we denote with kets and bras as

$$\gamma_N := \sum_{i=1}^{N} |u_{N,i}\rangle\langle u_{N,i}|. \tag{7.30}$$

Providing γ_N is equivalent to giving the space generated by the $u_{N,i}$ (its range) which is itself equivalent to the specification of Ψ_N, up to a phase. All the average values $\langle \Psi_N, A\Psi_N \rangle$ can be expressed using the operator γ_N (even if the explicit formulas can be quite complicated). Our quantum state is therefore completely characterized by the operator γ_N, which is called the **one-body density matrix** of Ψ_N [BLS94]. When $\lambda_{\ell\Omega,N} < \lambda_{\ell\Omega,N+1}$, the operator γ_N coincides with the spectral projector

$$\gamma_N = \mathbb{1}_{(-\infty,\lambda_{\ell\Omega,N}]} \left((-\Delta + V)_{|\ell\Omega}\right).$$

If $\lambda_{\ell\Omega,N} = \lambda_{\ell\Omega,N+1}$, the operator γ_N is not quite a spectral projector, but Corollary 7.6 and its proof suggest that the few problematic eigenvalues of the last level will again not affect the limit.

Corollary 7.8 (Convergence of γ_N) *Let $\rho > 0$ and ε_ρ be a real number such that $n(\varepsilon_\rho) = \rho$ where n is the integrated density of states defined in (7.25). Then γ_N converges to the spectral projector $\mathbb{1}_{(-\infty,\varepsilon_\rho]}(-\Delta + V)$ strongly, that is for all $g \in L^2(\mathbb{R}^d)$ we have*

$$\gamma_N\left(\mathbb{1}_{\ell\Omega}g\right) \to \mathbb{1}_{(-\infty,\varepsilon_\rho]}(-\Delta + V)g \qquad in \; L^2(\mathbb{R}^d) \tag{7.31}$$

where the function on the left is by convention extended by 0 outside of $\ell\Omega$.

The interpretation of this result is that our infinite system of electrons occupies all the energy states located below the **Fermi level** ε_ρ, somewhat like in Fig. 6.3 but with a continuous spectrum. This infinite system should be thought of as a Slater determinant as in (7.27) but with infinitely many variables, hence infinitely many particles. It can be mathematically represented by the spectral projector $\mathbb{1}_{(-\infty,\varepsilon_\rho]}(-\Delta + V)$. This projector is called in physics the **Fermi sea** because it describes an infinity of electrons whose density is $\mathscr{L}$-periodic, somewhat like waves, and equals

$$\rho_{\text{per}}^{(1)}(x) = \frac{1}{(2\pi)^d} \sum_{n\geq 1} \int_B \mathbb{1}_{(-\infty,\varepsilon_\rho]}\left(\lambda_n(\xi)\right) |u_{n,\xi}(x)|^2 \, d\xi.$$

It can be shown that the latter function is also the local limit of the density $\rho_{\Psi_N}^{(1)} = \sum_{i=1}^{N} |u_{N,i}|^2$ of Ψ_N introduced in (6.29).

Proof The proof is very similar to that of Corollary 7.6. Let $\eta > 0$. We know that $\varepsilon_\rho^- - \eta < \lambda_{\ell\Omega,N} < \varepsilon_\rho^+ + \eta$ for all $\eta > 0$ and N large enough. We can therefore write

$$\gamma_N = \mathbb{1}_{(-\infty,\varepsilon_\rho^- -\eta]}\left((-\Delta + V)_{|\ell\Omega}\right) + \delta_N$$

where δ_N is the orthogonal projector on the space generated by the $u_{N,i}$ with $i \leq N$ and $\lambda_{\ell\Omega,i} > \varepsilon_\rho^- - \eta$. This operator satisfies the inequality

$$\delta_N \leq \mathbb{1}_{(\varepsilon_\rho^- -\eta,\varepsilon_\rho^+ +\eta]}\left((-\Delta + V)_{|\ell\Omega}\right).$$

As these are orthogonal projectors, this implies

$$\|\delta_N \mathbb{1}_{\ell\Omega}g\| \leq \left\|\mathbb{1}_{(\varepsilon_\rho^- -\eta,\varepsilon_\rho^+ +\eta]}\left((-\Delta + V)_{|\ell\Omega}\right)\mathbb{1}_{\ell\Omega}g\right\|$$

and thus provides

$$\left\|\gamma_N \mathbb{1}_{\ell\Omega}g - \mathbb{1}_{(-\infty,\varepsilon_\rho^-]}(-\Delta + V)g\right\|$$

$$\leq \left\|\mathbb{1}_{(-\infty,\varepsilon_\rho^- -\eta]}\left((-\Delta + V)_{|\ell\Omega}\right)\mathbb{1}_{\ell\Omega}g - \mathbb{1}_{(-\infty,\varepsilon_\rho^- -\eta]}(-\Delta + V)g\right\|$$

$$+ \left\|\mathbb{1}_{(\varepsilon_\rho^- -\eta,\varepsilon_\rho^+ +\eta]}\left((-\Delta + V)_{|\ell\Omega}\right)\mathbb{1}_{\ell\Omega}g\right\| + \left\|\mathbb{1}_{(\varepsilon_\rho^- -\eta,\varepsilon_\rho^-]}(-\Delta + V)g\right\|.$$

The limit (7.23) in Theorem 7.4 implies

$$\limsup_{N\to\infty} \left\|\gamma_N\left(\mathbb{1}_{\ell\Omega}g\right) - \mathbb{1}_{(-\infty,\varepsilon_\rho]}(-\Delta + V)g\right\|$$

$$\leq \left\|\mathbb{1}_{(\varepsilon_\rho^- -\eta,\varepsilon_\rho^+ +\eta]}(-\Delta + V)g\right\| + \left\|\mathbb{1}_{(\varepsilon_\rho^- -\eta,\varepsilon_\rho^-]}(-\Delta + V)g\right\|.$$

According to the convergence (v) of Theorem 4.8 on functional calculus, the term on the right tends to 0 when $\eta \to 0$ since $-\Delta+V$ has no eigenvalue by Theorem 7.3.
□

In practice, there is often an integer number of electrons $K \in \mathbb{N}$ per unit cell C of the lattice $\mathscr{L}$. The density ρ is then not an arbitrary real number and equals $\rho = K/|C|$. Using (7.9) we see that the Fermi level $E_K := \varepsilon_{K/|C|}$ must satisfy

$$K = n(E_K) = \frac{1}{|B|}\int_B \sum_{n\geq 1} \mathbb{1}_{(-\infty, E_K]}\left(\lambda_n(\xi)\right) d\xi.$$

The situation is now as shown in Fig. 7.3. If the K-th Bloch band is completely below the $(K + 1)$-th band, that is

$$\max_{\xi \in \overline{B}} \lambda_K(\xi) < \min_{\xi \in \overline{B}} \lambda_{K+1}(\xi),$$

then E_K can be taken equal to any number in the interval between the two bands. This physically corresponds to an **insulator**. In Fig. 7.3, this is for example the case for $K = 1$ and $K = 3$ with the energies E_1 and E_3. On the other hand, if the image of λ_K intersects that of λ_{K+1}, the value of E_K will fall in the middle of the spectrum of $-\Delta + V$, which corresponds to a **metal** and is illustrated for $K = 2$ with the energy E_2 in Fig. 7.3.

In an insulator, the electrons completely occupy the first K Bloch bands and exciting one of them costs an energy at least equal to the height of the spectral gap above the K-th band. If the height of this spectral gap is large enough, the electrons will therefore be quite stable and will have little tendency to move around in the system. On the contrary, in a metal the electrons partially occupy the last bands and they can access unoccupied states with a tiny amount of energy. They are therefore much more mobile and this explains the very different conduction properties of metals compared to insulators. For metals, the shape of the spectrum in the vicinity of E_K also plays an important role. Graphene, a material in dimension $d = 2$ whose two bands have the shape of two inverted cones that touch exactly at their tip at the level E_K, has very different properties from other materials where the intersection is larger (typically a curve in dimension $d = 2$). The calculation of the eigenvalues $\lambda_n(\xi)$ is therefore important, as it allows one to deduce the macroscopic behavior of the system depending on the number K of electrons per unit cell.

The distinction between insulator and metal based on the spectrum of $-\Delta + V$ is due to Sommerfeld, Bloch and Bethe in the years 1927–1933 [Som28, Blo29, SB33] and it is of course too simplified. If the spectral gap is small (on the order of 1 electron-volt), the electrons will in practice be easily excited due to temperature effects and we then speak of a **semiconductor**. Moreover, a realistic model must obviously take into account the interactions between the electrons. As a fine understanding of the N-electron Hamiltonian seen in Chap. 6 is out of reach for infinite N, approximate models are used. In Kohn-Sham theory [CF23], one must solve a nonlinear equation in the form

$$\gamma = \mathbb{1}_{(-\infty, \varepsilon_\rho]}(-\Delta + V_\gamma)$$

where the periodic potential V_γ depends on the projector γ itself [CLL01, CDL08]. Once this potential is determined, the interpretation is similar to the one presented here.

Semiconductors play a central role in electronics. The most famous is of course silicon, the band spectrum of which is provided in Fig. 7.5. It is possible to alter the conductivity of semiconductors using **doping techniques**, which involve introducing defects into the lattice, so as to slightly decrease or increase the

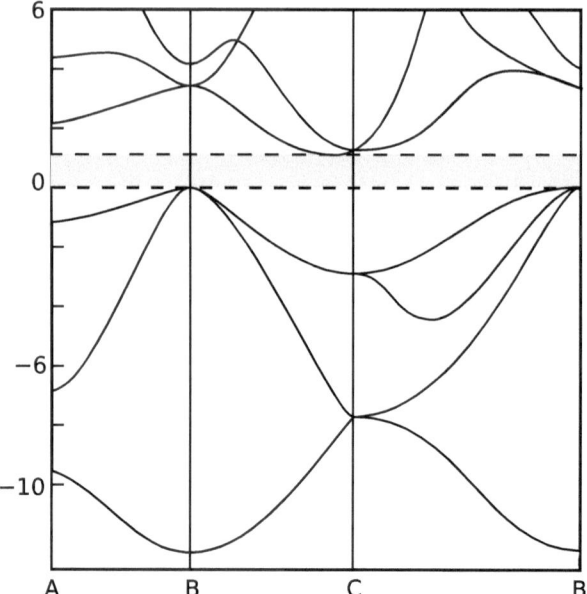

Fig. 7.5 Shape of the $\lambda_n(\xi)$'s for pure silicon, when ξ moves along a path $ABCB$ in the Brillouin zone, where A, B, C are three natural points of symmetry. The energy is expressed in electron-volts, and shifted vertically so that the Fermi level E_K is placed at 0. The spectral gap is indicated in gray. Adaptation of a figure from [CC74]. © from derivative work: Cepheiden, CC BY-SA 2.5/original:S-kei, public domain/Wikimedia Commons

spectral gap. Diodes and transistors operate from an assembly of differently doped semiconductors, in order to control the direction in which electrons can move.

The physical importance of the intersection of Bloch bands raises a natural theoretical question: Will the bands really intersect for a general potential V? Is it possible that the bands are all separated? It was shown [Sim76, MT76] that, generically, the bands are always all separated for a regular potential V in dimension $d = 1$. On the contrary, in dimension $d \geq 2$ there are always at least two bands that intersect [Kar97, Par08], as Bethe and Sommerfeld had conjectured [SB33].

7.4.3 Proof of Theorem 7.4

Step 1: Proof of the Limit (7.22) The proof follows the same strategy as that of Theorems 5.49 and 5.51. We only give the main ideas. By superposition, we can limit ourselves to dealing with the case of the Heaviside function $f = \mathbb{1}_{(-\infty, E]}$. Then, we partition the large domain $\ell\Omega$ with a smaller reference domain Ω' and use the Dirichlet conditions at the boundary of these for the upper bound and the Neumann conditions for the lower bound. This shows that it is sufficient to prove the theorem for a well-chosen reference domain Ω'.

In the case of the Laplacian, we had chosen for Ω' a cube because the Dirichlet and Neumann eigenvalues are explicit (Lemma 5.48). In order to use the structure of the lattice $\mathscr{L}$, we rather use the Wigner-Seitz unit cell C as a reference. If $\ell \in \mathbb{N}$, on $\Omega' = \ell C$ we can also consider the operator $-\Delta + V$ with periodic conditions at the boundary, because ℓC is the unit cell of the sub-lattice $\ell \mathscr{L} \subset \mathscr{L}$ and V is $\ell \mathscr{L}$-periodic. We denote $H_{\text{Dir/Neu/per}}(\ell)$ the three corresponding operators $-\Delta + V$ on ℓC, and $N_{\text{Dir/Neu/per}}(E, \ell)$ the number of their eigenvalues less than or equal to E. As $H_0^1(\ell C) \subset H_{\text{per}}^1(\ell C) \subset H^1(\ell C)$ and the quadratic forms are all equal, we have the inequality

$$H_{\text{Neu}}(\ell) \leq H_{\text{per}}(\ell) \leq H_{\text{Dir}}(\ell)$$

in the sense of Definition 5.3, so that

$$N_{\text{Dir}}(E, \ell) \leq N_{\text{per}}(E, \ell) \leq N_{\text{Neu}}(E, \ell).$$

It turns out that we know exactly the spectrum of $H_{\text{per}}(\ell)$, which is

$$\sigma\left(H_{\text{per}}(\ell)\right) = \left\{\lambda_n(\xi) \ : \ n \geq 1, \ \xi \in \mathscr{L}^*/\ell \cap \overline{B}\right\}.$$

That is, we must only retain the values of $\lambda_n(\xi)$ on the small lattice $\mathscr{L}^*/\ell$. To see this, it is enough to use that the eigenfunctions of $(-\Delta)_{\text{per},\xi} + V$ on C are solutions of the eigenvalue equation on all $\mathbb{R}^d$ in the sense of distributions, and are $\ell \mathscr{L}$-periodic when $\xi \in \mathscr{L}^*/\ell$, so they are eigenfunctions of $H_{\text{per}}(\ell)$. Moreover, we can verify that they form an orthonormal basis of $L^2(C)$. We leave the verification of these statements as an exercise. So, we simply have

$$\lim_{\ell \to \infty} \frac{N_{\text{per}}(E, \ell)}{\ell^d |C|} = \lim_{\ell \to \infty} \frac{1}{\ell^d |C|} \sum_{n \geq 1} \sum_{\xi \in B \cap \frac{\mathscr{L}^*}{\ell}} \mathbb{1}_{(-\infty, E]}(\lambda_n(\xi))$$

$$= \frac{1}{|B| \, |C|} \sum_{n \geq 1} \int_B \mathbb{1}_{(-\infty, E]}(\lambda_n(\xi)) \, d\xi$$

because it is a Riemann sum and the unit cell of $\mathscr{L}^*/\ell$ has volume $|B|/\ell^d$. The series in n is actually finite according to the divergence (7.20) of λ_n to infinity. The limit also uses the fact that the $\lambda_n(\xi)$'s cannot be constant on sets of positive measure. Thus, the theorem is proved for $N_{\text{per}}(E, \ell)$ and it just remains to prove that $N_{\text{Dir}}(E, \ell)$ and $N_{\text{Neu}}(E, \ell)$ have the same behavior to leading order:

$$N_{\text{Dir}}(E, \ell) = N_{\text{Neu}}(E, \ell) + o(\ell^d).$$

This rather technical result is obtained by comparing the resolvents of $H_{\text{Dir}}(\ell)$ and $H_{\text{Neu}}(\ell)$. We refer for example to [Nak01, DIM01] for this estimate.

Step 2: Proof of the Limit (7.23) Let us denote by $h_\ell := (-\Delta + V)_{|\ell\Omega}$ the operator with Dirichlet boundary conditions on $\ell\Omega$, and by $h = -\Delta + V$ the operator on $\mathbb{R}^d$. We have

$$h_\ell \geq \frac{(-\Delta)_{|\ell\Omega}}{2} + \frac{1}{2}(-\Delta + 2V)_{|\ell\Omega} \geq \frac{(-\Delta)_{|\ell\Omega}}{2} + \frac{1}{2}\lambda_1\left((-\Delta + 2V)_{|\ell\Omega}\right).$$

As the functions of $H_0^1(\ell\Omega)$ extended by 0 outside of $\ell\Omega$ are in $H^1(\mathbb{R}^d)$, the Courant-Fischer formula implies

$$\lambda_1\left((-\Delta + 2V)_{|\ell\Omega}\right) \geq \min \sigma(-\Delta + 2V),$$

where the term on the right is finite according to Theorem 7.1. We therefore obtain

$$h + C \geq \frac{-\Delta + 1}{2}, \qquad h_\ell + C \geq \frac{(-\Delta)_{|\ell\Omega} + 1}{2} \tag{7.32}$$

on, respectively, $L^2(\mathbb{R}^d)$ and $L^2(\ell\Omega)$, with an appropriate constant C.

Next, we prove the limit (7.23) for $f(x) = (x - z)^{-1}$ where z is any complex number with a non-zero imaginary part. That is, we show that

$$u_\ell := (h_\ell - z)^{-1}\mathbb{1}_{\ell\Omega}g \to (h - z)^{-1}g =: u \tag{7.33}$$

for all $g \in L^2(\mathbb{R}^d)$. By convention, we extend u_ℓ by 0 outside of $\ell\Omega$ and we then obtain a function from $H^1(\mathbb{R}^d)$, since $u_\ell \in H_0^1(\ell\Omega)$. As $\|(h_\ell - z)^{-1}\| \leq |\Im(z)|^{-1}$, we have

$$\int_{\mathbb{R}^d} |u_\ell|^2 = \int_{\ell\Omega} |u_\ell|^2 \leq \frac{1}{|\Im(z)|^2} \int_{\ell\Omega} |g|^2.$$

Thus, (u_ℓ) is a bounded sequence in $L^2(\mathbb{R}^d)$. We have $(h_\ell - z)u_\ell = \mathbb{1}_{\ell\Omega}g$ on $\ell\Omega$ and, by taking the scalar product against u_ℓ then the real part, we find

$$\int_{\ell\Omega}\left(|\nabla u_\ell|^2 + (V + C)|u_\ell|^2\right) = (C + \Re(z))\int_{\ell\Omega} |u_\ell|^2 + \Re \int_{\ell\Omega} \overline{u_\ell}g.$$

The term on the right is bounded since u_ℓ is bounded and the term on the left is lower bounded by (7.32). We have therefore proved that $\int_{\ell\Omega} |\nabla u_\ell|^2 = \int_{\mathbb{R}^d} |\nabla u_\ell|^2$ is bounded, that is, (u_ℓ) is bounded in $H^1(\mathbb{R}^d)$. After extracting a subsequence we can therefore assume that $u_\ell \rightharpoonup v$ weakly in $H^1(\mathbb{R}^d)$. By integrating the equation against a function $\varphi \in C_c^\infty(\mathbb{R}^d)$ we also find (as soon as the support of φ is included in $\ell\Omega$)

$$\int_{\mathbb{R}^d} \overline{u_\ell}(-\Delta\varphi + V\varphi - \bar{z}\varphi) = \int_{\ell\Omega} \overline{u_\ell}\varphi.$$

By letting $\ell \to \infty$ we obtain $(-\Delta + V - z)v = g$ in the sense of distributions on $\mathbb{R}^d$. Since $v \in H^1(\mathbb{R}^d)$ this shows that $v \in D(h)$ and $(h - z)v = g$. Thus $v = u = (h-z)^{-1}g$ and we have shown the weak convergence $u_\ell \rightharpoonup u$ in (7.33). It remains to show that the convergence is strong. For this we take a function $\eta \in C^\infty(\mathbb{R}^d, [0, 1])$ which is 0 on the unit ball B_1 and is constant equal to 1 outside the ball B_2 of radius 2. We set $\eta_R(x) = \eta(x/R)$ and we compute the imaginary part

$$0 = \Im\langle\eta_R u_\ell, g - (h_\ell - z)u_\ell\rangle$$

$$= \Im(z)\int_{\ell\Omega} \eta_R|u_\ell|^2 + \Im\int_{\ell\Omega}\eta_R\overline{u_\ell}g - \Im\int_{\ell\Omega}\overline{u_\ell}\nabla\eta_R\cdot\nabla u_\ell$$

where we have used that

$$\int_{\ell\Omega}\nabla(\eta_R\overline{u_\ell})\cdot\nabla u_\ell = \underbrace{\int_{\ell\Omega}\eta_R|\nabla u_\ell|^2}_{\in\mathbb{R}} + \int_{\ell\Omega}\overline{u_\ell}\nabla\eta_R\cdot\nabla u_\ell.$$

By dividing by $\Im(z) \neq 0$ and using the Cauchy-Schwarz inequality, this provides

$$\int_{\ell\Omega}\eta_R|u_\ell|^2 \leq \frac{1}{|\Im(z)|}\left(\|u_\ell\|_{L^2}\|\eta_R g\|_{L^2} + \frac{1}{R}\|u_\ell\|_{L^2}\|\nabla u_\ell\|_{L^2}\|\nabla\eta\|_{L^\infty}\right)$$

$$\leq \frac{C}{|\Im(z)|}\left(\|\eta_R g\|_{L^2} + \frac{1}{R}\right).$$

The norm of u_ℓ outside the ball B_{2R} is therefore small when R is large, uniformly with respect to ℓ. Moreover, we know that $u_\ell\mathbb{1}_{B_{2R}} \to u\mathbb{1}_{B_{2R}}$ strongly in $L^2(B_{2R})$ by the Rellich-Kondrachov Theorem A.18, since u_ℓ is bounded in $H^1(\mathbb{R}^d)$. By an $\varepsilon/2$ argument, this proves that $u_\ell \to u$ strongly in $L^2(\mathbb{R}^d)$, as announced in (7.33).

Now that we have shown the limit (7.23) for $f(x) = (x - z)^{-1}$ with $\Im(z) \neq 0$, the same convergence follows for all $f \in C_0^0(\mathbb{R})$ by density of the resolvent algebra in this space, as seen in Chap. 4 (for a similar argument, see Exercise 4.47). Since the operators considered are all bounded from below, the convergence also follows for any continuous function with support in an interval in the form $(-\infty, E]$.

The limit (7.23) for f piecewise continuous uses the fact that $-\Delta + V$ has no eigenvalues and we only explain it for $f(x) = \mathbb{1}_{(-\infty,E]}(x)$. Consider the two continuous functions

$$f_\eta(x) = \mathbb{1}_{(-\infty,E]}(x) + \frac{E + \eta - x}{\eta}\mathbb{1}_{(E,E+\eta]}(x)$$

and

$$\varphi_\eta(x) = \mathbb{1}_{[E,E+\eta]}(x) + \frac{E + 2\eta - x}{\eta}\mathbb{1}_{(E+\eta,E+2\eta]}(x) + \frac{x - E + \eta}{\eta}\mathbb{1}_{[E-\eta,E)}(x),$$

which are regularizations of $\mathbb{1}_{(-\infty,E]}$ and $\mathbb{1}_{[E,E+\eta]}$, respectively. Then we have

$$\big\| \mathbb{1}_{(-\infty,E]}(h_\ell)(\mathbb{1}_{\ell\Omega}g) - \mathbb{1}_{(-\infty,E]}(h)g \big\|$$
$$\leq \big\| f_\eta(h_\ell)(\mathbb{1}_{\ell\Omega}g) - f_\eta(h)g \big\| + \big\| \mathbb{1}_{(E,E+\eta]}(h_\ell)(\mathbb{1}_{\ell\Omega}g) \big\| + \big\| \mathbb{1}_{(E,E+\eta]}(h)g \big\|$$
$$\leq \big\| f_\eta(h_\ell)(\mathbb{1}_{\ell\Omega}g) - f_\eta(h)g \big\| + \big\| \varphi_\eta(h_\ell)(\mathbb{1}_{\ell\Omega}g) \big\| + \big\| \varphi_\eta(h)g \big\| .$$

In the last line we used the fact that

$$\big\| \mathbb{1}_{(E,E+\eta]}(A)v \big\|^2 = \big\langle v, \mathbb{1}_{(E,E+\eta]}(A)v \big\rangle \leq \big\langle v, \varphi_\eta(A)^2 v \big\rangle = \big\| \varphi_\eta(A)v \big\|^2$$

for any self-adjoint operator A and any vector v, because $\mathbb{1}_{(E,E+\eta]}^2 = \mathbb{1}_{(E,E+\eta]} \leq \varphi_\eta^2$. The limit for continuous functions therefore provides

$$\limsup_{\ell\to\infty} \big\| \mathbb{1}_{(-\infty,E]}(h_\ell)(\mathbb{1}_{\ell\Omega}g) - \mathbb{1}_{(-\infty,E]}(h)g \big\| \leq 2 \big\| \varphi_\eta(h)g \big\|$$

for all $\eta > 0$. As we have the convergence $\varphi_\eta(x) \to_{\eta\to0} \mathbb{1}_{\{E\}}(x)$ and $\mathbb{1}_{\{E\}}(h) = 0$ because there is no eigenvalue, we deduce from Theorem 4.8 (v) that $\varphi_\eta(h)g \to 0$ when $\eta \to 0$. This shows that

$$\lim_{\ell\to\infty} \big\| \mathbb{1}_{(-\infty,E]}(h_\ell)(\mathbb{1}_{\ell\Omega}g) - \mathbb{1}_{(-\infty,E]}(h)g \big\| = 0$$

and concludes the proof of Theorem 7.4.   □

Appendix A
Sobolev Spaces

Sobolev spaces play an essential role in the analysis of partial differential equations. They allow equations to be formulated in a "weak" sense which is however more restrictive than in the sense of distributions and is therefore closer to "strong" solutions. Moreover, as explained in Chap. 2, Sobolev spaces naturally appear as the domains where the momentum and the Laplacian are closed and self-adjoint. They are therefore unavoidable spaces in spectral theory.

We provide here the main tools without always providing the proofs and refer to the books [LL01, Eva10, Bre11, AF03, Gri85] for more details. While Sobolev spaces over the entire space $\mathbb{R}^d$ are quite simple objects (thanks to the use of the Fourier transform), the case of bounded domains is much more technical. The regularity of the boundary indeed plays an important role.

A.1 Definition

Let Ω be an open set of $\mathbb{R}^d$ and $k \geq 1$ an integer. The Sobolev space $H^k(\Omega)$ is defined by

$$H^k(\Omega) = \left\{ f \in L^2(\Omega) \text{ such that } \partial^\alpha f \in L^2(\Omega), \right.$$

$$\left. \text{for every multi-index } \alpha \text{ of length } |\alpha| \leq k \right\}. \qquad (A.1)$$

For $\alpha = (\alpha_1, \ldots, \alpha_d)$, recall that the partial derivative is $\partial^\alpha f = \partial_{x_1}^{\alpha_1} \cdots \partial_{x_d}^{\alpha_d} f$ and is interpreted in the sense of distributions in Ω, while the length of the multi-index α is $|\alpha| := \alpha_1 + \cdots + \alpha_d$. In the definition of $H^k(\Omega)$ we therefore require that all derivatives of f in the sense of distributions, of order less than k, identify with

© The Editor(s) (if applicable) and The Author(s), under exclusive license to Springer Nature Switzerland AG 2024
M. Lewin, *Spectral Theory and Quantum Mechanics*, Universitext,
https://doi.org/10.1007/978-3-031-66878-4

functions of $L^2(\Omega)$. It can be shown that $H^k(\Omega)$ is a Hilbert space, when it is equipped with the scalar product

$$\langle f, g\rangle_{H^k(\Omega)} := \sum_{|\alpha|\leq k} \langle \partial^\alpha f, \partial^\alpha g\rangle_{L^2(\Omega)}. \qquad (A.2)$$

The Sobolev space $W^{k,p}(\Omega)$ is defined similarly by replacing $L^2(\Omega)$ with $L^p(\Omega)$ everywhere. It is a Banach space but in this book we almost exclusively use the simplest case of the Hilbert space $H^k(\Omega)$.

The space $H_0^k(\Omega)$ is defined as the closure of $C_c^\infty(\Omega)$ for the norm of $H^k(\Omega)$. It is a closed subspace that is generally different from $H^k(\Omega)$. Intuitively, it contains the functions that vanish at the boundary (to a certain order depending on k), while the functions of $H^k(\Omega)$ can take any value at the boundary. For now, however, we have not yet discussed the boundary of Ω. The following result specifies that $H^k(\Omega)$ is itself the closure of $C^\infty(\Omega)$ for its norm.

Theorem A.1 (Meyers-Serrin) *Let $\Omega \subset \mathbb{R}^d$ be an open set. Then $C^\infty(\Omega)$ is dense in $H^k(\Omega)$ for the norm induced by the scalar product (A.2).*

The proof of the theorem is difficult when Ω is arbitrary, but simpler if Ω is sufficiently regular. We will give it later for $\Omega = (0, 1) \subset \mathbb{R}$ and $\Omega = \mathbb{R}^d$.

A.2   Sobolev Spaces on the Interval $I = (0, 1)$

The case of dimension $d = 1$ is easier. For convenience we work on the interval $I = (0, 1)$ but the arguments are similar on any finite interval. Consider first a function $f \in L^1(I)$ whose derivative in the sense of distributions f' is also in $L^1(I)$. Then the function $g(x) := \int_0^x f'(t)\, dt$ is continuous on the closed interval $[0, 1]$ (in particular it has a right limit at 0 and a left limit at 1). It satisfies $g' = f'$ in the sense of distributions on $I = (0, 1)$, which implies that $g - f$ is almost everywhere constant on I. In particular, f coincides almost everywhere with a continuous function and we can assume for simplicity that f is itself continuous. After evaluating at $x = 0^+$ we therefore have the relation

$$f(x) = f(0^+) + \int_0^x f'(t)\, dt.$$

In order to obtain an estimate involving only the L^1 norms of f and f', we use the above formula for $f(x)$ and $f(y)$ and subtract them to obtain

$$f(x) = f(y) + \int_y^x f'(t)\, dt \qquad (A.3)$$

for all $x, y \in (0, 1)$. Next we integrate over y and get

$$f(x) = \int_0^1 f(y)\,dy + \int_0^1 \int_y^x f'(t)\,dt\,dy.$$

Estimating $|\int_y^x f'(t)\,dt| \leq \int_0^1 |f'(t)|\,dt$, this provides the inequality

$$\max_{x\in[0,1]} |f(x)| \leq \|f\|_{L^1(I)} + \|f'\|_{L^1(I)}. \tag{A.4}$$

This proves the continuous embedding between the Banach spaces

$$W^{1,1}(I) \hookrightarrow C^0([0, 1]).$$

In particular, the boundary restriction map $f \in W^{1,1}(I) \mapsto (f(0^+), f(1^-)) \in \mathbb{C}^2$ is continuous, where we are of course always using the continuous representative of f.

If $f \in L^2(I)$ and $f' \in L^2(I)$ then these functions are *a fortiori* in $L^1(I)$ and f therefore also coincides almost everywhere with a continuous function on $[0, 1]$. But we can improve the estimate (A.4) by using that $|f|^2 \in L^1(I)$ and $(|f|^2)' = 2\Re(\overline{f}f') \in L^1(I)$. From (A.4) and the Cauchy-Schwarz inequality we obtain

$$\max_{x\in[0,1]} |f(x)| \leq \sqrt{2}\,\|f\|_{L^2(I)}^{\frac{1}{2}}\,\|f'\|_{L^2(I)}^{\frac{1}{2}} + \|f\|_{L^2(I)}. \tag{A.5}$$

Compared to (A.4), the derivative appears here with a power of $1/2$, which is useful in certain applications. In fact, f is even Hölder continuous with exponent $1/2$ since we can also write, from (A.3),

$$|f(x) - f(y)| \leq \left|\int_y^x f'(t)\,dt\right| \leq |x - y|^{1/2}\,\|f'\|_{L^2(I)}.$$

In summary, we have shown that the functions of $H^1(I)$ are $1/2$–Hölder continuous on $[0, 1]$. Similarly, the functions $f \in H^2(I)$ are of class $C^1([0, 1])$ and their derivative f' is $1/2$–Hölder. By iterating the argument, it is possible to prove the following lemma.

Lemma A.2 (Boundary Restriction for $H^k(I)$)   *Let* $I = (0, 1)$. *For* $k \geq 1$, *we have the continuous embedding*

$$H^k(I) \hookrightarrow C^{k-1}([0, 1]).$$

The boundary restriction map

$$f \in H^k(I) \mapsto \left(f(0^+), \ldots, f^{(k-1)}(0^+), f(1^-), \ldots, f^{(k-1)}(1^-) \right) \in \mathbb{C}^{2k}$$
$$\text{(A.6)}$$

is therefore continuous. On the right we use the $C^{k-1}([0, 1])$ representative of f.

By taking for f a polynomial of sufficiently high degree, it is clear that the boundary restriction map (A.6) is also surjective.

Exercise A.3 (Density of Smooth Functions) We fix $f \in H^1(I)$. Construct a sequence $f_n \in C^\infty([0, 1])$ such that $f_n \to f$ strongly in $H^1(I)$. If we further assume that $f(0^+) = f(1^-) = 0$, show that we can take $f_n \in C_c^\infty(I)$. Thus, you will have shown that

$$H_0^1(I) = \{ f \in H^1(I) \, : \, f(0^+) = f(1^-) = 0 \}$$

where it is recalled that $H_0^1(I)$ is by definition the closure of $C_c^\infty(I)$ in $H^1(I)$. Extend these results to $H^k(I)$ and $H_0^k(I)$.

Elliptic regularity often plays a central role. It stipulates that it is sufficient to verify that $f \in L^2(I)$ and that $f^{(k)} \in L^2(I)$ to immediately deduce that $f \in H^k(I)$. We state it here with $k = 2$ to simplify and leave the case of $H^k(I)$ as an exercise.

Lemma A.4 (Elliptic Regularity on $I = (0, 1)$) *Let $I = (0, 1)$ and $f \in L^2(I)$ such that $f'' \in L^2(I)$. Then $f' \in C^0([0, 1])$ with*

$$\max_{x \in [0,1]} |f'(x)| \leq 6\|f\|_{L^2(I)} + 2\|f''\|_{L^2(I)}. \qquad \text{(A.7)}$$

In particular, $f' \in L^2(I)$ and $f \in H^2(I)$.

Proof Since $f'' \in L^1(I)$ we know that there exists a constant C such that f' coincides almost everywhere with the continuous function $C + \int_0^x f''(t)\,dt$. In particular, f' belongs to $L^2(I)$. As a result, f is a C^1 function. To prove the inequality (A.7), we can for example start from the relation $f'(y) = f'(0^+) + \int_0^y f''(t)\,dt$, and integrate against $y(1 - y)$, which provides

$$\int_0^1 y(1 - y) \left(\int_0^y f''(t)\,dt \right) dy + \frac{f'(0^+)}{6} = -\int_0^1 (1 - 2y) f(y)\,dy.$$

By integrating by parts the term on the left, this can be rewritten as

$$f'(0^+) = -6 \int_0^1 (1 - 2y) f(y)\,dy + \int_0^1 \left(3y^2 - 2y^3 - 1 \right) f''(y)\,dy.$$

We have therefore obtained the formula

$$f'(y) = \int_0^y f''(t)\, dt - 6 \int_0^1 (1 - 2y) f(y)\, dy + \int_0^1 \left(3y^2 - 2y^3 - 1\right) f''(y)\, dy$$

from which the Cauchy-Schwarz inequality provides

$$\|f'\|_{L^\infty(I)} \le 2\|f''\|_{L^1(I)} + 6\|f\|_{L^1(I)} \le 2\|f''\|_{L^2(I)} + 6\|f\|_{L^2(I)},$$

as we wanted.   □

Exercise A.5 (Case of the Half-Line) Extend the results of this section to the case of the half-line $I = (0, \infty)$:

1. Show that $H^1(I) \hookrightarrow C_0^0([0, \infty))$ (the space of continuous functions that tend to 0 at $+\infty$), with the same estimate as (A.5).
2. Show that C^∞ functions are dense in $H^1(I)$ and that $H_0^1(I)$ is the space of functions $f \in H^1(I)$ such that $f(0^+) = 0$.
3. Show that if $f, f'' \in L^2(I)$ then $f' \in C_0^0([0, \infty))$ with the same estimate as (A.7).

A.3   Sobolev Spaces on $\mathbb{R}^d$

As we have

$$\widehat{\partial^\alpha f}(p) = (ip)^\alpha \widehat{f}(p) = i^{|\alpha|} \prod_{j=1}^d p_j^{\alpha_j} \widehat{f}(p),$$

we deduce from Parseval that a function f belongs to $H^k(\mathbb{R}^d)$ if and only if its Fourier transform $\widehat{f}$ is such that

$$p^\alpha \widehat{f}(p) \in L^2(\mathbb{R}^d) \qquad \text{for every multi-index } \alpha \text{ of length } |\alpha| \le k.$$

However, by Hölder's inequality we can estimate

$$|p^\alpha| \le \sum_{j=1}^d \frac{\alpha_j}{k} |p_j|^k + 1 - \frac{|\alpha|}{k} \le 1 + \sum_{j=1}^d |p_j|^k \le C_{d,k}(1 + |p|^2)^{\frac{k}{2}} \qquad (A.8)$$

with for example $C_{d,k} = \max(d^{1-k/2}, 1)$. This allows us to deduce the equivalent characterizations (and norms)

$$H^k(\mathbb{R}^d) = \left\{ f \in L^2(\mathbb{R}^d) : \partial_{x_j}^k f \in L^2(\mathbb{R}^d) \text{ for all } j = 1, \ldots, d \right\}$$

$$= \left\{ f \in L^2(\mathbb{R}^d) : \int_{\mathbb{R}^d} (1 + |p|^2)^k |\widehat{f}(p)|^2 \, dp < \infty \right\}. \tag{A.9}$$

In other words, we can restrict ourselves to the highest order derivatives. As $|p|^2 \widehat{f} = \widehat{-\Delta f}$, when k is even, we deduce by the same reasoning an important characterization that we state in the form of a lemma.

Lemma A.6 (Elliptic Regularity on $\mathbb{R}^d$) *For $k = 2\ell$ an even integer, we have*

$$H^{2\ell}(\mathbb{R}^d) = \left\{ f \in L^2(\mathbb{R}^d) : \Delta^\ell f \in L^2(\mathbb{R}^d) \right\}. \tag{A.10}$$

The advantage of the characterization (A.9) of $H^k(\mathbb{R}^d)$ is that it easily extends to the case of a non-integer exponent. We therefore set

$$\boxed{H^s(\mathbb{R}^d) = \left\{ f \in L^2(\mathbb{R}^d) : \int_{\mathbb{R}^d} (1 + |p|^2)^s |\widehat{f}(p)|^2 \, dp < \infty \right\}} \tag{A.11}$$

for all $s > 0$. While the Sobolev norm is easily expressed in space variables when $s = k$ is an integer, it does not seem obvious to write that of $H^s(\mathbb{R}^d)$ in terms of $f(x)$ when s is not an integer. The following theorem is then useful.

Theorem A.7 (Definition of $H^s(\mathbb{R}^d)$ in Space Variables for $0 < s < 1$) *Let $0 < s < 1$ and $f \in H^s(\mathbb{R})$ be any function. Then we have*

$$\int_{\mathbb{R}^d} |p|^{2s} |\widehat{f}(p)|^2 \, dp = C_{d,s} \int_{\mathbb{R}^d} \int_{\mathbb{R}^d} \frac{|f(x) - f(y)|^2}{|x - y|^{d+2s}} \, dx \, dy \tag{A.12}$$

where $C_{d,s} = \frac{2^{2s-1} s \Gamma(s)}{\pi^{d/2} \Gamma(1-s)}$.

Here Γ is the Euler Gamma function. When $0 < s < 1$, $f \in H^s(\mathbb{R}^d)$ if and only if $f \in L^2(\mathbb{R}^d)$ and the integral on the right of (A.12) converges. If $f \in H^s(\mathbb{R}^d)$ with $s \geq 1$, we start by writing $s = k + \sigma$ with $0 < \sigma < 1$ and we use the above characterization for all derivatives of order k of f, which must belong to $H^\sigma(\mathbb{R}^d)$.

Proof We use the following integral formula

$$|p|^{2s} = \frac{s}{\Gamma(1-s)} \int_0^\infty \left(1 - e^{-t|p|^2}\right) \frac{dt}{t^{1+s}}$$

which is obtained by a change of variable. Note that the integral on the right converges for $0 < s < 1$ since $1 - e^{-t|p|^2} = t|p|^2 + O(t^2)$ at 0, which reduces the singularity to t^{-s} at 0. The latter is integrable since $s < 1$. The integrability at infinity follows from the condition that $s > 0$. The value of the constant comes from the equality

$$\int_0^\infty \left(1 - e^{-t}\right) \frac{dt}{t^{1+s}} = \frac{1}{s} \int_0^\infty e^{-t} \frac{dt}{t^s} = \frac{\Gamma(1-s)}{s}.$$

We thus obtain

$$\int_{\mathbb{R}^d} |p|^{2s} |\widehat{f}(p)|^2 \, dp = \frac{s}{\Gamma(1-s)} \int_0^\infty \left(\int_{\mathbb{R}^d} (1 - e^{-t|p|^2}) |\widehat{f}(p)|^2 \, dp \right) \frac{dt}{t^{1+s}}.$$

The exchange of integrals is permitted by Fubini since the integrand is non-negative. We can easily express the integral on the right in space using the inverse Fourier transform of $e^{-t|p|^2}$ which is $(2t)^{-d/2} e^{-|x|^2/(4t)}$, and the fact that the Fourier transform of a product is a convolution. We find

$$\int_{\mathbb{R}^d} (1 - e^{-t|p|^2}) |\widehat{f}(p)|^2 \, dp = \frac{1}{2} (4\pi t)^{-d/2} \iint_{\mathbb{R}^d \times \mathbb{R}^d} e^{-\frac{|x-y|^2}{4t}} |f(x) - f(y)|^2 \, dx \, dy$$

since $(4\pi t)^{-d/2} \int_{\mathbb{R}^d} e^{-\frac{|x|^2}{4t}} \, dx = 1$. We have therefore proven that

$$\int_{\mathbb{R}^d} |p|^{2s} |\widehat{f}(p)|^2 \, dp$$

$$= \frac{s(4\pi)^{-\frac{d}{2}}}{2\Gamma(1-s)} \iint_{\mathbb{R}^d \times \mathbb{R}^d} \left(\int_0^\infty e^{-\frac{|x-y|^2}{4t}} \frac{dt}{t^{1+s+\frac{d}{2}}} \right) |f(x) - f(y)|^2 \, dx \, dy,$$

and as

$$\int_0^\infty e^{-\frac{|x-y|^2}{4t}} \frac{dt}{t^{1+s+\frac{d}{2}}} = \frac{1}{|x-y|^{d+2s}} \int_0^\infty e^{-\frac{1}{4t}} \frac{dt}{t^{1+s+\frac{d}{2}}} = \frac{2^{d+2s} \Gamma(s)}{|x-y|^{d+2s}},$$

we obtain the result.   □

The following theorem indicates that $H_0^s(\mathbb{R}^d) = H^s(\mathbb{R}^d)$ for all $s > 0$, that is, $C_c^\infty(\mathbb{R}^d)$ is dense in $H^s(\mathbb{R}^d)$ for the associated norm.

Theorem A.8 (Density of $C_c^\infty(\mathbb{R}^d)$) *The space $C_c^\infty(\mathbb{R}^d)$ is dense in $H^s(\mathbb{R}^d)$ for all $s > 0$.*

Proof We argue in two steps. First, we show the density of C^∞ functions (not necessarily with compact support) by regularization, before deducing the density of C^∞ functions with compact support by smoothly truncating. Let $f \in H^s(\mathbb{R}^d)$

and $\chi \in C_c^\infty(\mathbb{R}^d)$ such that $\int_{\mathbb{R}^d} \chi(x)\,dx = 1$. We define the convolution $f_n = f * \chi_n$ where $\chi_n(x) = n^d \chi(nx)$. The function f_n is C^∞ on $\mathbb{R}^d$ because we can place all the derivatives on χ_n in the convolution. It remains to show that $f_n \to f$ in $H^s(\mathbb{R}^d)$. We calculate in Fourier

$$\|f_n - f\|_{H^s(\mathbb{R}^d)}^2 = \int_{\mathbb{R}^d} (1 + |p|^{2s}) |\widehat{f}(p)|^2 \left(1 - (2\pi)^{\frac{d}{2}} \widehat{\chi}(p/n)\right)^2 dp$$

which tends to 0 by dominated convergence, because $(2\pi)^{\frac{d}{2}} \widehat{\chi}(p/n) \to (2\pi)^{\frac{d}{2}} \widehat{\chi}(0) = \int_{\mathbb{R}^d} \chi = 1$ for all $p \in \mathbb{R}^d$ and

$$\left| (2\pi)^{\frac{d}{2}} \widehat{\chi}(p/n) \right| \leq (2\pi)^{\frac{d}{2}} \|\widehat{\chi}\|_{L^\infty(\mathbb{R}^d)} \leq \|\chi\|_{L^1(\mathbb{R}^d)}.$$

Thus, C^∞ functions are dense in $H^s(\mathbb{R}^d)$ for all $s > 0$. We note that the function $f * \chi_n$ also belongs to all the $H^{s'}(\mathbb{R}^d)$ for $s' > 0$, since $|k|^\alpha \widehat{\chi}(k/n)$ is uniformly bounded for all $\alpha > 0$. So we have also shown that $H^{s'}(\mathbb{R}^d)$ is dense in $H^s(\mathbb{R}^d)$ for all $s' \geq s$.

We can now take $f \in H^{s'}(\mathbb{R}^d) \cap C^\infty(\mathbb{R}^d)$ for any $s' > s$ and approximate it in $H^s(\mathbb{R}^d)$ by a sequence of functions with compact support. We simply take $s' = m$, any integer greater than or equal to s. As the norm of $H^m(\mathbb{R}^d)$ controls that of $H^s(\mathbb{R}^d)$, we can even show convergence in $H^m(\mathbb{R}^d)$. This allows us to use the characterization of $H^m(\mathbb{R}^d)$ with spatial derivatives rather than Fourier. To simplify the presentation we will first assume that $m = 1$, that is, we take $f \in H^1(\mathbb{R}^d) \cap C^\infty(\mathbb{R}^d)$ which we approximate by a sequence of functions from $C_c^\infty(\mathbb{R}^d)$ in $H^1(\mathbb{R}^d)$. Let $\chi \in C_c^\infty(\mathbb{R}^d)$ such that, this time, $\chi(0) = 1$. We introduce the C^∞ function with compact support $f_n(x) := f(x)\chi(x/n)$. We then have

$$\|f - f_n\|_{L^2(\mathbb{R}^d)}^2 = \int_{\mathbb{R}^d} |f(x)|^2 (1 - \chi(x/n))^2 \, dx \longrightarrow 0$$

by dominated convergence. Moreover, in the sense of distributions, we have

$$\nabla f_n = \chi(x/n)\nabla f(x) + \frac{1}{n} f(x)\nabla\chi(x/n).$$

This shows by the triangle inequality that

$$\|\nabla f - \nabla f_n\|_{L^2(\mathbb{R}^d)} \leq \|\nabla f - \chi(\cdot/n)\nabla f\|_{L^2(\mathbb{R}^d)} + \frac{\|\nabla\chi\|_{L^\infty(\mathbb{R}^d)} \|f\|_{L^2(\mathbb{R}^d)}}{n}$$

which tends to 0 by the previous argument. For $m > 1$ we similarly calculate $\partial^\alpha f_n$ by Leibniz' formula. This provides the estimate

$$\|\partial^\alpha f - \partial^\alpha f_n\|_{L^2(\mathbb{R}^d)} \leq \|\partial^\alpha f - \chi(\cdot/n)\partial^\alpha f\|_{L^2(\mathbb{R}^d)} + \frac{C}{n}$$

which again tends to 0 for every multi-index α of length $|\alpha| \leq m$.   □

A.4   Trace, Lifting and Extension on a Bounded Open Set

We have already seen that it is possible to define the trace of a function $f \in H^1(I)$ at the boundary of the interval $I = (0, 1)$, this trace being just the vector composed of the two values $f(0^+)$ and $f(1^-)$ of the continuous representative of f. In higher dimensions, the situation is a bit more delicate. It is possible to define the trace $f_{|\partial\Omega}$ at the boundary of a domain Ω, but it is an object defined almost everywhere on $\partial\Omega$ because the functions $f \in H^1(\Omega)$ are not necessarily continuous.

Let us start by quickly discussing the half-space

$$\Omega = \{x = (x_1, \ldots, x_d) \in \mathbb{R}^d \;:\; x_d > 0\},$$

which can be reduced to the one-dimensional case studied in Sect. A.2 and serves as a basis for any domain. If a function f belongs to $H^1(\Omega)$, then we have by Fubini

$$\int_{\mathbb{R}^{d-1}} \left(\int_0^\infty |f(x_1, \ldots, x_d)|^2 \, dx_d + \int_0^\infty |\partial_{x_d} f(x_1, \ldots, x_d)|^2 \, dx_d \right)$$
$$dx_1 \cdots dx_{d-1} < \infty.$$

For almost all $x_1, \ldots, x_{d-1}$ the function $y \mapsto f(x_1, \ldots, x_{d-1}, y)$ belongs to the Sobolev space $H^1((0, \infty))$, so it coincides almost everywhere with a continuous function according to Exercise A.5. Its value at $x_d = 0$ is simply noted $f(x_1, \ldots, x_{d-1}, 0)$ or $f_{|\partial\Omega}(x_1, \ldots, x_{d-1})$, for $x_1, \ldots, x_{d-1}$ in the set of total measure in $\mathbb{R}^{d-1}$ considered. It verifies

$$|f(x_1, \ldots, x_{d-1}, 0)|^2 = -2\Re \int_0^\infty \overline{f(x_1, \ldots, x_{d-1}, y)} \partial_{x_d} f(x_1, \ldots, x_{d-1}, y) \, dy.$$

The function on the right is integrable over $\mathbb{R}^{d-1}$, so the same is true for the term on the left, with

$$\int_{\mathbb{R}^{d-1}} |f(x_1, \ldots, x_{d-1}, 0)|^2 dx_1 \cdots dx_{d-1} = -2\Re \int_\Omega \overline{f(x)} \partial_{x_d} f(x) \, dx.$$

The so-defined restriction to the boundary of Ω therefore belongs to $L^2(\mathbb{R}^{d-1})$ with the estimate

$$\| f_{|\partial\Omega} \|^2_{L^2(\mathbb{R}^{d-1})} \leq 2 \, \| f \|_{L^2(\Omega)} \, \| \partial_{x_d} f \|_{L^2(\Omega)}. \tag{A.13}$$

Note that only the derivative ∂_{x_d} in the direction normal to the boundary of Ω is necessary to control the restriction of f to the boundary.

Using the half-space case as a model, we can define the restriction of a function $f \in H^1(\Omega)$ to the boundary of any smooth-enough open set Ω, a restriction that belongs to $L^2(\partial\Omega)$. The idea is to reduce to small neighborhoods of the boundary using a partition of unity, and to use local maps to "flatten the boundary". The proof is not very difficult when the boundary of Ω is a sufficiently regular hypersurface. We will now state without proof results valid with fairly weak assumptions on the regularity of Ω.

We say that an open set $\Omega \subset \mathbb{R}^d$ has a *Lipschitz boundary* when $\partial\Omega$ is, in the neighborhood of any point, a hypersurface which is the graph of a Lipschitz function φ with Lipschitz constant less than or equal to M, where M is independent of the chosen point. We will say that Ω has a $C^{k,\alpha}$ *boundary* if the function φ is of class C^k with its kth derivatives α–Hölder continuous, and similar uniform estimates as before. Finally, we will say that Ω has a *piecewise Lipschitz (resp. $C^{k,\alpha}$) boundary* when it is the union of a finite number of Lipschitz (resp. $C^{k,\alpha}$) hypersurfaces. Unfortunately, irregular domains often occur in practice. The simplest example is the cube $(0, 1)^d$ which is composed of flat faces, thus has a piecewise C^k boundary for all $k \geq 1$. Note that the unit cell of a periodic lattice of $\mathbb{R}^d$ (Chap. 7) is always a polyhedron.

A piecewise Lipschitz boundary $\partial\Omega$ can be equipped with a surface measure on each of the hypersurfaces that compose it, which is just the deformation of the Lebesgue measure on $\mathbb{R}^{d-1}$. This allows one to define the space $L^2(\partial\Omega)$. The main theorem is the following.

Theorem A.9 (Trace) *Let Ω be a bounded open set of $\mathbb{R}^d$ whose boundary is piecewise Lipschitz. Then there exists a unique continuous map*

$$
\begin{aligned}
H^1(\Omega) &\rightarrow L^2(\partial\Omega)\\
f &\mapsto f_{|\partial\Omega}
\end{aligned}
\tag{A.14}
$$

which coincides with the restriction to the boundary when $f \in H^1(\Omega) \cap C^0(\overline{\Omega})$. Moreover, we have

$$
\|f\|^2_{L^2(\partial\Omega)} \leq C\|f\|_{L^2(\Omega)}\|f\|_{H^1(\Omega)}
\tag{A.15}
$$

where the constant C depends only on Ω. If Ω is of class $C^{k-1,1}$ with $k \geq 2$, there exists a unique continuous map

$$
\begin{aligned}
H^k(\Omega) &\rightarrow \quad\quad L^2(\partial\Omega)^k\\
f &\mapsto \left(f_{|\partial\Omega}, \cdots, \partial_n^{k-1} f_{|\partial\Omega}\right)
\end{aligned}
\tag{A.16}
$$

which coincides with the restriction to the boundary when $f \in H^k(\Omega) \cap C^k(\overline{\Omega})$,
where $\partial_n = n \cdot \nabla$ *denotes the derivative in the direction of the outgoing normal* n
on $\partial\Omega$, *with*

$$\sum_{j=0}^{k-1} \left\| \partial_n^j f_{|\partial\Omega} \right\|_{L^2(\partial\Omega)} \leq C \|f\|_{H^{k-1}(\Omega)} \|f\|_{H^k(\Omega)}.$$

We have $f \in H_0^k(\Omega)$ *(the closure of* $C_c^\infty(\Omega)$ *in* $H^k(\Omega)$*) if and only if* $f \in H^k(\Omega)$
and all its traces vanish: $f_{|\partial\Omega} = \cdots = \partial_n^{k-1} f_{|\partial\Omega} = 0$.

If Ω has a piecewise $C^{k-1,1}$ boundary, the second part of the result is the same
but the derivatives must be calculated outside the "edges", that is, the intersection of
the various hypersurfaces composing the boundary. The inequality (A.15) is the generalization of the same estimate (A.13) for the half-space. See for example [Gri85,
Thm. 1.5.1.10] for the proof.

The range of the trace map (A.14) is not the whole of $L^2(\partial\Omega)$, it is smaller (in
dimension $d \geq 2$). However, the image is dense because we can "lift" any given
sufficiently smooth boundary condition.

Theorem A.10 (Smooth Lifting) *Let* $k \geq 1$. *If* Ω *has a* $C^{k-1,1}$ *boundary and*
$g_j \in C^{k-1-j}(\partial\Omega)$ *for* $j = 0, \ldots, k-1$, *then there exists a function* $f \in H^k(\Omega)$
such that $\partial_n^j f_{|\Omega} = g_j$.

The result is the same if $\partial\Omega$ has a piecewise $C^{k-1,1}$ boundary, but then we must
assume that the support of the g_j does not intersect the edges.

The idea of the construction is quite simple. We first use partitions of unity
to reduce to the case where the g_j have small support. Then, on this support the
boundary $\partial\Omega$ is almost flat and everything again boils down to understanding the
half space. If $k = 2$ and $\Omega \subset \mathbb{R}^d$ is exactly equal to the half space $\{x_d > 0\}$ in the
neighborhood of the support of g_0 and g_1, then we can simply take

$$f(x_1, \ldots, x_d) = \Big(g_0(x_1, \ldots, x_{d-1}) - x_d g_1(x_1, \ldots, x_{d-1})\Big) \eta(x_1, \ldots, x_d)$$

where η is a C^∞ function with compact support, which equals 1 in the vicinity of
the boundary. The general case follows from the use of local charts.

It is possible to precisely identify the image of the restriction to the boundary.
By comparison with the formula obtained in the whole space in Theorem A.7, we
define the fractional Sobolev space on a domain Ω by its norm

$$\|f\|_{H^s(\Omega)}^2 = \iint_{\Omega\times\Omega} \frac{|f(x) - f(y)|^2}{|x-y|^{2s+2d}} \, dx \, dy$$

for $0 < s < 1$. If $s > 1$ we of course require that the derivatives of order $[s]$
all belong to $H^{s-[s]}$. It can then be proved that the range of $f \in H^k(\Omega) \mapsto$

$\partial_n^j f_{|\partial\Omega}$ is exactly the space $H^{k-j-1/2}(\partial\Omega)$ when Ω has a $C^{k-1,1}$ boundary (see for example [Gri85, Thm 1.5.1.2]). Of course, the map cannot be one-to-one because we can change the function f inside Ω at will without changing its values on the boundary. However, it is possible to construct a lifting that is continuous for the norms in question.

Theorem A.11 (Lifting) *Let Ω be an open set of class $C^{k-1,1}$ with $k \geq 1$. There exists a continuous map*

$$R : (g_0, \ldots, g_{k-1}) \in \prod_{j=0}^{k-1} H^{k-j-\frac{1}{2}}(\partial\Omega) \mapsto f \in H^k(\Omega)$$

such that $\partial_n^j f = g_j$ and

$$\|f\|_{H^k(\Omega)} \leq C \sum_{j=0}^{k-1} \|g_j\|_{H^{k-j-\frac{1}{2}}(\partial\Omega)} . \tag{A.17}$$

The function f can be constructed so that it is supported as close as desired to the boundary, but the constant C explodes as the desired distance decreases.

It is possible to use Theorem A.11 to extend a function from $H^k(\Omega)$ to a function of $H^k(\mathbb{R}^d)$: We apply the result on the complement $\mathbb{R}^d \setminus \Omega$ (or a neighborhood of $\partial\Omega$ in this complement), so that the restrictions to the boundary of the function on $\mathbb{R}^d \setminus \Omega$ all coincide with those of f. Using Green's formula, we can then verify that the compatibility of all the restrictions is sufficient for the function to be in $H^k(\mathbb{R}^d)$. However, this construction based on the continuity of traces and the extension outside of Ω requires strong assumptions on the domain Ω, since it is necessary to know the exact spaces in which the traces of f live. It is actually possible to extend a function from $H^k(\Omega)$ without using traces, which then requires very weak regularity assumptions on Ω (see for example [Gri85, Thm 1.4.3.1]).

Theorem A.12 (Extension) *Let $\Omega \subset \mathbb{R}^d$ be a bounded open set with a Lipschitz boundary and $k \geq 1$ an integer. Then there exists a continuous linear map*

$$f \in H^k(\Omega) \mapsto \tilde{f} \in H^k(\mathbb{R}^d)$$

such that $\tilde{f} \mathbb{1}_\Omega = f$ and

$$\|\tilde{f}\|_{H^k(\mathbb{R}^d)} \leq C\|f\|_{H^k(\Omega)} \tag{A.18}$$

This result is very important to be able to easily extend properties proven in $\mathbb{R}^d$ to a bounded domain. We will see an example in the next section. Once again, we can ask that the extension be supported as close as desired to Ω, but the constant C diverges as the distance tends to 0.

A.5   Sobolev Embedding and Rellich-Kondrachov Compactness

We now discuss the Sobolev embeddings, which specify that the functions of $H^k(\Omega)$ belong to certain $L^p(\Omega)$ with $p > 2$, the embedding being compact when Ω is bounded. But let us start with the case of the whole space.

Theorem A.13 (Sobolev Inequality)   *Let $d \geq 1$ and $0 < s < d/2$. There exists a constant $S_{d,s}$ such that, for any measurable function $f \in L^1_{loc}(\mathbb{R}^d) \cap S'$ so that the set $\{x \ : \ |f(x)| > M\}$ has a finite measure for all $M > 0$, we have*

$$\|f\|^2_{L^{p^*}(\mathbb{R}^d)} \leq S_{d,s} \int_{\mathbb{R}^d} |\xi|^{2s} |\widehat{f}(\xi)|^2 \, d\xi \tag{A.19}$$

with $p^* = \frac{2d}{d-2s}$.

The value of p^* is imposed by the invariance of the two terms of the inequality under space dilations.

Remark A.14   We have assumed that f is a tempered distribution in order to write its Fourier transform. In practice we often have $f \in L^2(\mathbb{R}^d)$, which immediately implies that $f \in S'$ and that the sets $\{|f| > M\}$ are of finite measure for all $M > 0$. However, as the inequality does not involve the L^2 norm of f, it is not very natural to state a theorem with this assumption. Obviously, the inequality (A.19) is void if the integral on the right diverges, that is, if $|\xi|^s \widehat{f}(\xi)$ is not in $L^2(\mathbb{R}^d)$.

If $f \in H^s(\mathbb{R}^d)$ then $f \in L^2(\mathbb{R}^d)$ and we can use Hölder's inequality, to deduce that $f \in L^p(\mathbb{R}^d)$ for all $2 \leq p \leq p^*$, with the *Gagliardo-Nirenberg inequality*:

$$\|f\|_{L^p(\mathbb{R}^d)} \leq \|f\|^\theta_{L^2(\mathbb{R}^d)} \|f\|^{1-\theta}_{L^{p^*}(\mathbb{R}^d)} \leq C \|f\|^\theta_{L^2(\mathbb{R}^d)} \left(\int_{\mathbb{R}^d} |\xi|^{2s} |\widehat{f}(\xi)|^2 \, d\xi \right)^{\frac{1-\theta}{2}}$$

$$\leq C \left(\int_{\mathbb{R}^d} (1 + |\xi|^2)^s |\widehat{f}(\xi)|^2 \, d\xi \right)^{\frac{1}{2}} = C \|f\|_{H^s(\mathbb{R}^d)}, \tag{A.20}$$

where $\frac{1}{p} = \frac{\theta}{2} + \frac{1-\theta}{p^*}$ and $C = S_{d,s}^{\frac{1-\theta}{2}}$. We provide a fairly simple proof of the Sobolev inequality (A.19) from [CX97].

Proof   Let f be as in the statement, such that $K^2 := \int |\xi|^{2s} |\widehat{f}(\xi)|^2 \, d\xi$ is finite. We first write

$$\int_{\mathbb{R}^d} |f(x)|^{\frac{2d}{d-2s}} \, dx = \frac{d-2s}{d+2s} \int_{\mathbb{R}^d} \int_0^\infty \lambda^{\frac{2d}{d-2s}-1} \mathbb{1}(|f(x)| \geq \lambda) \, d\lambda \, dx$$

$$= \frac{d-2s}{d+2s} \int_0^\infty \lambda^{\frac{2d}{d-2s}-1} |\{x \ : \ |f(x)| \geq \lambda\}| \, d\lambda. \tag{A.21}$$

We now need to estimate $|\{x \ : \ |f(x)| \geq \lambda\}|$ for all λ. We write $f = v + w$ with $\widehat{v}(\xi) = \widehat{f}(\xi)\mathbb{1}(|\xi| \leq a)$, where a is a parameter depending on λ that will be chosen later. We then use that

$$|\{|f(x)| \geq \lambda\}| \leq |\{|v(x)| \geq \lambda/2\}| + |\{|w(x)| \geq \lambda/2\}|$$

and we choose a such that $\|v\|_{L^\infty} < \lambda/2$, which gives $|\{|v(x)| \geq \lambda/2\}| = 0$. In fact, we have

$$\|v\|_{L^\infty} \leq (2\pi)^{-d} \int_{|\xi| \leq a} |\widehat{f}(\xi)|\, d\xi$$

$$\leq (2\pi)^{-d} \left(\int_{|\xi| \leq a} \frac{d\xi}{|\xi|^{2s}} \right)^{1/2} \left(\int_{|\xi| \leq a} |\xi|^{2s}|\widehat{f}(\xi)|^2 d\xi \right)^{1/2} \leq C\, K\, a^{\frac{d-2s}{2}},$$

$$(A.22)$$

for some constant C.[1] This suggests taking $a^{\frac{d-2s}{2}} = \lambda/(4CK)$, that is, $a = C(\lambda/K)^{\frac{2}{d-2s}}$. It remains to estimate the term involving w, for which we write

$$|\{|w(x)| \geq \lambda/2\}| \leq \frac{4}{\lambda^2} \int_{\mathbb{R}^d} |w|^2 \leq \frac{4}{\lambda^2} \int_{|\xi| \geq a} |\widehat{f}(\xi)|^2\, d\xi.$$

By inserting into (A.21), we obtain

$$\int_{\mathbb{R}^d} |f(x)|^{\frac{2d}{d-2s}}\, dx \leq C \int_0^\infty \lambda^{\frac{2d}{d-2s}-1} \frac{1}{\lambda^2} \int_{|\xi| \geq a} |\widehat{f}(\xi)|^2\, d\xi\, d\lambda$$

$$\leq C \int_0^\infty \lambda^{\frac{6s-d}{d-2s}} \int_{|\xi| \geq C'(\lambda/K)^{\frac{2}{d-2s}}} |\widehat{f}(\xi)|^2\, d\xi\, d\lambda$$

$$= C K^{\frac{4s}{d-2s}} \int |\xi|^{2s}|\widehat{f}(\xi)|^2\, d\xi = C K^{\frac{2d}{d-2s}}.$$

This concludes the proof of the Sobolev inequality (A.19).   □

For $s = 1$, we find that the functions of $H^1(\mathbb{R}^d)$ are in $L^2(\mathbb{R}^d) \cap L^{\frac{2d}{d-2}}(\mathbb{R}^d)$ when $d \geq 3$. In dimension 1 we can use the formula $|f(x)|^2 = 2\Re \int_{-\infty}^x \overline{f} f'$ as in Sect. A.2. This allows us to show that $H^1(\mathbb{R}) \hookrightarrow C_0^0(\mathbb{R})$, the space of continuous functions that tend to 0 at infinity, with the estimate

$$\|f\|_{L^\infty(\mathbb{R})}^2 \leq 2\|f\|_{L^2(\mathbb{R})} \|f'\|_{L^2(\mathbb{R})}.$$

$$(A.23)$$

---

[1] For simplicity, throughout the proof we call C a generic constant that can change from line to line. We do not compute the value of the constant our proof provides for the Sobolev inequality (A.19), since it is not the optimal one.

In particular, we have $H^1(\mathbb{R}) \hookrightarrow L^p(\mathbb{R})$ for all $2 \le p \le \infty$ by Hölder's inequality. In dimension two, there is no continuous embedding of $H^1(\mathbb{R}^2)$ into $L^\infty(\mathbb{R}^2)$, but the continuous embedding into $L^p(\mathbb{R}^2)$ is true for all $2 \le p < \infty$. The general theorem for $H^s(\mathbb{R}^d)$ is as follows.

Theorem A.15 (Sobolev Embedding on $\mathbb{R}^d$)

- *If $0 < s < d/2$, we have the continuous embedding*

$$H^s(\mathbb{R}^d) \hookrightarrow L^p(\mathbb{R}^d), \qquad \forall 2 \le p \le \frac{2d}{d-2s}. \tag{A.24}$$

- *If $s = d/2$, we have*

$$H^s(\mathbb{R}^d) \hookrightarrow L^p(\mathbb{R}^d), \qquad \forall 2 \le p < \infty. \tag{A.25}$$

- *If $s > d/2$, we have*

$$H^s(\mathbb{R}^d) \hookrightarrow C_0^{\ell,\theta}(\mathbb{R}^d) \tag{A.26}$$

where ℓ is the unique integer such that $0 \le \ell < s - d/2 < \ell + 1$ and $\theta = s - \ell - d/2$.

Here, $C_0^{\ell,\theta}(\mathbb{R}^d)$ is the space of functions of class C^ℓ, which tend to 0 at infinity as well as all their ℓ first derivatives, and which are such that the derivatives of order ℓ are Hölder continuous with exponent θ. The corresponding norm is given by

$$\|f\|_{C_0^{\ell,\theta}(\mathbb{R}^d)} := \sum_{|\alpha| \le \ell} \max_{x \in \mathbb{R}^d} |\partial^\alpha f(x)| + \sum_{|\alpha|=\ell} \sup_{x \ne y \in \mathbb{R}^d} \frac{|\partial^\alpha f(x) - \partial^\alpha f(y)|}{|x-y|^\theta}.$$

When $s \ge d/2$ we have the same Gagliardo-Nirenberg inequality as (A.20)

$$\|f\|_{L^p(\mathbb{R}^d)} \le C_{d,s,p} \|f\|_{L^2(\mathbb{R}^d)}^\theta \left(\int_{\mathbb{R}^d} |\xi|^{2s} |\widehat{f}(\xi)|^2 \, d\xi \right)^{\frac{1-\theta}{2}}, \qquad \theta = 1 - \frac{d(p-2)}{2sp}, \tag{A.27}$$

this time for all $2 \le p \le \infty$ if $s > d/2$ and all $2 \le p < \infty$ if $s = d/2$.[2] The form of the inequality and the value of θ are imposed by the homogeneity of each of the terms.

The "subcritical" embeddings of Theorem A.15 are not very difficult. Let us quickly explain how we can prove the Gagliardo-Nirenberg inequality (A.27) directly, for $p \ne 2d/(d-2s)$, without using the Sobolev inequality if $s < d/2$ as we did in (A.20). According to the characterization with the Fourier transform, a

---

[2] The statement of Theorem A.15 provides the inequality with the norm of H^s on the right. We can then obtain (A.27) by dilating the function f and optimizing with respect to the dilation coefficient.

function f belongs to $H^s(\mathbb{R}^d)$ if and only if it can be written in Fourier

$$\widehat{f}(\xi) = \frac{\widehat{g}(\xi)}{(1 + |\xi|^2)^{s/2}}$$

with $g \in L^2(\mathbb{R}^d)$. Moreover, the L^2 norm of g is equal to the H^s norm of f. In space variables, the formula can be written

$$f(x) = \int_{\mathbb{R}^d} h_s(x - y)g(y)\,dy$$

where h_s is the function such that

$$\widehat{h_s}(\xi) = \frac{(2\pi)^{d/2}}{(1 + |\xi|^2)^{s/2}}. \tag{A.28}$$

Thus, we are led to prove that if $g \in L^2(\mathbb{R}^d)$, then $h_s * g$ belongs to $L^p(\mathbb{R}^d)$ for $2 \le p \le 2d/(d - 2s)$ if $s < d/2$, for $2 \le p < \infty$ if $s = d/2$ and for $2 \le p \le \infty$ if $s > d/2$ (and also to smoother function spaces in this latter case). In the subcritical case, this simply follows from elementary properties of convolution, whereas in the case of $p = 2d/(d - 2s)$, it is the Hardy-Littlewood-Sobolev inequality (5.31) that is equivalent to the Sobolev inequality [LL01].

Lemma A.16 *The function* $h_s = (2\pi)^{d/2}\mathcal{F}^{-1}\{(1 + |\xi|^2)^{-s/2}\}$ *is positive, radial, decreasing, belongs to* $C^\infty(\mathbb{R}^d \setminus \{0\})$ *and satisfies*

$$h_s(x) \underset{|x|\to 0}{\sim} C_{d,s} \begin{cases} \dfrac{1}{|x|^{d-s}} & \text{for } s < d, \\[2mm] -\log|x| & \text{for } s = d, \\[2mm] 1 & \text{for } s > d. \end{cases}$$

Moreover, it decreases exponentially at infinity: $h_s(x) \le e^{-\alpha|x|}$ *for* $|x| \ge 1$ *and* $\alpha > 0$.

Proof Using the integral formula

$$\frac{1}{a^s} = \Gamma(s/2)^{-1} \int_0^\infty e^{-ta} t^{s-1}\,dt,$$

and the Fourier transform of Gaussians, we can express h_s in the following form

$$h_s(x) = \frac{2^{-d/2}}{\Gamma(s/2)} \frac{1}{|x|^{d-s}} \int_0^\infty e^{-t|x|^2 - \frac{1}{4t}} \frac{dt}{t^{1+\frac{d-s}{2}}}. \tag{A.29}$$

This formula shows that h_s is positive, radial, decreasing, belongs to $C^\infty(\mathbb{R}^d \setminus \{0\})$ and that $h_s(x) \sim C|x|^{s-d}$ when $x \to 0$, for $s < d$. For $s = d$ we can split the integral in two, for example by writing $[0, \infty) = [0, 1] \cup (1, \infty)$ and we find the behavior in $-\log|x|$ for the integral over $[1, \infty)$ by integrating by parts. If $s > d$ then $\xi \mapsto (1 + |\xi|^2)^{-s/2}$ belongs to $L^1(\mathbb{R}^d)$ and is strictly positive, which immediately implies that h_s is continuous and bounded, with $h_s(0) > 0$. For $s < d$, using that $e^{-\frac{1}{4t}} \leq 1$ we get $h_s(x) \leq C(d, s)|x|^{s-d}$ and therefore h_s is also bounded at infinity. Similarly, we can prove that $h_s(x) \leq |\log x| + C$ if $s = d$. To find a better estimate at infinity we can set $t = u/|x|$ and we obtain

$$h_s(x) = \frac{2^{-d/2}c(s)}{|x|^{\frac{d-s}{2}}} \int_0^\infty e^{-\left(u + \frac{1}{4u}\right)|x|} \frac{du}{u^{1 + \frac{d-s}{2}}}.$$

Since $u + 1/(4u) \geq 2\sqrt{u/4u} = \sqrt{2}$, we get for example

$$h_s(x) \leq 2^{\frac{s-d}{2}} e^{-|x|/\sqrt{2}} h_s(x/2)$$

which shows the exponential decay.   □

We deduce from the lemma that

$$h_s \in \begin{cases} L^p(\mathbb{R}^d), & 1 \leq p < \frac{d}{d-s}, & \text{for } s < d, \\[2mm] L^p(\mathbb{R}^d), & 1 \leq p < \infty, & \text{for } s = d, \\[2mm] L^1(\mathbb{R}^d) \cap L^\infty(\mathbb{R}^d), & & \text{for } s > d. \end{cases} \tag{A.30}$$

It then remains to use Young's inequality

$$\|f * g\|_{L^r(\mathbb{R}^d)} \leq \|f\|_{L^p(\mathbb{R}^d)} \|g\|_{L^q(\mathbb{R}^d)}, \qquad \frac{1}{p} + \frac{1}{q} = 1 + \frac{1}{r}, \tag{A.31}$$

to conclude the proof of inequality (A.27), hence of Theorem A.15, in all the subcritical cases. When $s > d/2$ we have $h_s, g \in L^2(\mathbb{R}^d)$ so that $f = h_s * g$ is in fact continuous and tends to 0 at infinity. When $s = d/2 + \ell + \theta$ with $\ell \geq 1$, we start by noticing that $f \in H^s(\mathbb{R}^d)$ if and only if the derivatives of order $\leq \ell$ of f are all in $H^{d/2+\theta}(\mathbb{R}^d)$. Applying the previous argument shows that these derivatives are bounded and continuous, so that f is in fact C^ℓ. We do not discuss the Hölder character here.

If Ω is a bounded open set, we can use Theorem A.12 that allows to extend any function from $H^s(\Omega)$ to $\mathbb{R}^d$ and we immediately deduce a result similar to Theorem A.15.

Theorem A.17 (Sobolev Embedding on Ω) *Let $\Omega \subset \mathbb{R}^d$ be a bounded open set whose boundary is Lipschitz and k an integer.*

- If $2k < d$, we have the continuous embedding

$$H^k(\Omega) \hookrightarrow L^{\frac{2d}{d-2k}}(\Omega).$$

- If $2k = d$, we have

$$H^k(\Omega) \hookrightarrow L^p(\Omega), \qquad \forall 2 \le p < \infty.$$

- If $2k > d$, we have

$$H^k(\Omega) \hookrightarrow C_b^{\ell,\theta}(\overline{\Omega})$$

where ℓ is the unique integer such that $0 \le \ell < k - d/2 < \ell + 1$ and $\theta = k - \ell - d/2$.

In the case of a bounded open set, the embedding into $L^2(\Omega)$ is actually compact, which is important information for passing to the strong limit locally when we have a sequence that converges weakly in a Sobolev space.

Theorem A.18 (Rellich-Kondrachov) *Let Ω be a bounded open set. The embedding $H^s(\mathbb{R}^d) \hookrightarrow L^2(\Omega)$ is compact for all $s > 0$.*

By Hölder's inequality, we immediately deduce that the embedding $H^s(\mathbb{R}^d) \hookrightarrow L^p(\Omega)$ is compact for all $2 \le p < 2d/(d-2s)$ when $0 < s < d/2$ and for $2 \le p < \infty$ in the other cases. Similarly, we have the compact embedding $H^k(\Omega) \hookrightarrow L^p(\Omega)$ when Ω is bounded and k is an integer. We now give the proof of Theorem A.18.

Proof Let Ω be a bounded open set and (f_n) a sequence of functions from $H^s(\mathbb{R}^d)$, which converges weakly to $f \in H^s(\mathbb{R}^d)$, $f_n \rightharpoonup f$. We must show that $\mathbb{1}_\Omega f_n \to \mathbb{1}_\Omega f$ strongly in $L^2(\Omega)$. Following the discussion after Theorem A.15, we can write $f_n = h_s * g_n$ where (g_n) is a bounded sequence in $L^2(\mathbb{R}^d)$. In fact, we have $g_n \rightharpoonup g$ weakly in $L^2(\mathbb{R}^d)$ where $\widehat{g}(\xi) = (1 + |\xi|^2)^{s/2} \widehat{f}(\xi)$. We must therefore finally show that if $g_n \rightharpoonup g$ weakly in $L^2(\mathbb{R}^d)$, then $\mathbb{1}_\Omega(h_s * g_n) \to \mathbb{1}_\Omega(h_s * g)$ strongly in $L^2(\Omega)$. Another way to state this result is to say that the operator $g \mapsto \mathbb{1}_\Omega(h_s * g)$ is compact on $L^2(\mathbb{R}^d)$. The result is much more general and follows immediately from the following lemma since $h_s \in L^1(\mathbb{R}^d)$ according to Lemma A.16. $\quad\square$

Lemma A.19 *Let $F \in L^\infty(\mathbb{R}^d)$ with $F \to 0$ at infinity, and $G \in L^1(\mathbb{R}^d)$. Then the operator $K_{F,G} : g \mapsto F(G * g)$ is compact on $L^2(\mathbb{R}^d)$.*

Proof (of the Lemma) By Young's inequality (A.31) we have $\|G * g\|_{L^2} \le \|G\|_{L^1} \|g\|_{L^2}$. Thus $\|K_{F,G}g\|_{L^2} \le \|F\|_{L^\infty} \|G\|_{L^1} \|g\|_{L^2}$, which means that $K_{F,G}$ is controlled in operator norm by $\|K_{F,G}\| \le \|F\|_{L^\infty} \|G\|_{L^1}$. If we have sequences $F_n \to F$ in L^∞ and $G_n \to G$ in L^1, we deduce by the same type of inequality that $\|K_{F_n,G_n} - K_{F,G}\| \to 0$ in operator norm. But we know that a norm limit of a sequence of compact operators is always compact. We therefore deduce that it is sufficient to prove the lemma for $G \in C_c^\infty(\mathbb{R}^d)$ and $F \in L_c^\infty(\mathbb{R}^d)$ (any function in

L^∞ that tends to 0 at infinity can be approached by a sequence of bounded functions with compact support). Now, we have

$$\int_{\mathbb{R}^d} G(x - y)g_n(y)\, dy \rightarrow \int_{\mathbb{R}^d} G(x - y)g(y)\, dy$$

for almost every x, by the definition of weak convergence $g_n \rightharpoonup g$, under the only condition that $G \in L^2$. Moreover, $\|G * g_n\|_{L^\infty} \leq \|G\|_{L^2} \|g_n\|_{L^2}$ by the Cauchy-Schwarz inequality and thus $|(G * g_n)F| \leq C|F|$. This domination is in $L^2(\mathbb{R}^d)$ because $F \in L^\infty_c(\mathbb{R}^d)$. By the dominated convergence theorem, we deduce that $(G * g_n)F \rightarrow (G * g)F$ strongly in $L^2(\mathbb{R}^d)$.   □

For a more general result of the same type as Lemma A.19, see Theorem 5.24.

A.6 Elliptic Regularity on a Bounded Open Set*

We have already mentioned the elliptic regularity theorem on $\Omega = I = (0, 1)$ and on $\Omega = \mathbb{R}^d$, which states that if $f \in L^2(\Omega)$ and $\Delta f \in L^2(\Omega)$ then $f \in H^2(\Omega)$. This result plays an important role when we seek to identify the domain of the adjoint of the Laplacian or its perturbations. We want to briefly discuss here the elliptic regularity theorem on any domain Ω in dimension $d \geq 2$, which is much more difficult and often eluded in works on the subject. In fact, on a domain Ω not equal to $\mathbb{R}^d$ in dimension $d \geq 2$ it is **false** that if f and Δf are both in $L^2(\Omega)$ then necessarily $f \in H^2(\Omega)$. For example, we recall that the function $f(x) = |x - x_0|^{-1}$ satisfies $-\Delta f = 4\pi \delta_{x_0}$ in dimension $d = 3$. If $\Omega \neq \mathbb{R}^3$ we can then place x_0 on the boundary $\partial\Omega$, so that we find $\Delta f = 0$ in the sense of distributions in Ω (this notion uses the functions $C_c^\infty(\Omega)$ which vanish at the boundary and do not see the singular measure δ_{x_0}). Thus, $\Delta f \in L^2(\Omega)$. Moreover,

$$\int_\Omega |f(x)|^2\, dx = \int_\Omega \frac{dx}{|x - x_0|^2} < \infty$$

if Ω is bounded, because the function $1/|x|$ is square integrable in the vicinity of the origin in dimension $d = 3$. But

$$\nabla f(x) = -\frac{x - x_0}{|x - x_0|^3}$$

is not square integrable in the vicinity of x_0. It is then sufficient that Ω contains a large enough set $A \subset \Omega$, such that $x_0 \in \overline{A}$, with the property that

$$\int_A |\nabla f(x)|^2\, dx = \int_A \frac{dx}{|x - x_0|^4} = +\infty$$

and we deduce that we do not even have $f \in H^1(\Omega)$. The integral diverges if A is for example a cone with tip x_0 and positive angle and it is therefore sufficient to be able to place such a cone at least one point of the boundary (which is feasible for all regular open sets) to deduce that the elliptic regularity theorem cannot be true in an open set in dimension $d = 3$. This example (generalizable by the same argument in any dimension $d \geq 2$) shows that it is imperative to have information on the value of f at the boundary, in order to deduce that $f \in H^2(\Omega)$. This greatly complicates the study, since it is therefore necessary to start by giving a meaning to the restriction of f to the boundary, with the only information that f and Δf are in $L^2(\Omega)$.

It turns out that every $f \in L^2(\Omega)$ such that $\Delta f \in L^2(\Omega)$ has well-defined restrictions to the boundary, but these are distributions on $\partial\Omega$. For example, if Ω has a Lipschitz boundary, then $f_{|\partial\Omega}$ belongs to the space $H^{-1/2}(\partial\Omega)$ which is by definition the dual of $H^{1/2}(\partial\Omega)$ (the image of the restriction map to the boundary). Similarly, $\partial_n f_{|\partial\Omega}$ belongs to the space $H^{-3/2}(\partial\Omega)$ when Ω is $C^{1,1}$. These assertions can be demonstrated based on Green's formula which, therefore, makes sense in these spaces [LM68]:

$$\int_\Omega g\,(-\Delta\overline{f}) - \int_\Omega \overline{f}(-\Delta g) = {}_{H^{-\frac{1}{2}}(\partial\Omega)}\big\langle f, \partial_n g\big\rangle_{H^{\frac{1}{2}}(\partial\Omega)} - {}_{H^{-\frac{3}{2}}(\partial\Omega)}\big\langle \partial_n f, g\big\rangle_{H^{\frac{3}{2}}(\partial\Omega)},$$
(A.32)

for all $g \in H^2(\Omega)$ and $f \in L^2(\Omega)$ such that $\Delta f \in L^2(\Omega)$. When the boundary of Ω is only piecewise $C^{1,1}$, the traces $f_{|\partial\Omega}$ and $\partial_n f_{|\partial\Omega}$ are well defined in $H^{-1/2}$ and $H^{-3/2}$ on each of the hypersurfaces in question [LM68, Gri85].

The following theorem provides the desired elliptic regularity for the Robin boundary condition, which requires that the normal derivative $\partial_n f_{|\partial\Omega}$ at the boundary be proportional to the restriction at the boundary $f_{|\partial\Omega}$.

Theorem A.20 (Elliptic Regularity) *Let Ω be a bounded open set whose boundary is piecewise C^2, such that Ω is convex in a neighborhood of the singularities of its boundary. Also let $0 \leq \theta < 1$. Then, there exists a constant $C(\Omega, \theta)$ (depending only on Ω and θ) such that for any function $f \in H^2(\Omega)$ satisfying*

$$\cos(\pi\theta)\, f_{|\partial\Omega} + \sin(\pi\theta)\, \partial_n f_{|\partial\Omega} = 0, \qquad (A.33)$$

almost everywhere on $\partial\Omega$, we have the estimate

$$\|f\|_{H^2(\Omega)} \leq C(\Omega, \theta)\left(\|f\|_{L^2(\Omega)} + \|\Delta f\|_{L^2(\Omega)}\right). \qquad (A.34)$$

Furthermore, if $f \in L^2(\Omega)$ is such that $\Delta f \in L^2(\Omega)$ and satisfies the Robin condition (A.33) in $H^{-3/2}$ on each of the hypersurfaces composing $\partial\Omega$, then $f \in H^2(\Omega)$ and satisfies the inequality (A.34).

This theorem specifies that the information that $f, \Delta f \in L^2(\Omega)$, combined with the Robin condition in the sense of distributions, is sufficient to ensure that $f \in H^2(\Omega)$, that is, to control all the other derivatives $\partial_{x_i} f$ and $\partial_{x_i}\partial_{x_j} f$ with $i \neq j$. The

Robin condition includes that of Dirichlet ($\theta = 0$) and Neumann ($\theta = 1/2$). There are many other boundary conditions than that of Robin for which the theorem is true, but we do not discuss them here and refer the interested reader to [LM68].

The proof of Theorem A.20 is long and technical. It is generally divided into two steps, the first being to obtain estimates on the derivatives inside Ω (the boundary condition then plays no role), and the second, more difficult, dedicated to regularity at the boundary. In most works on the subject, the second part of the theorem is solely stated for the Dirichlet condition, with the additional assumption that $f \in H_0^1(\Omega)$ (see for example [Bre11, Thm. 9.25]). The complete proof of Theorem A.20 can be found in [LM68].

Let us emphasize the new assumption that Ω is strictly convex in the neighborhood of the singularities of its boundary. Without this assumption, elliptic regularity may be false. This is for example the case of a polygon including a re-entrant angle in dimension $d = 2$ [Gri85, Dau88].

Appendix B
Problems

B.1 Fractional Hardy Inequality, Pseudo-Relativistic Hydrogen Atom

Part 1: Fractional Hardy Inequality

We have proved Hardy's inequality

$$\frac{1}{|x|^2} \leq \left(\frac{2}{d-2}\right)^2 (-\Delta) \tag{B.1}$$

in Exercise 1.27, which is to be interpreted in the sense of quadratic forms as in Definition 5.3. The constant $4/(d-2)^2$ is the smallest possible. We study here the more general inequality

$$\frac{1}{|x|^s} \leq C_{d,s}(-\Delta)^{\frac{s}{2}} \tag{B.2}$$

for all $0 < s < d$ in any dimension $d \geq 1$, which can also be written

$$\int_{\mathbb{R}^d} \frac{|u(x)|^2}{|x|^s} \, dx \leq C_{d,s} \left\| (-\Delta)^{\frac{s}{4}} u \right\|_{L^2(\mathbb{R}^d)}^2 = C_{d,s} \int_{\mathbb{R}^d} |p|^s |\widehat{u}(p)|^2 \, dp$$

for all $u \in H^s(\mathbb{R}^d)$. Inequality (B.2) appears in various forms in the literature. For $s = 4$ in dimensions $d \geq 5$ it is called **Rellich's inequality** [RB69] while for $s = 1$ in dimension $d = 3$, it is the **Kato** [Kat66, Eq. (5.33)] or **Kato-Herbst inequality** [Her77]. The best constant $C_{d,s}$ was found for all $0 < s < d$ in [Her77, Bec95, Yaf99]. We start by providing a fairly simple proof with a too large constant, before finding the optimal constant.

M. Lewin, *Spectral Theory and Quantum Mechanics*, Universitext, https://doi.org/10.1007/978-3-031-66878-4

1. Let $(A, D(A))$ and $(B, D(B))$ be two non-negative self-adjoint operators on a separable Hilbert space $\mathfrak{H}$, with form domains $Q(A) = D(A^{1/2})$ and $Q(B) = D(B^{1/2})$. We assume that $0 \leq A \leq B$, which means that $Q(B) \subset Q(A)$ and $q_A \leq q_B$ on $Q(B)$ (Definition 5.3).

   a. Show the formula

$$x^s = C_1(s) \int_0^\infty \frac{x \, dt}{(t+x)t^{1-s}}$$

   for all $x > 0$ and all $0 < s < 1$, with a constant $C_1(s) > 0$ to be determined.
   b. Deduce that $A^s \leq B^s$ for all $0 \leq s \leq 1$.
   c. Show the inequality

$$\int_{\mathbb{R}^d} \frac{|u(x)|^2}{|x|^s} \, dx \leq \left(\frac{2}{d-2}\right)^s \left\|(-\Delta)^{\frac{s}{4}} u\right\|_{L^2(\mathbb{R}^d)}^2 \tag{B.3}$$

   for all $u \in H^s(\mathbb{R}^d)$ and all $0 \leq s \leq 2$, in dimension $d \geq 3$.

The previous proof only works in dimension $d \geq 3$ for $s \leq 2$ and it does not provide the best constant. We now turn to a proof that comes from [HLP88, Chap. IX] and is more or less due to Hilbert. Similar arguments have been used more recently in [FLS07, Fra09, LS10b].

2. Let $d \geq 1$ and $0 < s < d$. We introduce the function

$$I_{d,s}(a) = \int_{\mathbb{R}^d} \frac{dq}{|q|^a |q - \omega|^{d-s}} \tag{B.4}$$

for some ω in the unit sphere $\mathbb{S}^{d-1}$, with $s < a < d$.

   a. Verify that $a \mapsto I_{d,s}(a)$ is well defined and C^∞ on the interval (s, d), that it diverges at s and d, and that it does not depend on the chosen vector $\omega \in \mathbb{S}^{d-1}$.
   b. By switching to radial coordinates $y = r\omega'$ and setting $u = 1/r$, show that $I_{d,s}(a) = I_{d,s}(d + s - a)$. In other words, $a \mapsto I_{d,s}(a)$ is even with respect to the center $(d+s)/2$ of its interval of definition (s, d).
   c. Show that $a \mapsto I_{d,s}(a)$ is a convex function, and therefore reaches its minimum at $a = \frac{s+d}{2}$. We call

$$I_{d,s} = \min_{s<a<d} I_{d,s}(a) = \int_{\mathbb{R}^d} \frac{dq}{|q|^{\frac{s+d}{2}} |q - \omega|^{d-s}}$$

   its minimum value.

3. Using the relation

$$\frac{1}{r^s} = C_2(s) \int_0^\infty e^{-tr^2} t^{\frac{s}{2}-1} \, dt$$

for a certain constant $C_2(s)$, show that the Fourier transform of $1/|x|^s$ is

$$\widehat{\frac{1}{|x|^s}}(k) = \frac{2^{\frac{d}{2}-s}\Gamma(\frac{d-s}{2})}{\Gamma(\frac{s}{2})} \frac{1}{|k|^{d-s}}.$$

Deduce that

$$\int_{\mathbb{R}^d} \frac{|u(x)|^2}{|x|^s} dx = \frac{\Gamma(\frac{d-s}{2})}{2^s \pi^{\frac{d}{2}} \Gamma(\frac{s}{2})} \int\int_{\mathbb{R}^d \times \mathbb{R}^d} \frac{\overline{\widehat{u}(p)}\widehat{u}(q)}{|p-q|^{d-s}} dp\, dq$$

for all $u \in C_c^\infty(\mathbb{R}^d)$.
4. By writing

$$\Re\left(\overline{\widehat{u}(p)}\widehat{u}(q)\right) = \frac{|p|^a|\widehat{u}(p)|^2}{2|q|^a} + \frac{|p|^a|\widehat{u}(p)|^2}{2|q|^a} - \frac{\left||p|^a\widehat{u}(p) - |q|^a\widehat{u}(q)\right|^2}{2|p|^a|q|^a},$$

show the inequality (B.2) with the constant

$$C_{d,s} = \frac{\Gamma(\frac{d-s}{2})}{2^s \pi^{\frac{d}{2}} \Gamma(\frac{s}{2})} I_{d,s} = 2^{-s} \frac{\Gamma(\frac{d-s}{4})^2}{\Gamma(\frac{d+s}{4})^2}.$$

5. Using the sequence $\widehat{u}_n(p) = |p|^{-\frac{s+d}{2}} \mathbb{1}(1 \le |p| \le n)$, prove that $C_{d,s}$ is optimal.
6. Verify that we recover the standard Hardy inequality (B.1) for $s = d - 2$ in dimension $d \ge 3$. Show then that $C_{3,1} = \pi/2$, which provides the **Kato-Herbst inequality**

$$\int_{\mathbb{R}^3} \frac{|u(x)|^2}{|x|} dx \le \frac{\pi}{2} \left\|(-\Delta)^{\frac{1}{4}} u\right\|_{L^2(\mathbb{R}^3)}^2 \tag{B.5}$$

for all $u \in H^{1/2}(\mathbb{R}^3)$, where $\pi/2$ is the best possible constant.

Part 2: The Pseudo-Relativistic Hydrogen Atom

In this part, we are in dimension $d = 3$. Following [Her77], we want to define and study the **pseudo-relativistic hydrogen atom** described by the operator

$$H_c = \sqrt{-c^2\Delta + c^4} - c^2 - \frac{1}{|x|}$$

where c is the speed of light. We are working here in the system of atomic units $m = \hbar = e^2/(4\pi\varepsilon_0) = 1$ in which the physical value of c is $c \simeq 137$. We call it a "pseudo"-relativistic model because the operator $\sqrt{-c^2\Delta + c^4}$ is non-local, which contradicts the principles of special relativity. A better model involves the **Dirac operator** [Tha92] instead of $\sqrt{-c^2\Delta + c^4}$.

7. In classical relativistic mechanics, the energy of a particle with momentum $p \in \mathbb{R}^3$ and mass $m > 0$ is given by the formula

$$E_{m,c}^{\text{kin}}(p) = \sqrt{c^2|p|^2 + m^2c^4} - mc^2.$$

a. We define the velocity by $v = \nabla_p E_{m,c}^{\text{kin}}(p)$. Verify that $|v| < c$.

b. Show that $E_{m,c}^{\text{kin}}(p) = \frac{|p|^2}{2m} + O(c^{-2})_{c\to+\infty}$ for any fixed $p \in \mathbb{R}^3$.

In the following, we take again $m = 1$.

8. For $c > 0$, we introduce the operator $T_c = \sqrt{-c^2\Delta + c^4} - c^2$ on the domain $D(T_c) = H^1(\mathbb{R}^3)$, which is the multiplication operator by the function $k \mapsto \sqrt{c^2|k|^2 + c^4} - c^2$ in Fourier.

a. Show that T_c is self-adjoint. What is its spectrum? What is its form domain $Q(T_c)$?

b. Prove that for all $\psi \in H^1(\mathbb{R}^3)$

$$\langle T_c\psi, \psi \rangle \le \frac{1}{2}\int_{\mathbb{R}^3} |\nabla\psi(x)|^2 dx, \quad \lim_{c\to\infty} \langle T_c\psi, \psi \rangle = \frac{1}{2}\int_{\mathbb{R}^3} |\nabla\psi(x)|^2 dx. \tag{B.6}$$

9. We now consider the operator $H_c = T_c - \frac{1}{|x|}$. Using Hardy's inequality (1.65), show that H_c is well defined on the domain $D(H_c) = H^1(\mathbb{R}^3)$ for all $c > 0$ and that it is self-adjoint when $c > 2$.

10. We consider the associated quadratic form

$$\mathcal{E}_c(\psi) = \left\| (-c^2\Delta + c^4)^{\frac{1}{4}}\psi \right\|^2 - c^2 \int_{\mathbb{R}^3} |\psi(x)|^2 \, dx - \int_{\mathbb{R}^3} \frac{|\psi(x)|^2}{|x|} \, dx.$$

a. Using the Kato-Herbst inequality (B.5), show that $\mathcal{E}_c$ is well defined and continuous on $H^{1/2}(\mathbb{R}^3)$ for all $c > 0$.

b. Show that

$$\inf_{\substack{\psi \in H^{1/2}(\mathbb{R}^3), \\ \int_{\mathbb{R}^3} |\psi|^2 = 1}} \mathcal{E}_c(\psi) \begin{cases} \ge -c^2 & \text{for } c \ge \pi/2, \\ = -\infty & \text{for } c < \pi/2. \end{cases}$$

c. How can we define the operator $T_c - 1/|x|$ when $\pi/2 < c \leq 2$?

11. We now derive some properties of the spectrum of $H_c = T_c - 1/|x|$, assuming for simplicity that $c > 2$ (which is the physical case). We therefore take $D(H_c) = H^1(\mathbb{R}^d)$.

a. Show that $\mathbb{R}_+ \subset \sigma_{ess}(H_c)$.
b. Let $\lambda \in \sigma_{ess}(H_c)$ and consider a sequence $(\psi_n) \subset H^1(\mathbb{R}^3)$ such that $\|\psi_n\|_{L^2(\mathbb{R}^3)} = 1$, $\psi_n \rightharpoonup 0$ weakly in $L^2(\mathbb{R}^3)$ and $(H_c - \lambda)\psi_n \to 0$ strongly in $L^2(\mathbb{R}^d)$. Show that (ψ_n) is bounded in $H^1(\mathbb{R}^3)$ and deduce that

$$\lim_{n \to \infty} \int_{\mathbb{R}^3} \frac{|\psi_n(x)|^2}{|x|} \, dx = 0.$$

Conclude that $\lambda \geq 0$, so that $\sigma_{ess}(H_c) = \mathbb{R}_+$.
c. Show that $T_c - 1/|x|$ has infinitely many strictly negative eigenvalues that accumulate at 0.

12. We now study the non-relativistic limit $c \to \infty$ of the first eigenvalue λ_1^c of the operator $T_c - 1/|x|$ (similar arguments can be used to deal with the next eigenvalues). We call $\lambda_1 = -1/2$ the first eigenvalue of $-\Delta/2 - 1/|x|$, which we recall is non-degenerate, with associated eigenfunction $\psi_0 = \pi^{-1/2}e^{-|x|}$.

a. Show that $\lambda_1^c \leq \lambda_1$ for all $c > 2$.
b. Let $\kappa > 0$ be fixed. Show that for large c we have

$$\sqrt{c^4 + c^2|p|^2} - c^2 \geq \kappa|p| - C_1(\kappa),$$

$$\left(\sqrt{c^4 + c^2|p|^2} - c^2\right)^2 \geq \kappa^2|p|^2 - C_2(\kappa)$$

for all $p \in \mathbb{R}^3$ with constants $C_j(\kappa)$ independent of c.
c. Deduce that λ_1^c is bounded in the limit $c \to \infty$.
d. Let (c_n) be a sequence such that $2 < c_n \to \infty$ and $(\psi_n) \subset H^1(\mathbb{R}^3)$ such that $\|\psi_n\| = 1$ and

$$(T_{c_n} - 1/|x|)\psi_n = \lambda_1^{c_n}\psi_n. \tag{B.7}$$

Show that (ψ_n) is bounded in $H^1(\mathbb{R}^3)$. Extracting a subsequence if necessary, we can suppose that $\psi_n \rightharpoonup \psi$ weakly in $H^1(\mathbb{R}^3)$ and strongly in $L^2_{loc}(\mathbb{R}^3)$, and that $\lambda_1^{c_n} \to \lambda_1'$. Show then that

$$\lim_{n \to \infty} \int_{\mathbb{R}^3} \frac{|\psi_n(x)|^2}{|x|} \, dx = \int_{\mathbb{R}^3} \frac{|\psi(x)|^2}{|x|} \, dx.$$

e. Prove that

$$\liminf_{n\to\infty} \mathcal{E}_{c_n}(\psi_n) \geq \frac{1}{2} \int_{\mathbb{R}^3} |\nabla \psi(x)|^2 \, dx - \int_{\mathbb{R}^3} \frac{|\psi(x)|^2}{|x|} \, dx.$$

Deduce that $\lambda'_1 = \lambda_1$, that $\int_{\mathbb{R}^3} |\psi|^2 = 1$ and that $\psi(x) = e^{i\theta} \psi_0 \in H^2(\mathbb{R}^3)$. Conclude that $\psi_n \to e^{i\theta} \psi_0$ strongly in $H^1(\mathbb{R}^3)$.

Remark B.1 We have seen here that the pseudo-relativistic hydrogen atom does not make sense for $c < \pi/2$, which is not a problem as the physical value is $c \simeq 137$. However, in the same model with a nucleus comprising Z protons, the external potential $1/|x|$ is replaced by $Z/|x|$, so that this theory is unstable for $Z > (2/\pi)c \simeq 87$; it is unable to describe the elements of the periodic table (with one electron) beyond $Z = 87$. A better relativistic model is based on the Dirac operator [Tha92] since the latter has the more physical constraint $Z \leq 137$ [ELS19, Sec. 1]. Its generalization to atoms containing several electrons is still very poorly understood, however.

B.2   The Radial Laplacian

In this problem we study the restriction of the Laplacian on $L^2(\mathbb{R}^d)$ to the subspace of radial functions, an operator that we then identify with a differential operator on $L^2(\mathbb{R}_+)$.

Part 1: Restriction of a Self-adjoint Operator

Let $(A, D(A))$ be a self-adjoint operator on a separable Hilbert space $\mathfrak{H}$. We recall that $\rho(A) = \mathbb{C} \setminus \sigma(A)$ is its resolvent set.

1. Let $\mathcal{V}$ be a closed subspace of $\mathfrak{H}$, such that $(A-z)^{-1}\mathcal{V} \subset \mathcal{V}$ for one $z \in \rho(A) \subset \mathbb{C}$. Show that $(A-z')^{-1}\mathcal{V} \subset \mathcal{V}$ for all z' belonging to the open disk of center z and radius $r = \|(A-z)^{-1}\|^{-1}$ in the complex plane.
2. Suppose there exists $z \in \rho(A)$ such that $(A-z)^{-1}\mathcal{V} \subset \mathcal{V}$ and $(A-\bar{z})^{-1}\mathcal{V} \subset \mathcal{V}$. Show then that $(A-z')^{-1}\mathcal{V} \subset \mathcal{V}$ for all $z' \in \rho(A)$.
3. Deduce the equivalence of the propositions

   (*i*) $(A-z)^{-1}\mathcal{V} \subset \mathcal{V}$ and $(A-\bar{z})^{-1}\mathcal{V} \subset \mathcal{V}$ for one $z \in \rho(A)$;
   (*ii*) $(A-z)^{-1}\mathcal{V} \subset \mathcal{V}$ for all $z \in \rho(A)$;
   (*iii*) $f(A)\mathcal{V} \subset \mathcal{V}$ for every bounded continuous function f on $\mathbb{R}$.

   We say that $\mathcal{V}$ is an invariant subspace (Exercise 4.45).
4. Let $\mathcal{V}$ be a closed subspace of $\mathfrak{H}$ satisfying one of the equivalent conditions of the previous question and such that in addition $D(A) \cap \mathcal{V}$ is dense in $\mathcal{V}$. On

the Hilbert space $\mathcal{V}$, equipped with the scalar product of $\mathfrak{H}$, we set $D(B) = D(A) \cap \mathcal{V}$ and $Bf = Af$. Show that B takes values in $\mathcal{V}$ and that $(B, D(B))$ is a self-adjoint operator on $\mathcal{V}$, with $\sigma(B) \subset \sigma(A)$.

Part 2: Radial Laplacian

In this section we work in dimension $d \geq 2$. We recall that a measurable function $f : \mathbb{R}^d \to \mathbb{R}$ is *radial* when for every orthogonal matrix $U \in SO(d)$ we have $f(Ux) = f(x)$ for almost every x. This implies $f(x) = f(|x|e_1)$ where e_1 is the first vector of the canonical basis of $\mathbb{R}^d$, that is, f depends only on the Euclidean norm $|x|$ of x. In the following we denote by $L_r^2(\mathbb{R}^d)$ the subspace of radial functions in $L^2(\mathbb{R}^d)$ and

$$H_r^s(\mathbb{R}^d) = H^s(\mathbb{R}^d) \cap L_r^2(\mathbb{R}^d) = \left\{ f \in L_r^2(\mathbb{R}^d) \ : \ \int_{\mathbb{R}^d} (1 + |k|^2)^s |\widehat{f}(k)|^2 \, dk < \infty \right\}$$

the corresponding Sobolev spaces, for $s \geq 0$. If $s = 0$, we simply have $H_r^0(\mathbb{R}^d) = L_r^2(\mathbb{R}^d)$. Similarly, we call $C_{c,r}^\infty(\mathbb{R}^d)$ the space of radial C^∞ functions with compact support.

5. Show that $H_r^s(\mathbb{R}^d)$ is a closed subspace of $H^s(\mathbb{R}^d)$, for all $s \geq 0$.
6. Show that $C_{c,r}^\infty(\mathbb{R}^d)$ is dense in $H_r^s(\mathbb{R}^d)$, for all $s \geq 0$.
7. For $U \in SO(d)$ and $f \in L^1(\mathbb{R}^d) \cap L^2(\mathbb{R}^d)$, calculate the Fourier transform of $x \mapsto f(Ux)$. Deduce that $\widehat{f} \in L_r^2(\mathbb{R}^d)$ if and only if $f \in L_r^2(\mathbb{R}^d)$.
8. Show that the restriction $-\Delta_r$ of the operator $-\Delta$ to $D(-\Delta_r) = H_r^2(\mathbb{R}^d)$ defines a self-adjoint operator on the Hilbert space $\mathcal{V} = L_r^2(\mathbb{R}^d)$.
9. We introduce the minimal operator $-\Delta_{r,\min}$ defined on $D(-\Delta_{r,\min}) = C_{c,r}^\infty(\mathbb{R}^d) \subset L_r^2(\mathbb{R}^d)$. Show that its closure in $L_r^2(\mathbb{R}^d)$ is $-\Delta_r$, that is, the latter is essentially self-adjoint on $C_{c,r}^\infty(\mathbb{R}^d)$.
10. Show that $\sigma(-\Delta_r) = \mathbb{R}_+$.

Part 3: 3D Radial Laplacian as an Operator on the Half Line

We now consider the space $L^2(I)$ with $I = (0, +\infty)$ (equipped with the Lebesgue measure). We recall that the functions $u \in H^1(I)$ all have a continuous representative that has a limit at 0^+ and tends to 0 at infinity (Exercise A.5). We finally recall that if u and u'' (understood in the sense of distributions on $I = (0, +\infty)$) are both in $L^2(I)$ then we automatically have $u' \in L^2(I)$, that is, $u \in H^2(I)$.

We introduce the Laplacian operator on the half-line with Dirichlet boundary condition

$$Lu := -u'', \qquad D(L) = \left\{ u \in H^2(I) \ : \ u(0^+) = 0 \right\}.$$

11. Show that L is self-adjoint on its domain $D(L)$.

We now study the operator $-\Delta_r$ from Part B, but only in dimension $d = 3$. We introduce the operator $\mathscr{U} : L_r^2(\mathbb{R}^3) \to L^2(I)$ defined by $(\mathscr{U} f)(r) = \sqrt{4\pi} \, r \, f(re_1)$.

12. Show that $\mathscr{U}$ is an isomorphism of Hilbert spaces.

13. Let $f \in C_{c,r}^\infty(\mathbb{R}^3)$ and $u = \mathscr{U} f \in L^2(I)$. Show first that $\nabla f(0) = 0$, then that $u \in C^\infty([0, +\infty))$ and finally that $u(0^+) = u''(0^+) = 0$. Prove the relation

$$\Delta f(x) = \frac{u''(|x|)}{|x|\sqrt{4\pi}} \tag{B.8}$$

for all $x \in \mathbb{R}^d \setminus \{0\}$ and deduce that

$$\int_0^\infty |u''(r)|^2 \, dr = \int_{\mathbb{R}^3} |\Delta f(x)|^2 \, dx,$$

$$\int_0^\infty \overline{u(r)} u''(r) \, dr = \int_{\mathbb{R}^3} \overline{f(x)} \, \Delta f(x) \, dx. \tag{B.9}$$

14. Show that if $f \in H_r^2(\mathbb{R}^3)$ then $\mathscr{U} f \in D(L)$ and that the formulas (B.8) and (B.9) remain true.

15. Show that $\mathscr{U} H_r^2(\mathbb{R}^3) = D(L)$ and $\mathscr{U}(-\Delta_r)\mathscr{U}^{-1} = L$.

16. Let V be a radial function in the form $V(x) = v(|x|)$, with $v \in C_0^0(\mathbb{R}_+)$ (continuous and tending to 0 at infinity).

a. Show that the operator $-\Delta + V$ is self-adjoint on $H^2(\mathbb{R}^3)$, that its spectrum is bounded below and that its essential spectrum is $\sigma_{\text{ess}}(-\Delta + V) = \mathbb{R}_+$.

b. Assuming that $-\Delta + V$ has a smallest eigenvalue $\lambda_1 < 0$. Show that λ_1 is non-degenerate and that the corresponding eigenfunction f is radial and strictly positive (up to a phase). Deduce that $u = \sqrt{4\pi} \, r \, f(re_1)$ solves the differential equation

$$\begin{cases} -u''(r) + v(r)u(r) = \lambda_1 \, u(r), \\ u(0^+) = 0. \end{cases}$$

B.3 The Delta Potential

In this problem we define and study the operator

$$\boxed{H_\alpha := -\Delta + \alpha\delta_0.} \qquad (B.10)$$

where δ_0 is the Dirac delta. This operator must be an extension of the minimal Laplacian

$$H^{\min} := -\Delta, \qquad D(H^{\min}) = C_c^\infty(\mathbb{R}^d \setminus \{0\}). \qquad (B.11)$$

We therefore need to determine which self-adjoint extension of $H^{\min}$ could represent H_α, if it exists. One such extension is the usual Laplacian

$$H_0 = -\Delta, \qquad D(H_0) = H^2(\mathbb{R}^d) \qquad (B.12)$$

which, by convention, we associate with $\alpha = 0$. The latter is self-adjoint and its spectrum is $\sigma(H_0) = \mathbb{R}_+$ by Theorem 2.36. We will show that H_α is well defined in dimension $d = 1$ for all $\alpha \in \mathbb{R}$ and, with more work, in dimensions $d \in \{2, 3\}$ for $\alpha < 0$. However, H_α does not exist in dimension $d \geq 4$, except of course for $\alpha = 0$.

Let us recall that the functions of the Sobolev space $H^2(\mathbb{R}^d)$ are all continuous and tend to 0 at infinity, when $d \in \{1, 2, 3\}$. The linear map $u \in H^2(\mathbb{R}^d) \mapsto u(0) \in \mathbb{C}$ is then continuous. In dimension $d = 1$, the functions of $H^2(\mathbb{R})$ are even C^1 and $u \in H^2(\mathbb{R}) \mapsto u'(0) \in \mathbb{C}$ is also continuous. Let us also recall that the functions of the Sobolev space $H^1(\mathbb{R})$ are all continuous and tend to 0 at infinity, this being true only in dimension $d = 1$. The space $C_c^\infty(\mathbb{R}^d)$ is dense in $H^1(\mathbb{R}^d)$ and in $H^2(\mathbb{R}^d)$ for all $d \geq 1$, according to Theorem A.8.

Part 1: H_α Does Not Exist for $\alpha \neq 0$ in Dimension $d \geq 4$

Let $\eta \in C^\infty(\mathbb{R}^d)$ be any function such that $\eta \equiv 1$ outside the ball of radius 2 and which vanishes on the ball of radius 1. We set $\eta_\varepsilon(x) = \eta(x/\varepsilon)$.

1. Let $\varphi \in C_c^\infty(\mathbb{R}^d)$. Calculate the Laplacian of $\eta_\varepsilon\varphi$ and deduce that $\eta_\varepsilon\varphi \to \varphi$ strongly in $H^2(\mathbb{R}^d)$ when $\varepsilon \to 0$, in dimensions $d \geq 5$.
2. Let $\varphi \in C_c^\infty(\mathbb{R}^4)$, in dimension $d = 4$. Calculate the Laplacian of $|x|^\tau\varphi(x)$ and deduce that $|x|^\tau\varphi \to \varphi$ strongly in $H^2(\mathbb{R}^4)$.
3. Show that for fixed $\tau > 0$, $\eta_\varepsilon|x|^\tau\varphi \to |x|^\tau\varphi$ strongly in $H^2(\mathbb{R}^4)$.
4. Deduce that $C_c^\infty(\mathbb{R}^d \setminus \{0\})$ is dense in $H^2(\mathbb{R}^d)$ in dimensions $d \geq 4$.
5. Show that the closure of the minimal Laplacian $H^{\min}$ in (B.11) is the usual Laplacian H_0 in (B.12) and conclude that $H^{\min}$ has no other self-adjoint extension than H_0, in dimension $d \geq 4$.

Part 2: H_α in Dimension $d = 1$

In this part we are in dimension $d = 1$.

6. Show that the closure of $H^{\min}$ is the operator $H_{0,0}u = -u''$ defined on $D(H_{0,0}) = \{u \in H^2(\mathbb{R}) : u(0) = u'(0) = 0\}$. You can use, without re-proving it, that $C_c^\infty(\mathbb{R} \setminus \{0\})$ is dense in this space, for the norm of $H^2(\mathbb{R})$.

7. Show that the adjoint of $H_{0,0}$ is the operator

$$H^{\max}u = -u''_{|(-\infty,0)} - u''_{|(0,+\infty)}$$

defined on the domain $D(H^{\max}) = L^2(\mathbb{R}) \cap H^2((-\infty, 0)) \cap H^2((0, +\infty))$ which contains the functions of $L^2(\mathbb{R})$ such that its second derivatives $u''_{|(-\infty,0)}$ and $u''_{|(0,+\infty)}$ calculated in the sense of distributions on the intervals $(-\infty, 0)$ and $(0, +\infty)$, belong respectively to $L^2((-\infty, 0))$ and $L^2((0, +\infty))$.

The functions of $D(H^{\max})$ are in $C^1(\mathbb{R} \setminus \{0\})$ and admit a limit as well as their derivative to the left and right of 0. We denote these limits by $u(0^-)$, $u(0^+)$, $u'(0^-)$ and $u'(0^+)$. The derivative in the sense of distributions of $u \in D(H^{\max})$ over the whole of $\mathbb{R}$ is then given by

$$u'' = u''_{|(-\infty,0)} + u''_{|(0,+\infty)} + \big(u(0^+) - u(0^-)\big)\delta'_0 + \big(u'(0^+) - u'(0^-)\big)\delta_0.$$

In our case we want $H_\alpha u$ to belong to $L^2(\mathbb{R})$, which requires the delta parts to cancel out. This suggests defining the operator H_α on the domain

$$D(H_\alpha) = \Big\{u \in L^2(\mathbb{R}) \cap H^2((-\infty, 0)) \cap H^2((0, +\infty)) : u(0^+) = u(0^-),$$
$$u'(0^+) - u'(0^-) = \alpha u(0)\Big\} \qquad \text{(B.13)}$$

by

$$\boxed{H_\alpha u = -u'' + \alpha u(0)\delta_0 = -u''_{|(-\infty,0)} - u''_{|(0,+\infty)}.}$$

In other words, the functions of the domain are continuous over all $\mathbb{R}$ and their derivative has a jump that is proportional to $u(0)$. This jump is used to cancel the term $\alpha u(0)\delta_0$ so that $H_\alpha u$ (interpreted in the sense of distributions on $\mathbb{R}$) is equal to the opposite of the second derivative of u on $\mathbb{R} \setminus \{0\}$.

8. Show that the operator H_α thus defined is self-adjoint. Verify that for $\alpha = 0$, the operator H_0 is nothing but the usual Laplacian defined on all $H^2(\mathbb{R})$, as we wanted.

9. Show that the quadratic form of H_α is given by

$$\langle u, H_\alpha u \rangle = \int_{\mathbb{R}} |u'(x)|^2 \, dx + \alpha |u(0)|^2. \tag{B.14}$$

What is the form domain $Q(H_\alpha)$? How could we define H_α from the quadratic form (B.14)?

10. Show that the spectrum of H_α is included in $[0, +\infty)$ when $\alpha \geq 0$.

11. By constructing a Weyl sequence $u_n \rightharpoonup 0$, show that the essential spectrum of H_α is $\sigma_{\text{ess}}(H_\alpha) = [0, +\infty)$ for all $\alpha \in \mathbb{R}$, and therefore that $\sigma(H_\alpha) = [0, +\infty)$ when $\alpha \geq 0$.

12. Show that if $\alpha < 0$, the spectrum of H_α is

$$\sigma(H_\alpha) = \{\lambda(\alpha)\} \cup [0, +\infty[, \qquad \lambda(\alpha) = -\frac{\alpha^2}{4}$$

with a simple eigenvalue, of eigenvector $f_\alpha(x) = e^{\frac{\alpha}{2}|x|}$.

13. Show that $D(H_\alpha) = H^2(\mathbb{R}) \cap H_0^1(\mathbb{R}) + f_\alpha \mathbb{C} = \{u \in H^2(\mathbb{R}) : u(0) = 0\} + f_\alpha \mathbb{C}$ and that for $u_0 \in H^2(\mathbb{R}) \cap H_0^1(\mathbb{R})$ and $\beta \in \mathbb{C}$,

$$H_\alpha(u_0 + \beta f_\alpha) = -\Delta u_0 - \frac{\alpha^2}{4} \beta f_\alpha. \tag{B.15}$$

Part 3: H_α in Dimensions $d = 2, 3$

In this part we work in dimension $d \in \{2, 3\}$. The construction of H_α is more difficult and described in [FB61, KS95, Sim95, AK00, AGHKH04]. In particular, we cannot rely on the quadratic form (B.14) because $u(0)$ is not well defined in $H^1(\mathbb{R}^d)$.

In dimension $d = 1$, for $\alpha < 0$ we found the unique eigenvector f_α, which satisfies

$$-f_\alpha'' + \frac{\alpha^2}{4} f_\alpha = -\alpha \delta_0. \tag{B.16}$$

since $f_\alpha(0) = 1$. If we hope to find a negative eigenvalue $E < 0$ in dimension $d \geq 2$ and write the similar equation

$$(-\Delta + E)f = b(2\pi)^{d/2} \delta_0$$

we find that $\widehat{f}(k) = b(|k|^2 + E)^{-1}$. Hence f is proportional to the function h_2 considered before in (A.28). The problem is that, according to Lemma A.16, this function must diverge at the origin when $b \neq 0$. It is therefore impossible to have b proportional to $f(0)$, as we would like. Some renormalization is necessary. The idea is to base the construction of the operator H_α on the function f, by similarity with Formula (B.15). At the very end we will explain the link with the delta potential.

In the following, we will parameterize H_α by its smallest eigenvalue $\lambda = -E < 0$ rather than by α and as there is no confusion we will write H_E. For $E > 0$, let Y_E be the function whose Fourier transform is

$$\widehat{Y_E}(k) = \frac{1}{|k|^2 + E}$$

and which therefore solves the equation in the sense of distributions $-\Delta Y_E + E Y_E = (2\pi)^{d/2}\delta_0$.

14. Show that $Y_E \in L^2(\mathbb{R}^d)$ but that $Y_E \notin H^1(\mathbb{R}^d)$, in dimension $d \in \{2, 3\}$.
15. Show that the closure of the operator $H^{\min}$ defined in (B.11) is the operator

$$H_{0,0}u = -\Delta u, \qquad D(H_{0,0}) = \{u \in H^2(\mathbb{R}^d) \ : \ u(0) = 0\}.$$

   You can use, without re-proving it, that $C_c^\infty(\mathbb{R}^d \setminus \{0\})$ is dense in this space, for the norm of $H^2(\mathbb{R}^d)$, in dimension $d \in \{2, 3\}$.
16. Let $D(H_E) = \{u \in L^2(\mathbb{R}^d) \ : \ \exists \beta \in \mathbb{C}, \ u - \beta Y_E \in D(H_{0,0})\}$. Show that if $u \in D(H_E)$, then β is uniquely determined.
17. Let $u_0 \in D(H_{0,0})$, that is $u_0 \in H^2(\mathbb{R}^d)$ with $u_0(0) = 0$. Show that $-\int_{\mathbb{R}^d} Y_E \Delta u_0 = -E \int_{\mathbb{R}^d} Y_E u_0$.
18. For $u = u_0 + \beta Y_E \in D(H_E)$ with $u_0 \in D(H_{0,0})$ and $\beta \in \mathbb{C}$, we now set

$$H_E(u_0 + \beta Y_E) := -\Delta u_0 - E \beta Y_E. \tag{B.17}$$

   Show that H_E is a symmetric operator on $D(H_E)$, and that $-E$ is an eigenvalue of H_E, with eigenfunction Y_E.
19. Let $w \in L^2(\mathbb{R}^d)$ and $C > 0$ any constant such that $C \neq E$. By passing to Fourier, show that there exists $\beta \in \mathbb{C}$ and $u_0 \in D(H_{0,0})$ such that $(H_E + C)(u_0 + \beta Y_E) = w$. Conclude that H_E is self-adjoint on $D(H_E)$.
20. Show that $\sigma(H_E) = \{-E\} \cup [0, +\infty)$.
21. Now that we have constructed the operator H_E, we still have to explain the link with the delta potential. The idea is to regularize the problem. Although one could regularize the potential, we find it more convenient to instead add a fourth-order derivative in the form $\varepsilon(\Delta)^2 - \Delta + \alpha \delta_0$. Namely, we introduce the quadratic form defined on $H^2(\mathbb{R}^d)$ by

$$q_{\alpha,\varepsilon}(u) := \varepsilon \int_{\mathbb{R}^d} |\Delta u(x)|^2 \, dx + \int_{\mathbb{R}^d} |\nabla u(x)|^2 \, dx + \alpha |u(0)|^2.$$

a. Show that $q_{\alpha,\varepsilon}$ is well defined and bounded from below on $H^2(\mathbb{R}^d)$ in dimension $d \in \{2, 3\}$, for any $\alpha \in \mathbb{R}$.

b. Show that the corresponding self-adjoint operator $H_{\alpha,\varepsilon}$ admits negative eigenvalues only for $\alpha < 0$. Then show that $\sigma(H_{\alpha,\varepsilon}) = \{-E(\alpha,\varepsilon)\} \cup [0, +\infty)$ where $E(\alpha, \varepsilon) > 0$ is the unique solution of the implicit equation

$$\alpha \int_{\mathbb{R}^d} \frac{dk}{\varepsilon |k|^4 + |k|^2 + E(\alpha,\varepsilon)} = -(2\pi)^d,$$

with the unique associated eigenfunction

$$\widehat{Y_{\alpha,\varepsilon}}(k) = \frac{1}{\varepsilon |k|^4 + |k|^2 + E(\alpha,\varepsilon)}.$$

c. Study the limit $\varepsilon \to 0$.

B.4 On the Finiteness of the Discrete Spectrum

We have stated in Theorem 5.47 that a Schrödinger operator $-\Delta + V$ always has a finite number of negative eigenvalues in dimension $d \geq 3$ when $V_- \in L^{d/2}(\mathbb{R}^d)$, a number that can be estimated by a multiple of $\|V_-\|_{L^{d/2}(\mathbb{R}^d)}^{d/2}$. On the contrary, in dimensions $d \in \{1, 2\}$, the operator $-\Delta + V$ always has a negative eigenvalue, regardless of the size of V, if for example $V < 0$ everywhere (Proposition 1.17). Here we show some results on the number of eigenvalues with additional assumptions on V or with a worse estimate than in Theorem 5.47.

Part 1: Number of Negative Eigenvalues Generated by a Deep Local Well

We are in dimension $d \geq 1$. Let

$$V \in L^p(\mathbb{R}^d, \mathbb{R}) + L^\infty_\varepsilon(\mathbb{R}^d, \mathbb{R}), \qquad \text{with} \quad \begin{cases} p = 1 & \text{if } d = 1, \\ p > 1 & \text{if } d = 2, \\ p = \frac{d}{2} & \text{if } d \geq 3. \end{cases} \qquad (\text{B.18})$$

We consider the Friedrichs self-adjoint realization of $-\Delta + V$ on $\mathbb{R}^d$, given by Corollary 3.20. We recall that its essential spectrum is

$$\sigma_{\text{ess}}(-\Delta + V) = [0, +\infty)$$

by Corollary 5.38. We prove here that there exists a constant C_d depending only on the dimension d such that if we have for some $R > 0$ and $x_0 \in \mathbb{R}^d$

$$V(x) < -\frac{C_d k^{2/d}}{R^2} \qquad \text{on } B(x_0, R), \tag{B.19}$$

then $-\Delta + V$ admits at least k negative eigenvalues. The interpretation is that a deep well can create an arbitrary number of eigenvalues (but still finite, as we will see below).

Let λ_k^D denote the k-th eigenvalue of the Dirichlet Laplacian (counted with multiplicity) on the unit ball $B(0, 1)$ and recall that $\lambda_k^D \to +\infty$ by Theorem 5.31.

1. Using the Dirichlet eigenfunctions in $B(x_0, R)$, prove that if $V(x) < -\lambda_k^D/R^2$ on $B(x_0, R)$, then $-\Delta + V$ admits at least k negative eigenvalues in $\mathbb{R}^d$.
2. Give an upper bound on λ_k^D using Lemma 5.48 and conclude the proof of (B.19).

Part 2: Case of a Potential with Compact Support

In this part, we assume in addition to (B.18) that V has **compact support** and we show that there are then finitely many negative eigenvalues. The interpretation is that a deep local well can only generate a finite discrete spectrum. In order to create infinitely many eigenvalues, the potential V should be strictly negative on an unbounded set. In fact it should also not go to 0 too fast. In Theorem 5.46 we gave an explicit condition on the decay of V at infinity that implies the existence of an infinite discrete spectrum.

In the following we choose a number $R > 0$ large enough so that the support of V is strictly included in the ball B_R of radius R centered at the origin and this time involve the Neumann Laplacian of that ball.

3. Show that the quadratic form $\int_{B_R} |\nabla u(x)|^2 dx + \int_{B_R} V(x)|u(x)|^2 dx$ is bounded below and closed on $Q = H^1(B_R)$ in the Hilbert space $\mathfrak{H} = L^2(B_R)$. Deduce that it is associated with a unique self-adjoint operator H_R and give its domain.
4. Show that H_R has a compact resolvent and deduce that it has finitely many negative eigenvalues.
5. Deduce from the inequality

$$\int_{\mathbb{R}^d} |\nabla u(x)|^2 \, dx + \int_{\mathbb{R}^d} V(x)|u(x)|^2 \, dx \geq \int_{B_R} |\nabla u(x)|^2 \, dx + \int_{B_R} V(x)|u(x)|^2 \, dx$$

that there exists a finite-dimensional space $W \subset L^2(\mathbb{R}^d)$ such that $q_{-\Delta+V}(v) \geq 0$ for all $v \in H^1(\mathbb{R}^d) \cap W^\perp$.
6. Conclude that $-\Delta + V$ has finitely many negative eigenvalues.

The result of this part remains true if we assume that V tends to 0 fast enough at infinity, instead of having compact support. For example, it is enough that

$$V_-(x) = \begin{cases} o\left(|x|^{-2}\right) & \text{if } d \neq 2, \\ o\left((|x|\log|x|)^{-2}\right) & \text{if } d = 2, \end{cases}$$

see [Gla66, Chap. IV]. The proof is essentially based on a Hardy-type inequality, as in Exercise 1.27 but on $\mathbb{R}^d \setminus B_R$ with the Neumann condition on ∂B_R, in order to control the quadratic form outside of B_R.

Part 3: Case of $V \in L^{d/2}(\mathbb{R}^d)$ in Dimension $d \geq 3$

We now assume that $V \in L^{d/2}(\mathbb{R}^d)$ and that $d \geq 3$. We show here an estimate on the number of negative eigenvalues of (the Friedrichs realization of) the operator $-\Delta + V$, discovered independently in 1961 by Birman [Bir61] and Schwinger [Sch61]. It can also be read in [RS78, Thm. XIII.10] and [FLW22, Prop. 4.27]. This estimate is worse than the Cwikel-Lieb-Rozenblum (CLR) inequality seen in Theorem 5.47.

7. Using the Courant-Fischer principle (5.42), show that the number of negative eigenvalues of $-\Delta + V$ can only increase when replacing $V(x)$ by its negative part $-V(x)_-$, where $a_- := \max(-a, 0)$. In this way, we can always assume that $V \leq 0$, which we do throughout the rest.

8. We then consider the Birman-Schwinger operator as introduced in Sect. 5.6.3

$$K_E = |V(x)|^{\frac{1}{2}}(-\Delta + E)^{-1}|V(x)|^{\frac{1}{2}}.$$

Using the same arguments as for Theorem 5.24, verify that K_E is well defined on the domain $D(K_E) = C_c^\infty(\mathbb{R}^d) \subset L^2(\mathbb{R}^d)$, for all $E > 0$, and that it is non-negative and symmetric.

9. Show that K_E has an integral kernel $k_E(x, y) \geq 0$, that is, such that $(K_E u)(x) = \int_{\mathbb{R}^d} k_E(x, y) u(y) \, dy$ for all $u \in C_c^\infty(\mathbb{R}^d)$ (give its explicit formula), and calculate its limit $k_0 = \lim_{E \to 0^+} k_E$.

10. The Hardy-Littlewood-Sobolev inequality [LL01] stipulates that

$$\left\| f * |x|^{-s} \right\|_{L^p(\mathbb{R}^d)} \leq C \left\| f \right\|_{L^q(\mathbb{R}^d)} \tag{B.20}$$

when $1 < p, q < \infty$ and $1 + \frac{1}{p} = \frac{1}{q} + \frac{s}{d}$. This inequality turns out to be equivalent to the Sobolev inequality seen in Theorem A.13. Show that K_E is bounded on $D(K_E) = L_c^\infty(\mathbb{R}^d)$:

$$\forall u \in L_c^\infty(\mathbb{R}^d), \qquad \|K_E u\|_{L^2(\mathbb{R}^d)} \leq C \|u\|_{L^2(\mathbb{R}^d)},$$

where in addition C does not depend on $E \in \mathbb{R}^+$. Deduce that K_E is essentially self-adjoint for all $E \geq 0$, and that its unique extension, still denoted K_E, is defined on all $L^2(\mathbb{R}^d)$. Show that the self-adjoint operator K_E thus obtained is compact and positive on $L^2(\mathbb{R}^d)$. In the following, we denote by $\lambda_j(E)$ its eigenvalues (repeated according to their multiplicity), in *decreasing* order.

11. Using the Courant-Fischer principle, show that the eigenvalues $\lambda_j(E)$ are all continuous and non-increasing with respect to E on $\mathbb{R}^+$.

12. Show that $\lambda \leq 0$ is an eigenvalue of the Friedrichs realization of $-\Delta + V$, if and only if 1 is an eigenvalue of $K_{-\lambda}$, with the same multiplicities.

13. Deduce that the number of negative eigenvalues of $-\Delta + V$ is less than the number of eigenvalues ≥ 1 of K_0:

$$\operatorname{rank}\left(\mathbb{1}_{(-\infty,0]}(-\Delta + V)\right) \leq \operatorname{rank}\left(\mathbb{1}_{[1,+\infty)}(K_0)\right).$$

14. Let A be a self-adjoint compact non-negative operator on $L^2(\mathbb{R}^d)$, which is given by a non-negative integral kernel $k(x, y) \geq 0$ belonging to $L^1_{\mathrm{loc}}(\mathbb{R}^d \times \mathbb{R}^d)$. Show that for all integers $n \geq 1$,

$$\int_{(\mathbb{R}^d)^n} k(x_1, x_2)k(x_2, x_3) \cdots k(x_n, x_1)\, dx_1 \cdots dx_n$$

$$= \sum_{j \geq 1} \lambda_j(A)^n \geq \operatorname{rank}\left(\mathbb{1}_{[1;+\infty)}(A)\right).$$

Calculate the term on the left for $k = k_0$ (the kernel of the operator K_E for $E = 0$).

15. In dimension $d = 3$, verify that we can take $n = 2$ (the operator K_0 is Hilbert-Schmidt) and obtain

$$\operatorname{rank}\left(\mathbb{1}_{(-\infty,0]}(-\Delta + V)\right) \leq C \, \|V_-\|^2_{L^{3/2}(\mathbb{R}^3)} \tag{B.21}$$

using (B.20). In the CLR inequality, the norm on the right is to the power $3/2$ instead of 2.

16. Let $d \geq 1$, $0 < s < d$ and n an integer such that $d/(d-s) < n < 2d/(d-s)$. Show the multi-linear Hardy-Littlewood-Sobolev inequality

$$\iint_{(\mathbb{R}^d)^n} \frac{f(x_1)f(x_2) \cdots f(x_n)}{|x_1 - x_2|^s |x_2 - x_3|^s \cdots |x_{n-1} - x_n|^s |x_n - x_1|^s}\, dx_1 \cdots dx_n$$

$$\leq C \int_{\mathbb{R}^d} \int_{\mathbb{R}^d} \frac{f(x)^{n/2} f(y)^{n/2}}{|x - y|^{ns + (2-n)d}}\, dx\, dy \leq C' \, \|f\|^n_{L^{\frac{d}{d-s}}(\mathbb{R}^d)}, \tag{B.22}$$

for all $f \in L^{\frac{d}{d-s}}(\mathbb{R}^d, \mathbb{R}^+)$. For this, you can write the left integrand as a product of n functions in the form

$$\frac{\sqrt{f(x_1)f(x_2)}}{|x_2 - x_3|^{\frac{s}{n-1}} \cdots |x_{n-1} - x_n|^{\frac{s}{n-1}}|x_n - x_1|^{\frac{s}{n-1}}} \times$$

$$\times \frac{\sqrt{f(x_2)f(x_3)}}{|x_1 - x_2|^{\frac{s}{n-1}}|x_3 - x_4|^{\frac{s}{n-1}} \cdots |x_{n-1} - x_n|^{\frac{s}{n-1}}|x_n - x_1|^{\frac{s}{n-1}}} \times \cdots$$

$$\cdots \times \frac{\sqrt{f(x_n)f(x_1)}}{|x_1 - x_2|^{\frac{s}{n-1}}|x_2 - x_3|^{\frac{s}{n-1}} \cdots |x_{n-1} - x_n|^{\frac{s}{n-1}}}$$

and use Hölder's inequality, followed by the fact that

$$\underbrace{\frac{1}{|x|^{\frac{sn}{n-1}}} * \cdots * \frac{1}{|x|^{\frac{sn}{n-1}}}}_{n-1 \text{ convolutions}} = \frac{C}{|x|^{ns+(2-n)d}}.$$

Verify that the assumptions on d, s and n guarantee the convergence of all the integrals.

17. Using the inequality (B.22) with $s = d - 2$ and $d \geq 3$, show that

$$\text{range}\left(\mathbb{1}_{(-\infty,0]}(-\Delta + V)\right) \leq C \int_{\mathbb{R}^d} \int_{\mathbb{R}^d} \frac{V(x)^{\frac{n}{2}} V(y)^{\frac{n}{2}}}{|x - y|^{2(d-n)}} \, dx \, dy \leq C' \|V\|^n_{L^{\frac{d}{2}}(\mathbb{R}^d)} \tag{B.23}$$

for any integer n strictly between $d/2$ and d. In the CLR inequality of Theorem 5.47, the norm on the right is to the power $d/2$ instead of n.

B.5 Perron-Frobenius Theory and Phase Transitions in Statistical Physics

Here we study a particular class of operators, which preserve the positivity of functions. The central theorem of this theory, called Perron-Frobenius (or Krein-Rutman in a more abstract framework) is a tool used in multiple branches of mathematics: in probability theory (for the ergodicity of Markov chains), for dynamical systems, and of course in spectral theory and mathematical physics in general. We will see a famous application to statistical physics in Part 3. We also refer to [RS78, Sec. 12] and [Sim15, Sec. 7.6] for a presentation of the Perron-Frobenius theory for operators.

Part 1: Case of Bounded Operators

We work in a Hilbert space in the form $\mathfrak{H} = L^2(\Omega)$ where Ω is a non-empty open set of $\mathbb{R}^d$, $d \geq 1$. We recall that a function $f \in \mathfrak{H}$ is *non-negative* ($f \geq 0$) if $f(x) \in [0, \infty)$ almost everywhere and that f is (strictly) *positive* ($f > 0$) if f is non-negative and the set $\{f \equiv 0\}$ has zero Lebesgue measure.

Let A be a bounded operator on $\mathfrak{H}$. We say that A is **positivity-preserving** if we have $Af \geq 0$ for all $0 \leq f \in \mathfrak{H}$. We say that A is **positivity-improving** if we have $Af > 0$ for all $0 \leq f \in \mathfrak{H}$ with $f \neq 0$.

1. Show that if A is positivity-preserving, then it commutes with complex conjugation, in the sense that $\overline{Af} = A\overline{f}$ for all $f \in \mathfrak{H}$.
2. Show that a bounded operator A on $\mathfrak{H} = L^2(\Omega)$ is positivity-preserving if and only if $\langle f, Ag \rangle \geq 0$ for all $0 \leq f, g \in \mathfrak{H}$. Show then that A is positivity-improving if and only if $\langle f, Ag \rangle > 0$ for all $0 \leq f, g \in \mathfrak{H} \setminus \{0\}$. Also show that A^* is positivity-preserving or improving like A.
3. Show that a bounded operator A on $\mathfrak{H} = L^2(\Omega)$ is positivity-preserving if and only if $|\langle f, Ag \rangle| \leq \langle |f|, A|g| \rangle$ for all complex-valued $f, g \in \mathfrak{H}$, where $|f|$ denotes the modulus of f. For this you can consider the scalar product $\langle f, (Ag) * \chi_n \rangle$ where $\chi_n(x) = n^d \chi(nx)$ and $\chi \in C_c^\infty(\mathbb{R}^d, \mathbb{R}_+)$ with $\int_{\mathbb{R}^d} \chi(x)\, dx = 1$.
4. Now let A be a **bounded self-adjoint** operator that is positivity-preserving.

   a. Recall why $\{-\|A\|, \|A\|\} \cap \sigma(A) \neq \emptyset$.
   b. Recall that

$$\|A\| = \sup_{\substack{f \in \mathfrak{H} \\ \int_B |f|^2 d\mu = 1}} |\langle f, Af \rangle|, \qquad \max \sigma(A) = \sup_{\substack{f \in \mathfrak{H} \\ \int_B |f|^2 d\mu = 1}} \langle f, Af \rangle,$$

   with equality if and only if f is a corresponding eigenfunction. Deduce that $\|A\| \in \sigma(A)$.
   c. Suppose that f is a real eigenvector for $\|A\|$ or $-\|A\|$. Show then that $|f|$ is an eigenvector for $\|A\|$.
   d. Now suppose that A is positivity-improving. Show that $-\|A\|$ cannot be an eigenvalue. If $\|A\|$ is an eigenvalue, show that it must be simple, with a strictly positive eigenfunction.
   e. Suppose that A is positivity-improving and that $\sigma_{\text{ess}}(A) \subset [-\|A\| + \varepsilon, \|A\| - \varepsilon]$ for some $\varepsilon > 0$. Show that for all non-negative functions $g, h \in \mathfrak{H} \setminus \{0\}$, we have

$$\log\langle g, A^n h \rangle = n \log \|A\| + \log \langle g, f_0 \rangle + \log \langle f_0, h \rangle + o(1)_{n \to \infty}, \qquad \text{(B.24)}$$

   where f_0 is the unique positive eigenvector associated with the eigenvalue $\|A\|$.

We have shown that any positivity-improving self-adjoint operator has a maximum eigenvalue that is always simple, when it exists. The historical theorem was shown for matrices with positive coefficients by Perron and Frobenius at the beginning of the twentieth century. Then, Krein and Rutman found a generalization of this result, on any Banach space, by replacing the set of positive functions with an abstract convex cone. Self-adjointness is therefore not necessary in this theory. The simplicity of the maximum eigenvalue plays a central role in the behavior of A^n as $n \to \infty$, as we saw in the last question. Indeed, (B.24) means that A^n is given to leading order by $\|A\|^n$ multiplied by the orthogonal projector P_0 onto the corresponding eigenvector f_0.

Part 2: Case of Unbounded Self-adjoint Operators

In this part we generalize the previous definitions to the case of unbounded operators by examining the resolvent or the heat kernel.

5. Let A be a self-adjoint operator on its domain $D(A) \subset \mathfrak{H}$, which we assume to be bounded from below. Show the equivalence

   (i) e^{-tA} is positivity-preserving for all $t > 0$;
   (ii) $(A + C)^{-1}$ is positivity-preserving for all $C > -\min \sigma(A)$.

   In this case we say that A is positivity-preserving.
6. We assume that $(A + C_0)^{-1}$ is positivity-preserving for a $C_0 > -\min \sigma(A)$. Deduce that the same is true for $(A + C)^{-1}$ for all $C > -\min \sigma(A)$.
7. We assume that $(A + C_0)^{-1}$ is positivity-improving for a $C_0 > -\min \sigma(A)$. Deduce that if $\min \sigma(A)$ is an eigenvalue, then it is necessarily simple, with a strictly positive associated eigenfunction.
8. Let $V \in L^p(\mathbb{R}^d, \mathbb{R}) + L^\infty(\mathbb{R}^d)$ with $p = \max(2, d/2)$ in dimensions $d \neq 4$ and $p > 2$ in dimension $d = 4$. We further assume that V is bounded from above. Drawing inspiration from the proof of Theorem 1.18 in Sect. 1.6, show that $H = -\Delta + V$ is positivity-improving, which provides new insight into this result.

Part 3: Case of Hilbert-Schmidt Operators

We still consider the Hilbert space $\mathfrak{H} = L^2(\Omega)$ with Ω a non-empty open set of $\mathbb{R}^d$. We are given a function $a \in L^2(\Omega \times \Omega)$ and we define the operator

$$(Af)(x) := \int_\Omega a(x, y) f(y) \, dy.$$

We study here under which conditions on the function a the associated operator A is positivity-preserving or improving. The function a is called the integral kernel of the operator A.

9. Recall why A is a compact operator that satisfies $\|A\| \leq \|a\|_{L^2(\Omega \times \Omega)}$.
10. Show that A is self-adjoint if and only if we have $a(x, y) = \overline{a(y, x)}$ almost everywhere on $\Omega \times \Omega$.
11. Show that A is positivity-preserving if and only if $a(x, y) \geq 0$ almost everywhere on $\Omega \times \Omega$.
12. Show that if $a(x, y) > 0$ almost everywhere on $\Omega \times \Omega$, then A is positivity-improving.
13. We take $\Omega = (0, 1) \subset \mathbb{R}$ and we define

$$a(x, y) = \mathbb{1}_D(x - y), \qquad D = \bigcup_{n \geq 1} \left(r_n - \frac{\eta}{2^n}, r_n + \frac{\eta}{2^n} \right)$$

where $\eta > 0$ and $\{r_n\}$ is an enumeration of all the rationals in $[-1, 1]$.

a. Verify that we can choose η small enough so that $0 < |D| < 1$, then that $|D \cap I| > 0$ for every open interval $I \subset (-1, 1)$.
b. Show that for all Borel sets $E, F \subset (0, 1)$, we have

$$\int_E \int_F a(x, y) \, dx \, dy = \int_D (\mathbb{1}_E * \mathbb{1}_{-F}) (x) \, dx$$

where $-F = \{-x \; : \; x \in F\}$. Deduce that this integral is always strictly positive when E and F are both of non-zero measure.
c. Conclude that the operator A associated with the integral kernel a is positivity-improving, even though a is not strictly positive.

Hilbert-Schmidt operators are those that resemble matrices the most, because they are given by an "integral kernel" $a(x, y)$ which is the continuous version of a matrix A_{jk}. We have shown here that such operators are positivity-preserving if and only if this kernel is a non-negative function. This is therefore the equivalent of the matrices with non-negative coefficients considered by Perron and Frobenius. However, it is entirely possible that the operator A is positivity-improving while the function a is not strictly positive, which differs from the case of matrices.

Part 4: Application: Absence of Phase Transition in Dimension $d = 1$

Any material undergoes phase transitions when temperature and pressure vary. For example, water turns into ice at $0\,°C$ and evaporates at $100\,°C$, under normal pressure conditions. These phase changes occur at particular values of temperature and pressure, which can be represented by curves in the (T, P) plane, called a *phase diagram*. Outside of these curves, the system's observables are all very smooth functions of T and P.

Phase transitions are typical of the space dimension 3 and water would never turn into ice in dimensions $d = 1$ and $d = 2$! We will show here a simplified version of a theorem from 1950 due to Van Hove [vH50] (see also [Rue99]), which specifies that the free enthalpy is a real analytic function on the quadrant $\{(T, P) \in (0, \infty)^2\}$, for a one-dimensional classical system with a compactly supported interaction. Thus, there are no usual phase transitions in dimension 1. A similar (though weaker) result was shown by Mermin and Wagner in dimension 2 [MW66, FP81, FP86]. There are also similar results for quantum systems but the proof is more difficult.

We consider a system of N classical particles in the interval $C_L = (-L/2, L/2)$, labeled in ascending order

$$-\frac{L}{2} \leq x_1 \leq x_2 \leq \cdots \leq x_N \leq \frac{L}{2}.$$

We assume that they interact by pairs through an interaction potential w, so that the classical Hamiltonian of the system is

$$E(x_1, p_1, \ldots, x_N, p_N) = \sum_{j=1}^{N} \frac{|p_j|^2}{2} + \sum_{1 \leq j < k \leq N} w(x_j - x_k)$$

as seen at the beginning of Chap. 6. The equilibrium state of the system at temperature $T = \beta^{-1} > 0$ is given by the probability on the phase space $(C_L \times \mathbb{R})^N$

$$\mu(x_1, p_1 \ldots, x_N, p_N) = \frac{e^{-\beta E(x_1, p_1 \ldots, x_N, p_N)} \mathbb{1}(x_1 \leq \cdots \leq x_N)}{\int_{(C_L \times \mathbb{R})^N} e^{-\beta E} \mathbb{1}(x_1 \leq \cdots \leq x_N)}$$

called the **Gibbs measure**. When $T = \beta^{-1} \to 0$, this measure concentrates on the set of minimizers of E. The corresponding free energy is the sum of the energy $\int E\mu$ and the entropy $T \int \mu \log \mu$. It equals

$$F_{\text{tot}}(\beta, L, N) = -\beta^{-1} \log \left(\int_{(C_L \times \mathbb{R})^N} e^{-\beta E} \mathbb{1}(x_1 \leq \cdots \leq x_N) \right)$$

$$= -\frac{N}{2\beta} \log \left(\frac{2\pi}{\beta} \right)$$

$$- \beta^{-1} \log \left(\int_{-\frac{L}{2} \leq x_1 \leq \cdots \leq x_N \leq \frac{L}{2}} e^{-\beta \sum_{1 \leq j < k \leq N} w(x_j - x_k)} dx_1 \cdots dx_N \right).$$

It is the second term that interests us here. The first term comes from the integration of the momenta and we can eliminate it without loss of generality because it is already analytic in β. We still need to apply a pressure P on our system. Remembering that P is the variable dual to the volume, this amounts to considering $-\beta^{-1} \log \Delta(\beta, P, N)$ where

$$\Delta(\beta, P, N) = \int_0^\infty e^{-PL} \int_{-\frac{L}{2} \le x_1 \le \cdots \le x_N \le \frac{L}{2}} e^{-\beta \sum_{1 \le j < k \le N} w(x_j - x_k)} dx_1 \cdots dx_N \, dL.$$

The following theorem is due to Van Hove [vH50].

Theorem B.2 (Absence of Phase Transition in 1D) *We assume that w is an even function satisfying*

$$w(x) \begin{cases} = +\infty & \text{if } |x| < R_1 \\ \in (-C, C) & \text{if } R_1 \le |x| \le R_2 \\ = 0 & \text{if } |x| \ge R_2. \end{cases} \tag{B.25}$$

with $0 < R_1 < R_2$. Then the thermodynamic limit $g(\beta, P) = \lim_{N \to \infty} \frac{\log \Delta(\beta, P, N)}{N}$ exists and it is real analytic on the quadrant $\{(\beta, P) \in (0, \infty)^2\}$.

Assumption (B.25) means that the particles have a "hard core" that prevents them from getting closer than a distance R_1, and on the other hand, they no longer see each other at all when they are at a distance greater than R_2. The interaction is just assumed to be bounded at intermediate distances. The same result is true with weaker assumptions on w but the proof is more difficult, see for example [Dob74, CCO83]. A rather fast decay at infinity is however necessary, transitions can exist if w decays slower than $1/|x|^2$ at infinity [Dys69].

Here we will only show Theorem B.2 in the case where R_2 is not too large and refer to [vH50] and [Rue99, Thm. 5.6.7] for the general case.

14. We assume that $R_2 < 2R_1$, so that each particle only interacts with its nearest neighbors. By introducing the new variables $y_1 = x_1 + L/2$, $y_2 = x_2 - x_1, \ldots$, $y_N = x_N - x_{N-1}$, calculate $\Delta(\beta, P, N)$ and deduce that

$$g(\beta, P) = \lim_{N \to \infty} \frac{\log \Delta(\beta, P, N)}{N} = \log \left(\int_{R_1}^\infty e^{-\beta w(y) - Py} \, dy \right).$$

Show that this is a real analytic function on $\{(\beta, P) \in (0, \infty)^2\}$.

15. We introduce the function

$$a_{\beta, P}(x, y) = \exp \left(-\beta w(x + y) - \frac{\beta}{2} w(x) - \frac{\beta}{2} w(y) - \frac{Px}{2} - \frac{Py}{2} \right)$$

and the operator $A_{\beta, P}$ whose integral kernel is $a_{\beta, P}(x, y)$, on $\mathcal{H} = L^2((R_1, \infty))$. Show that $A_{\beta, P}$ is self-adjoint, compact, and positivity-improving. Deduce that $\|A_{\beta,P}\|$ is a simple eigenvalue, with eigenvector $f_{\beta,P} > 0$.

16. We now assume that $R_2 < 3R_1$ so that each particle only interacts with its nearest neighbor and the next one (to its right and left). We also set

$$h_{\beta,P}(x) = \exp\left(-\beta \frac{w(x)}{2} - P\frac{x}{2}\right)$$

which is in $L^2\big((R_1, \infty)\big)$. Show that

$$\Delta(\beta, P, N) = \frac{1}{P^2}\Big\langle h_{\beta,P}, \left(A_{\beta,P}\right)^{N-2}h_{\beta,P}\Big\rangle_{L^2\big((R_1,\infty)\big)}$$

and deduce that

$$\lim_{N\to\infty} \frac{\log \Delta(\beta, P, N)}{N} = \log \|A_{\beta,P}\|.$$

17. Using similar arguments as in the proof of Theorem 5.7, show that $g(\beta, P) = \log \|A_{\beta,P}\|$ is real analytic on the quadrant $\{(\beta, P) \in (0, \infty)^2\}$.

18. We finally determine the law of the gap between two particles in the bulk of the system. We therefore give ourselves an index $n \in [2, N - 1]$, a test function $\eta \in C_c^\infty(\mathbb{R}, \mathbb{R}_+)$ and we study the average

$$\mathscr{L}_{\beta,P,N,n}(\eta) :=$$

$$\frac{1}{\Delta(\beta, P, N)} \int_0^\infty e^{-PL}$$

$$\int_{-\frac{L}{2}\le x_1 \le \cdots \le x_N \le \frac{L}{2}} \eta(x_n - x_{n-1})e^{-\beta \sum_{1\le j<k\le N} w(x_j-x_k)}dx_1 \cdots dx_N \, dL.$$

Express again $\mathscr{L}_{\beta,P,N,n}(\eta)$ in terms of $A_{\beta,P}$, $h_{\beta,P}$, η and deduce that

$$\lim_{\substack{N\to\infty \\ n\to\infty \\ N-n\to\infty}} \mathscr{L}_{\beta,P,N,n}(\eta) = \int_{\mathbb{R}} \eta(t) f_{\beta,P}(t)^2 \, dt.$$

Thus, the law of the gaps in the bulk is simply $f_{\beta,P}^2$.

References

[AF03] R. A. ADAMS AND J. J. F. FOURNIER, *Sobolev spaces*, vol. 140 of Pure and Applied Mathematics (Amsterdam), Elsevier/Academic Press, Amsterdam, second ed., 2003.

[AG74] W. O. AMREIN AND V. GEORGESCU, *On the characterization of bound states and scattering states in quantum mechanics*, Helv. Phys. Acta, 46 (1973/74), pp. 635–658.

[AGHKH04] S. ALBEVERIO, F. GESZTESY, R. HOEGH-KROHN, AND H. HOLDEN, *Solvable Models in Quantum Mechanics*, American Mathematical Soc., second ed., 2004. with an appendix by P. Exner.

[AK00] S. ALBEVERIO AND P. KURASOV, *Singular Perturbations of Differential Operators: Solvable Schrödinger-type Operators*, Lecture note series/London mathematical society, Cambridge University Press, 2000.

[AS82] M. AIZENMAN AND B. SIMON, *Brownian motion and Harnack inequality for Schrödinger operators*, Commun. Pure Appl. Math., 35 (1982), pp. 209–273.

[Atk73] F. V. ATKINSON, *On some results of Everitt and Giertz*, Proc. R. Soc. Edinb., Sect. A, Math., 71 (1973), pp. 151–158.

[AW15] M. AIZENMAN AND S. WARZEL, *Random operators*, vol. 168 of Graduate Studies in Mathematics, American Mathematical Society, Providence, RI, 2015.

[Bac91] V. BACH, *Ionization energies of bosonic Coulomb systems*, Lett. Math. Phys., 21 (1991), pp. 139–149.

[Bec95] W. BECKNER, *Pitt's inequality and the uncertainty principle*, Proc. Amer. Math. Soc., 123 (1995), pp. 1897–1905.

[Bha97] R. BHATIA, *Matrix analysis*, vol. 169, Springer, 1997.

[BHJ26] M. BORN, W. HEISENBERG, AND P. JORDAN, *Zur Quantenmechanik. II*, Z. Phys., 35 (1926), pp. 557–615.

[Bir61] M. V. BIRMAN, *On the spectrum of singular boundary-value problems*, Mat. Sb. (N.S.), 55 (97) (1961), pp. 125–174.

[BJ25] M. BORN AND P. JORDAN, *Zur Quantenmechanik*, Z. Phys., 34 (1925), pp. 858–888.

[BL83] R. BENGURIA AND E. H. LIEB, *Proof of the stability of highly negative ions in the absence of the Pauli principle*, Phys. Rev. Lett., 50 (1983), pp. 1771–1774.

[BL07] L. BOULTON AND M. LEVITIN, *On approximation of the eigenvalues of perturbed periodic Schrödinger operators*, J. Phys. A, 40 (2007), pp. 9319–9329.

[BLLS93] V. BACH, R. LEWIS, E. H. LIEB, AND H. SIEDENTOP, *On the number of bound states of a bosonic N-particle Coulomb system*, Math. Z., 214 (1993), pp. 441–459.

[Blo29] F. BLOCH, *Über die Quantenmechanik der Elektronen in Kristallgittern*, Z. Phys., 52 (1929), pp. 555–600.

M. Lewin, *Spectral Theory and Quantum Mechanics*, Universitext, https://doi.org/10.1007/978-3-031-66878-4

[BLS94] V. BACH, E. H. LIEB, AND J. P. SOLOVEJ, *Generalized Hartree-Fock theory and the Hubbard model*, J. Statist. Phys., 76 (1994), pp. 3–89.

[BO27]   M. BORN AND R. OPPENHEIMER, *Quantum theory of molecules*, Ann. Physics, 84 (1927), pp. 457–484.

[Bor26] M. BORN, *Zur Quantenmechanik der Stoßvorgänge. (Vorläufige Mitteilung.)*, Z. Phys., 37 (1926), pp. 863–867.

[BR02a] O. BRATELLI AND D. W. ROBINSON, *Operator Algebras and Quantum Statistical Mechanics. 1: C*- and W*-Algebras. Symmetry Groups. Decomposition of States*, Texts and Monographs in Physics, Springer, 2nd ed., 2002.

[BR02b] ———, *Operator Algebras and Quantum Statistical Mechanics 2: Equilibrium States. Models in Quantum Statistical Mechanics*, Texts and Monographs in Physics, Springer, 2nd ed., 2002.

[Bre11] H. BREZIS, *Functional analysis, Sobolev spaces and partial differential equations*, Universitext, Springer, New York, 2011.

[BS70]   M. V. BIRMAN AND M. Z. SOLOMJAK, *The principal term of the spectral asymptotics for "non-smooth" elliptic problems*, Funkcional. Anal. i Priložen., 4 (1970), pp. 1–13. English translation in Functional Anal. Appl. 4 (1970), pp. 265–275.

[BS98]   M. S. BIRMAN AND T. A. SUSLINA, *Absolute continuity of a two-dimensional periodic magnetic Hamiltonian with discontinuous vector potential*, Algebra i Analiz, 10 (1998), pp. 1–36.

[BS99]   ———, *A periodic magnetic Hamiltonian with a variable metric. The problem of absolute continuity*, Algebra i Analiz, 11 (1999), pp. 1–40.

[CC74]   J. R. CHELIKOWSKY AND M. L. COHEN, *Electronic structure of silicon*, Phys. Rev. B, 10 (1974), pp. 5095–5107.

[CCO83] M. CAMPANINO, D. CAPOCACCIA, AND E. OLIVIERI, *Analyticity for one-dimensional systems with long-range superstable interactions*, J. Stat. Phys., 33 (1983), pp. 437–476.

[CDL08] É. CANCÈS, A. DELEURENCE, AND M. LEWIN, *A new approach to the modelling of local defects in crystals: the reduced Hartree-Fock case*, Commun. Math. Phys., 281 (2008), pp. 129–177.

[CEM12] É. CANCÈS, V. EHRLACHER, AND Y. MADAY, *Periodic schrödinger operators with local defects and spectral pollution*, SIAM J. Numer. Anal., 50 (2012), pp. 3016–3035.

[CF23]   É. CANCÈS AND G. FRIESECKE, eds., *Density Functional Theory: Modeling, Mathematical Analysis, Computational Methods, and Applications*, no. 1 in Springer series on Molecular Modeling and Simulation, Springer International Publishing, 2023.

[CH53]   R. COURANT AND D. HILBERT, *Methods of mathematical physics. Vol. I*, 1953.

[Cie70] Z. CIESIELSKI, *On the spectrum of the Laplace operator*, Comment. Math. Prace Mat., 14 (1970), pp. 41–50.

[CLL98] I. CATTO, C. LE BRIS, AND P.-L. LIONS, *The mathematical theory of thermodynamic limits: Thomas-Fermi type models*, Oxford Mathematical Monographs, The Clarendon Press Oxford University Press, New York, 1998.

[CLL01] ———, *On the thermodynamic limit for Hartree-Fock type models*, Ann. Inst. H. Poincaré Anal. Non Linéaire, 18 (2001), pp. 687–760.

[Cou20] R. COURANT, *Über die Eigenwerte bei den Differentialgleichungen der mathematischen Physik*, Math. Z., 7 (1920), pp. 1–57.

[Cwi77] M. CWIKEL, *Weak type estimates for singular values and the number of bound states of Schrödinger operators*, Ann. of Math., 106 (1977), pp. 93–100.

[CX97]   J.-Y. CHEMIN AND C.-J. XU, *Inclusions de Sobolev en calcul de Weyl-Hörmander et champs de vecteurs sous-elliptiques*, Ann. Sci. École Norm. Sup. (4), 30 (1997), pp. 719–751.

[Dau88] M. DAUGE, *Elliptic boundary value problems on corner domains*, vol. 1341 of Lecture Notes in Mathematics, Springer-Verlag, Berlin, 1988.

[Dav83] E. B. DAVIES, *Some norm bounds and quadratic form inequalities for Schrödinger operators*, J. Oper. Theory, 9 (1983), pp. 147–162.

[Dav95] ———, *Spectral theory and differential operators*, vol. 42 of Cambridge Studies in Advanced Mathematics, Cambridge University Press, Cambridge, 1995.

[Die81] J. DIEUDONNÉ, *History of functional analysis*, vol. 49 of North-Holland Mathematics Studies, North-Holland Publishing Co., Amsterdam-New York, 1981.

[DIM01] S.-I. DOI, A. IWATSUKA, AND T. MINE, *The uniqueness of the integrated density of states for the Schrödinger operators with magnetic fields*, Math. Z., 237 (2001), pp. 335–371.

[Dir27] P. A. M. DIRAC, *The physical interpretation of the quantum dynamics*, Proceedings Royal Soc. London (A), 113 (1927), pp. 621–641.

[DL92] R. DAUTRAY AND J.-L. LIONS, *Mathematical Analysis and Numerical Methods for Science and Technology*, vol. 5 of Evolution Problems I, Springer-Verlag, Berlin, 1992.

[Dob74] R. L. DOBRUŠIN, *Analyticity of the correlation functions for one-dimensional classical systems with power law decay of the potential*, Mat. Sb. (N.S.), 23 (1974), p. 13.

[DP04] E. B. DAVIES AND M. PLUM, *Spectral pollution*, IMA J. Numer. Anal., 24 (2004), pp. 417–438.

[Dys69] F. J. DYSON, *Existence of a phase-transition in a one-dimensional Ising ferromagnet*, Comm. Math. Phys., 12 (1969), pp. 91–107.

[EG74] W. N. EVERITT AND M. GIERTZ, *Inequalities and separation for certain ordinary differential operators*, Proc. Lond. Math. Soc. (3), 28 (1974), pp. 352–372.

[EG78] ———, *Inequalities and separation for Schrödinger type operators in $L_2(\mathbf{R}^n)$*, Proc. Roy. Soc. Edinburgh Sect. A, 79 (1977/78), pp. 257–265.

[Ehr27] P. EHRENFEST, *Bemerkung über die angenäherte gültigkeit der klassischen mechanik innerhalb der quantenmechanik*, Z. Phys., 45 (1927), pp. 455–457.

[ELS19] M. J. ESTEBAN, M. LEWIN, AND É. SÉRÉ, *Domains for Dirac-Coulomb min-max levels*, Rev. Mat. Iberoam., 35 (2019), pp. 877–924.

[Eng88] B.-G. ENGLERT, *Semiclassical Theory of Atoms*, vol. 300 of Lecture Notes in Physics, Springer Verlag, Berlin, 1988.

[Ens78] V. ENSS, *Asymptotic completeness for quantum mechanical potential scattering. I. Short range potentials*, Commun. Math. Phys., 61 (1978), pp. 285–291.

[Eva10] L. C. EVANS, *Partial differential equations*, vol. 19 of Graduate Studies in Mathematics, American Mathematical Society, Providence, RI, second ed., 2010.

[EZ78] W. D. EVANS AND A. ZETTL, *Dirichlet and separation results for Schrödinger-type operators*, Proc. R. Soc. Edinb., Sect. A, Math., 80 (1978), pp. 151–162.

[Fan49] K. FAN, *On a theorem of weyl concerning eigenvalues of linear transformations. i*, Proc. Nat. Acad. Sci. U. S. A., 35 (1949), pp. 652–655.

[FB61] L. D. FADDEEV AND F. A. BEREZIN, *A remark on schrödinger's equation with a singular potential*, Dokl. Akad. Nauk SSSR, 137 (1961).

[Fer27] E. FERMI, *Un metodo statistico per la determinazione di alcune priorieta dell'atome*, Rend. Accad. Naz. Lincei, 6 (1927), pp. 602–607.

[FGP10] S. FRABBONI, G. C. GAZZADI, AND G. POZZI, *Ion and electron beam nanofabrication of the which-way double-slit experiment in a transmission electron microscope*, Appl. Phys. Lett., 97 (2010), p. 263101.

[Fis05] E. FISCHER, *Über quadratische Formen mit reellen Koeffizienten*, Monatsh. Math. Phys., 16 (1905), pp. 234–249.

[FLLØ16] S. FOURNAIS, J. LAMPART, M. LEWIN, AND T. ØSTERGAARD SØRENSEN, *Coulomb potentials and Taylor expansions in Time-Dependent Density Functional Theory*, Phys. Rev. A, 93 (2016), p. 062510.

[FLLS22] R. L. FRANK, A. LAPTEV, M. LEWIN, AND R. SEIRINGER, eds., *The Physics and Mathematics of Elliott Lieb: The 90th Anniversary Volume (2 books)*, EMS Press, 2022.

[Flo83] G. FLOQUET, *Sur les équations différentielles linéaires à coefficients périodiques*, Ann. Sci. École Norm. Sup., 2e série, 12 (1883), pp. 47–88.

[FLS07] R. FRANK, E. LIEB, AND R. SEIRINGER, *Hardy-Lieb-Thirring inequalities for fractional Schrödinger operators*, J. Amer. Math. Soc., 21 (2007), pp. 925–950.

[FLS18] S. FOURNAIS, M. LEWIN, AND J. P. SOLOVEJ, *The semi-classical limit of large fermionic systems*, Calc. Var. Partial Differ. Equ., (2018), pp. 57–105.

[FLW22] R. FRANK, A. LAPTEV, AND T. WEIDL, *Schrödinger operators: Eigenvalues and Lieb-Thirring inequalities*, Cambridge Studies in Advanced Mathematics, Cambridge University Press, Cambridge, 2022.

[FP81] J. FRÖHLICH AND C.-E. PFISTER, *On the absence of spontaneous symmetry breaking and of crystalline ordering in two-dimensional systems*, Comm. Math. Phys., 81 (1981), pp. 277–298.

[FP86] ———, *Absence of crystalline ordering in two dimensions*, Comm. Math. Phys., 104 (1986), pp. 697–700.

[Fra09] R. L. FRANK, *A Simple Proof of Hardy-Lieb-Thirring Inequalities*, Comm. Math. Phys., 290 (2009), pp. 789–800.

[Fra23] ———, *The Lieb-Thirring inequalities: recent results and open problems*, in Nine mathematical challenges—an elucidation, vol. 104 of Proc. Sympos. Pure Math., Amer. Math. Soc., Providence, RI, 2023, pp. 45–86.

[Fri34] K. FRIEDRICHS, *Spektraltheorie halbbeschränkter Operatoren und Anwendung auf die Spektralzerlegung von Differentialoperatoren. I, II*, Math. Ann., 109 (1934), pp. 465–487, 685–713.

[FS90] C. L. FEFFERMAN AND L. A. SECO, *On the energy of a large atom*, Bull. Amer. Math. Soc. (N.S.), 23 (1990), pp. 525–530.

[Gie00] F. GIERES, *Mathematical surprises and Dirac's formalism in quantum mechanics*, Rep. Prog. Phys., 63 (2000), p. 1893.

[GK12] M. J. GANDER AND F. KWOK, *Chladni figures and the Tacoma Bridge: motivating PDE eigenvalue problems via vibrating plates*, SIAM Rev., 54 (2012), pp. 573–596.

[Gla66] I. M. GLAZMAN, *Direct methods of qualitative spectral analysis of singular differential operators*, Israel Program for Scientific Translations, Jerusalem, 1965; Daniel Davey & Co., Inc., New York, 1966. Translated from the Russian by the IPST staff.

[Gon20] D. GONTIER, *Edge states in ordinary differential equations for dislocations*, J. Math. Phys., 61 (2020), pp. 043507, 21.

[Gri85] P. GRISVARD, *Elliptic problems in nonsmooth domains*, vol. 24 of Monographs and Studies in Mathematics, Pitman (Advanced Publishing Program), Boston, MA, 1985.

[GWW92] C. GORDON, D. WEBB, AND S. WOLPERT, *Isospectral plane domains and surfaces via Riemannian orbifolds*, Invent. Math., 110 (1992), pp. 1–22.

[Hal13] B. C. HALL, *Quantum theory for mathematicians*, vol. 267 of Graduate Texts in Mathematics, Springer, New York, 2013.

[Hei25] W. HEISENBERG, *Über quantentheoretische Umdeutung kinematischer und mechanischer Beziehungen*, Z. Phys., 33 (1925), pp. 879–893.

[Her77] I. W. HERBST, *Spectral theory of the operator* $(p^2+m^2)^{1/2} - Ze^2/r$, Commun. Math. Phys., 53 (1977), pp. 285–294.

[Hil02] D. HILBERT, *Mathematical problems*, Bull. Am. Math. Soc., 8 (1902), pp. 437–479. Lecture delivered before the international congress of mathematicians at Paris in 1900. Translated by *Mary Winston Newson*.

[Hil06] ———, *Grundzüge einer allgemeinen Theorie der linearen Integralgleichungen. Vierte Mitteilung*, Nachr. Ges. Wiss. Göttingen, Math.-Phys. Kl., 1906 (1906), pp. 157–227.

[HLP88] G. H. HARDY, J. E. LITTLEWOOD, AND G. PÓLYA, *Inequalities*, Cambridge Mathematical Library, Cambridge University Press, 1988.

[HÖ12] D. HAHN AND M. ÖZISIK, *Heat Conduction*, Wiley, 2012.

[HU30] E. A. HYLLERAAS AND B. UNDHEIM, *Numerische berechnung der 2 S-terme von ortho- und par- helium*, Z. Phys., 65 (1930), pp. 759–772.

[Hun66] W. HUNZIKER, *On the spectra of Schrödinger multiparticle Hamiltonians*, Helv. Phys. Acta, 39 (1966), pp. 451–462.

[HvN27] D. HILBERT, J. VON NEUMANN, AND L. W. NORDHEIM, *Über die Grundlagen der Quantenmechanik*, Math. Ann., 98 (1927), pp. 1–30.

[ILS96] A. IANTCHENKO, E. H. LIEB, AND H. SIEDENTOP, *Proof of a conjecture about atomic and molecular cores related to Scott's correction*, J. Reine Angew. Math., 472 (1996), pp. 177–195.

[Ivr16] V. IVRII, *100 years of Weyl's law*, Bull. Math. Sci., 6 (2016), pp. 379–452.

[Jor27] P. JORDAN, *Über eine neue Begründung der Quantenmechanik. I*, Z. Phys., 40 (1927), pp. 809–838.

[Kac66] M. KAC, *Can one hear the shape of a drum?*, Amer. Math. Monthly, 73 (1966), pp. 1–23.

[Kar97] Y. E. KARPESHINA, *Perturbation theory for the Schrödinger operator with a periodic potential*, vol. 1663 of Lecture Notes in Mathematics, Springer-Verlag, Berlin, 1997.

[Kat51] T. KATO, *Fundamental properties of Hamiltonian operators of Schrödinger type*, Trans. Amer. Math. Soc., 70 (1951), pp. 195–221.

[Kat66] ———, *Perturbation theory for linear operators*, Springer, second ed., 1966.

[KS95] A. KISELEV AND B. SIMON, *Rank one perturbations with infinitesimal coupling*, J. Funct. Anal., 130 (1995), pp. 345–356.

[Lal19] F. LALOË, *Do we really understand quantum mechanics? 2nd revised edition*, Cambridge: Cambridge University Press, 2nd revised edition ed., 2019.

[Lew10] M. LEWIN, *Variational Methods in Quantum Mechanics*. Unpublished lecture notes (University of Cergy-Pontoise), 2010.

[Lew11] ———, *Geometric methods for nonlinear many-body quantum systems*, J. Funct. Anal., 260 (2011), pp. 3535–3595.

[Lew17] ———, *Éléments de théorie spectrale : le Laplacien sur un ouvert borné*. Notes de cours de Master 2, 2017.

[Lew22] ———, *Théorie spectrale et mécanique quantique*, Mathématiques et Applications, Springer International Publishing, 2022.

[Lew23] ———, *Mean-field limits for quantum systems and nonlinear Gibbs measures*, in International Congress of Mathematicians 2022, D. Beliaev and S. Smirnov, eds., vol. 5, EMS Press, December 2023, pp. 3800–3821.

[Lie76] E. H. LIEB, *The stability of matter*, Rev. Mod. Phys., 48 (1976), pp. 553–569.

[Lie79] ———, *The $n^{5/3}$ law for bosons*, Phys. Lett. A, 70 (1979), pp. 71–73.

[Lie80] ———, *The number of bound states of one-body Schrödinger operators and the Weyl problem*, in Geometry of the Laplace operator, Proc. Sympos. Pure Math., XXXVI, Amer. Math. Soc., Providence, R.I., 1980, pp. 241–252. (Proc. Sympos. Pure Math., Univ. Hawaii, Honolulu, Hawaii, 1979).

[Lie81] ———, *Thomas-Fermi and related theories of atoms and molecules*, Rev. Mod. Phys., 53 (1981), pp. 603–641.

[Lie84] ———, *Bound on the maximum negative ionization of atoms and molecules*, Phys. Rev. A, 29 (1984), pp. 3018–3028.

[Lie90] ———, *The stability of matter: from atoms to stars*, Bull. Amer. Math. Soc. (N.S.), 22 (1990), pp. 1–49.

[LL01] E. H. LIEB AND M. LOSS, *Analysis*, vol. 14 of Graduate Studies in Mathematics, American Mathematical Society, Providence, RI, 2nd ed., 2001.

[LL13] E. LENZMANN AND M. LEWIN, *Dynamical ionization bounds for atoms*, Analysis & PDE, 6 (2013), pp. 1183–1211.

[LM54] P. D. LAX AND A. N. MILGRAM, *Parabolic equations*, Ann. Math. Stud., 33 (1954), pp. 167–190.

[LM68] J.-L. LIONS AND E. MAGENES, *Problèmes aux limites non homogènes et applications. Vol. 1*, Travaux et Recherches Mathématiques, No. 17, Dunod, Paris, 1968.

[LM77] J. M. LEINAAS AND J. MYRHEIM, *On the theory of identical particles*, Nuovo Cimento B Serie, 37 (1977), pp. 1–23.

[LNR14] M. LEWIN, P. T. NAM, AND N. ROUGERIE, *Derivation of Hartree's theory for generic mean-field Bose systems*, Adv. Math., 254 (2014), pp. 570–621.

[LS77] E. H. LIEB AND B. SIMON, *The Thomas-Fermi theory of atoms, molecules and solids*, Adv. Math., 23 (1977), pp. 22–116.

[LS10a] M. LEWIN AND É. SÉRÉ, *Spectral pollution and how to avoid it (with applications to Dirac and periodic Schrödinger operators)*, Proc. London Math. Soc., 100 (2010), pp. 864–900.

[LS10b] E. H. LIEB AND R. SEIRINGER, *The Stability of Matter in Quantum Mechanics*, Cambridge Univ. Press, 2010.

[LSST88] E. H. LIEB, I. M. SIGAL, B. SIMON, AND W. THIRRING, *Approximate neutrality of large-Z ions*, Commun. Math. Phys., 116 (1988), pp. 635–644.

[LSSY05] E. H. LIEB, R. SEIRINGER, J. P. SOLOVEJ, AND J. YNGVASON, *The mathematics of the Bose gas and its condensation*, Oberwolfach Seminars, Birkhäuser, 2005.

[LT75] E. H. LIEB AND W. E. THIRRING, *Bound on kinetic energy of fermions which proves stability of matter*, Phys. Rev. Lett., 35 (1975), pp. 687–689.

[LT76] ———, *Inequalities for the moments of the eigenvalues of the Schrödinger hamiltonian and their relation to Sobolev inequalities*, Studies in Mathematical Physics, Princeton University Press, 1976, pp. 269–303.

[Mac33] J. K. L. MACDONALD, *Successive approximations by the Rayleigh-Ritz variation method*, Phys. Rev., 43 (1933), pp. 830–833.

[Mac92] N. MACRAE, *John von Neumann: The Scientific Genius Who Pioneered the Modern Computer, Game Theory, Nuclear Deterrence, and Much More*, Pantheon Press, 1992.

[MT76] H. P. MCKEAN AND E. TRUBOWITZ, *Hill's operator and hyperelliptic function theory in the presence of infinitely many branch points*, Comm. Pure Appl. Math., 29 (1976), pp. 143–226.

[MTWB10] N. T. MAITRA, T. N. TODOROV, C. WOODWARD, AND K. BURKE, *Density-potential mapping in time-dependent density-functional theory*, Phys. Rev. A, 81 (2010), p. 042525.

[MW66] N. D. MERMIN AND H. WAGNER, *Absence of ferromagnetism or antiferromagnetism in one- or two-dimensional isotropic heisenberg models*, Phys. Rev. Lett., 17 (1966), pp. 1133–1136.

[Nak01] S. NAKAMURA, *A remark on the Dirichlet-Neumann decoupling and the integrated density of states*, J. Funct. Anal., 179 (2001), pp. 136–152.

[Nam12] P. T. NAM, *New bounds on the maximum ionization of atoms*, Commun. Math. Phys., 312 (2012), pp. 427–445.

[NS05] Y. NETRUSOV AND Y. SAFAROV, *Weyl asymptotic formula for the Laplacian on domains with rough boundaries*, Comm. Math. Phys., 253 (2005), pp. 481–509.

[Oka82] N. OKAZAWA, *On the perturbation of linear operators in Banach and Hilbert spaces*, J. Math. Soc. Japan, 34 (1982), pp. 677–701.

[PA09] P. PYYKKÖ AND M. ATSUMI, *Molecular single-bond covalent radii for elements 1–118*, Chemistry – A European Journal, 15 (2009), pp. 186–197.

[Par08] L. PARNOVSKI, *Bethe-Sommerfeld conjecture*, Ann. Henri Poincaré, 9 (2008), pp. 457–508.

[Poh65] S. I. POHOZAEV, *On the eigenfunctions of the equation* $\Delta u + \lambda f(u) = 0$, Dokl. Akad. Nauk SSSR, 165 (1965), pp. 36–39.

[Poi90] H. POINCARÉ, *Sur les équations aux dérivées partielles de la physique mathématique*, Amer. J. Math., 12 (1890), pp. 211–294.

[RB69] F. RELLICH AND J. BERKOWITZ, *Perturbation Theory of Eigenvalue Problems*, Gordon and Breach, New York, 1969.

[Rei70] C. REID, *Hilbert*, no. IX, Berlin-Heidelberg-New York: Springer-Verlag, 1970. With an appreciation of Hilbert's mathematical work by Hermann Weyl.

[Rel37] F. RELLICH, *Störungstheorie der Spektralzerlegung. IV.*, Math. Ann., 116 (1937), pp. 555–570.

[Rie13] F. RIESZ, *Les systèmes d'équations linéaires à une infinité d'inconnues*, vol. VI + 182 S. 8° of Collection Borel, Gauthier-Villars, Paris, 1913.

[Rit09] W. RITZ, *Über eine neue methode zur lösung gewisser variationsprobleme der mathematischen physik.*, J. Reine Angew. Math., 135 (1909), pp. 1–61.

[Rou16] N. ROUGERIE, *Théorèmes de de Finetti, limites de champ moyen et condensation de Bose-Einstein*, Spartacus-idh, Paris, 2016. Cours Peccot au Collège de France (2014).

[Roz72] G. V. ROZENBLUM, *Distribution of the discrete spectrum of singular differential operators*, Dokl. Akad. Nauk SSSR, 202 (1972), pp. 1012–1015.

[RS72] M. REED AND B. SIMON, *Methods of Modern Mathematical Physics. I. Functional analysis*, Academic Press, 1972.

[RS75] ——, *Methods of Modern Mathematical Physics. II. Fourier analysis, self-adjointness*, Academic Press, New York, 1975.

[RS78] ——, *Methods of Modern Mathematical Physics. IV. Analysis of operators*, Academic Press, New York, 1978.

[RS79] ——, *Methods of Modern Mathematical Physics. III. Scattering theory*, Academic Press, New York, 1979.

[Rue69] D. RUELLE, *A remark on bound states in potential-scattering theory*, Nuovo Cimento A (10), 61 (1969), pp. 655–662.

[Rue99] ——, *Statistical mechanics. Rigorous results*, Singapore: World Scientific. London: Imperial College Press, 1999.

[Rus82] M. B. RUSKAI, *Absence of discrete spectrum in highly negative ions: II. Extension to fermions*, Commun. Math. Phys., 85 (1982), pp. 325–327.

[SB33] A. SOMMERFELD AND H. BETHE, *Elektronentheorie der Metalle*, Springer Berlin Heidelberg, Berlin, Heidelberg, 1933, pp. 333–622.

[Sch26] E. SCHRÖDINGER, *Quantisierung als Eigenwertproblem. I*, Ann. der Phys. (4), 79 (1926), pp. 361–374.

[Sch61] J. SCHWINGER, *On the bound states of a given potential*, Proc. Nat. Acad. Sci. U.S.A., 47 (1961), pp. 122–129.

[Sch19] A. SCHIRRMACHER, *Establishing quantum physics in Göttingen. David Hilbert, Max Born, and Peter Debye in context, 1900–1926*, Cham: Springer, 2019.

[Sco52] J. SCOTT, *The binding energy of the Thomas-Fermi atom*, Lond. Edinb. Dubl. Phil. Mag., 43 (1952), pp. 859–867.

[See80] R. SEELEY, *An estimate near the boundary for the spectral function of the Laplace operator*, Amer. J. Math., 102 (1980), pp. 869–902.

[She01] Z. SHEN, *On absolute continuity of the periodic Schrödinger operators*, Internat. Math. Res. Notices, (2001), pp. 1–31.

[Sig82] I. M. SIGAL, *Geometric methods in the quantum many-body problem. Non existence of very negative ions*, Commun. Math. Phys., 85 (1982), pp. 309–324.

[Sig84] ——, *How many electrons can a nucleus bind?*, Ann. Phys., 157 (1984), pp. 307–320.

[Sim76] B. SIMON, *On the genericity of nonvanishing instability intervals in Hill's equation*, Ann. Inst. H. Poincaré Sect. A (N.S.), 24 (1976), pp. 91–93.

[Sim95] ——, *Spectral analysis of rank one perturbations and applications*, in Mathematical Quantum Theory II: Schrödinger Operators, J. Feldman, R. Froese, and L. Rosen, eds., vol. 8, Centre de Recherche Mathématiques, CRM Proceedings and Lecture Notes, 1995, p. 109.

[Sim15] ——, *Operator theory*, vol. Part 4 of A Comprehensive Course in Analysis, American Mathematical Society, Providence, RI, 2015.

[SM09] T. SAUER AND U. MAJER, eds., *David Hilbert's lectures on the foundations of physics, 1915-1927. Relativity, quantum theory and epistemology. In collaboration with Arne Schirrmacher and Heinz-Jürgen Schmidt*, Berlin: Springer, 2009.

[Sol90] J. P. SOLOVEJ, *Asymptotics for bosonic atoms*, Lett. Math. Phys., 20 (1990), pp. 165–172.

[Sol16] ——, *A new look at Thomas–Fermi theory*, Mol. Phys., 114 (2016), pp. 1036–1040.

[Som28] A. SOMMERFELD, *Zur Elektronentheorie der Metalle auf Grund der Fermischen Statistik*, Zeitschrift fur Physik, 47 (1928), pp. 1–32.

[Sto29] M. H. STONE, *Linear transformation in Hilbert space. I: Geometrical aspects. II: Analytical aspects*, Proc. Natl. Acad. Sci. USA, 15 (1929).

[Sto30] ———, *Linear transformations in Hilbert space. III: Operational methods and group theory*, Proc. Natl. Acad. Sci. USA, 16 (1930), pp. 172–175.

[Sto32a] ———, *Linear transformations in Hilbert space and their applications to analysis*, vol. 15, American Mathematical Society (AMS), Providence, RI, 1932.

[Sto32b] ———, *On one-parameter unitary groups in Hilbert space*, Ann. Math. (2), 33 (1932), pp. 643–648.

[Str71] J. STRUTT (LORD RAYLEIGH), *Some general theorems relating to vibrations*, Proc. London Math. Soc., s1-4 (1871), pp. 357–368.

[SW87] H. SIEDENTOP AND R. WEIKARD, *Upper bound on the ground state energy of atoms that proves Scott's conjecture*, Phys. Lett. A, 120 (1987), pp. 341–342.

[TEM+89] A. TONOMURA, J. ENDO, T. MATSUDA, T. KAWASAKI, AND H. EZAWA, *Demonstration of single-electron buildup of an interference pattern*, Amer. J. Phys., 57 (1989), pp. 117–120.

[Tes09] G. TESCHL, *Mathematical Methods in Quantum Mechanics; With Applications to Schrödinger Operators*, vol. 99 of Graduate Studies in Mathematics, Amer. Math. Soc, Providence, RI, 2009.

[Tha92] B. THALLER, *The Dirac equation*, Texts and Monographs in Physics, Springer-Verlag, Berlin, 1992.

[Tho27] L. H. THOMAS, *The calculation of atomic fields*, Proc. Camb. Philos. Soc., 23 (1927), pp. 542–548.

[Tho73] L. E. THOMAS, *Time dependent approach to scattering from impurities in a crystal*, Commun. Math. Phys., 33 (1973), pp. 335–343.

[Tru73] N. S. TRUDINGER, *Linear elliptic operators with measurable coefficients*, Ann. Sc. Norm. Super. Pisa, Sci. Fis. Mat., III. Ser., 27 (1973), pp. 265–308.

[Tru77] ———, *Maximum principles for linear, non-uniformly elliptic operators with measurable coefficients*, Math. Z., 156 (1977), pp. 291–301.

[Van64] C. VAN WINTER, *Theory of finite systems of particles. I. The Green function*, Mat.-Fys. Skr. Danske Vid. Selsk., 2 (1964), p. 60 pp.

[vH50] L. VAN HOVE, *Sur l'intégrale de configuration pour les systèmes de particules à une dimension*, Physica, 16 (1950), pp. 137–143.

[von27a] J. VON NEUMANN, *Mathematische Begründung der Quantenmechanik*, Nachr. Ges. Wiss. Göttingen, Math.-Phys. Kl., 1927 (1927), pp. 1–57.

[von27b] ———, *Thermodynamik quantenmechanischer Gesamtheiten*, Nachr. Ges. Wiss. Göttingen, Math.-Phys. Kl., 1927 (1927), pp. 276–291.

[von27c] ———, *Wahrscheinlichkeitstheoretischer Aufbau der Quantenmechanik*, Nachr. Ges. Wiss. Göttingen, Math.-Phys. Kl., 1927 (1927), pp. 245–272.

[von29a] ———, *Zur Algebra der Funktionaloperationen und Theorie der normalen Operatoren*, Math. Ann., 102 (1929), pp. 370–427.

[von29b] ———, *Zur Theorie der unbeschränkten Matrizen*, J. Reine Angew. Math., 161 (1929), pp. 208–236.

[von30] ———, *Allgemeine Eigenwerttheorie Hermitescher Funktionaloperatoren*, Math. Ann., 102 (1930), pp. 49–131.

[von32a] ———, *Mathematishe Grundlagen der Quantenmechanik*, Springer Verlag (Berlin), 1932.

[von32b] ———, *Uber Einen Satz Von Herrn M. H. Stone*, Ann. Math. (2), 33 (1932), p. 567.

[VZ77] S. VUGALTER AND G. M. ZHISLIN, *Finiteness of a discrete spectrum of many-particle hamiltonians in symmetry spaces (coordinate and momentum representations)*, Teoret. Mat. Fiz., 32 (1977), pp. 70–87.

[Web69] H. WEBER, *Über die Integration der partiellen Differentialgleichung $\frac{\partial^2 u}{\partial x^2} + \frac{\partial^2 u}{\partial y^2} + k^2 u = 0$*, Math. Ann., 1 (1869), pp. 1–36.

[Wei96] T. WEIDL, *On the Lieb-Thirring constants $L_{\gamma,1}$ for $\gamma \geq 1/2$*, Comm. Math. Phys., 178 (1996), pp. 135–146.

[Wey12] H. WEYL, *Das asymptotische verteilungsgesetz der eigenwerte linearer partieller differentialgleichungen (mit einer anwendung auf die theorie der hohlraumstrahlung)*, Math. Ann., 71 (1912), pp. 441–479.

[Wey13] ———, *Über die Randwertaufgabe der Strahlungstheorie und asymptotische Spektralgesetze*, J. Reine Angew. Math., 143 (1913), pp. 177–202.

[Yaf76] D. R. YAFAEV, *On the point spectrum in the quantum-mechanical many-body problem*, Math. USSR Izv., 40 (1976), pp. 861–896. English translation.

[Yaf99] ———, *Sharp constants in the Hardy-Rellich inequalities*, J. Funct. Anal., 168 (1999), pp. 121–144.

[Zhi60] G. M. ZHISLIN, *Discussion of the spectrum of Schrödinger operators for systems of many particles. (in Russian)*, Trudy Moskov. Mat. Obšč., 9 (1960), pp. 81–120.

[Zhi71] ———, *On the finiteness of the discrete spectrum of the energy operator of negative atomic and molecular ions*, Teoret. Mat. Fiz., 21 (1971), pp. 332–341.

[ZS65] G. M. ZHISLIN AND A. G. SIGALOV, *The spectrum of the energy operator for atoms with fixed nuclei on subspaces corresponding to irreducible representations of the group of permutations*, Izv. Akad. Nauk SSSR Ser. Mat., 29 (1965), pp. 835–860.

Index

A
Adjoint, 56
Angular momentum, 155
Atoms and molecules
    Hamiltonian, 231
    spectrum, 243

B
Beams, equation, 223
Birman-Schwinger, 210, 317
Bloch, bands, 262
Bloch-Floquet, 256, 257
Bose-Einstein, condensate, 230, 239
Bosonic atoms, 247
Bosons, 229, 232, 238
Boundary
    Lipschitz, 290
    piecewise Lipschitz, 290
Boundary conditions
    Born-von Kármán, 73, 79, 83, 98, 257
    Dirichlet, 79, 83, 99
    Neumann, 79, 83, 99
    periodic, 79, 81, 83
    Robin, 79, 99, 205, 225
Boundary restriction, 283
Bounded from below, operator, 95

C
Cauchy, formula, 140
Cell, Voronoi or Wigner-Seitz, 253
Chladni, figures, 116

Closed
    operator, 54
    quadratic form, 92
Closure
    of a quadratic form, 92
    of an operator, 54
Coercive, operator, 95
Commutator
    counter-examples, 82
Commuting operators
    definition, 161
    diagonalization, 163
    joint spectrum, 163
Compact
    operator, 185
    relatively, 197
Convergence of operators
    in norm, 165
    strong, 151, 165
    weak, 165
Core, of a self-adjoint operator, 88
Coulomb potential, 3
Courant-Fischer, formula, 201
Cwikel-Lieb-Rozenblum, inequality, 209
Cyclic
    subspace, 135
    vector, 135, 165

D
Deficiency indices, 83
Delta potential, 311
Density of states, 268